JACQUES COUSTEAU

THE OCEAN WORLD

JACQUES

THE OCE

HARRY N. ABRAMS, INC.

COUSTEAU

AN WORLD

PUBLISHERS, NEW YORK

Third printing 1981

Editor: Edith Pavese

Designer: Bruce Blair

Rights and Reproductions: Barbara Lyons

Library of Congress Cataloging in Publication Data
Cousteau, Jacques-Yves
　Jacques Cousteau/The Ocean World
　Text based on the author's 20-vol. series, first
published 1972–1974 by World Pub., New York.
　Includes index.
　1. Marine biology. 2. Ocean. I. Title.
QH91.15.C652　　　574.92　　　77-20197
ISBN 0-8109-0777-1

Library of Congress Catalogue Card Number: 77-20197

Printed and bound in Japan

CONTENTS

INTRODUCTION

The Third Infinity

Twenty-seven years ago, I wanted to keep an eye permanently open into the oceans, so I equipped the bow of my new ship *Calypso* with an underwater observation chamber. At that time, I was convinced that the oceans were immense, teeming with life, rich in resources of all kinds; during the long crossings in the Indian Ocean or in the Atlantic, I spent many hours, day and night, looking through my undersea portholes, dreaming of Captain Nemo in the *Nautilus*. But soon I had to face the evidence: the blue waters of the open sea appeared to be, most of the time, a discouraging desert. Like the deserts on land, it was far from dead, but the live ingredient, plankton, was thinly spread, like haze, barely visible and monotonous. Then, exceptionally, areas turned into meeting places; close to shores and reefs, around floating weeds or wrecks, fish would gather and make a spectacular display of vitality and beauty. Years of diving have revealed to me that the same situation occurs on the bottom of the sea. On the floor as in midwater, endless deserts are spotted with rare but exuberant oases.

The "oasis theory" was to help me understand that the ocean, huge as it may be when measured at human scale, is a thin layer of water covering most of our planet—a very small world in fact—extremely fragile and at our mercy. Yet, it is in the primordial ocean that life originated, approximately 3.5 billion years ago, in the shape of very simple cells. The most recent explorations of outer space have demonstrated that our planet Earth—our Water planet, I should write—is the only planet in the solar system to be endowed with appreciable quantities of liquid water. Life, born in water, must be at least as rare as water in the universe, and as such must be revered, under any of its forms, as a miracle. Diving in the open sea, I have encountered salps, barrel-shaped jelly creatures, linked to each other to form gracious living chains one hundred feet long, that must have inspired the legend of the divine belt worn by Venus; the sight of a Medusa, a delicate transparent dome of pulsating crystal suggested to me irresistibly that life is organized water and that *life identifies itself with the water system of our planet*.

Another time, diving in the Red Sea, in the archipelago off Suakin, I found the water so clear that sharks and barracudas seemed to be suspended in a three-dimensional nothingness, around exuberant coral reef communities, medleys of colorful patches, bustling swimmers, drifters, crawlers, and dwellers; the reef was actually noisy: choruses made of faint cracks, groans, and hisses were playing a vibrant symphony to the miracles of life and death. There I learned that *variety is the essence of a healthy life system*.

In the Sea of Oman, north and northeast of the coast of Somali, I encountered ocean-borne intelligence—whales, spermwhales, orcas, and dolphins. The marine mammals were displaying their power, their speed, their smartness; but the more I observed the creatures of the sea, the more I could relate them to those that live in our dry world. The behavior of fish, squid, birds, or whales was governed by the same basic motivations as that of snakes, insects, or apes. To me, the *unity of life* was in beautiful evidence.

Assumption Island, isolated in the Indian Ocean, north of Madagascar, was one of the richest undersea sanctuaries I ever visited; but when I came back thirteen years later, I found out that extensive damage had already been done to the coral fringe of the island, either by overfishing or by pollution. The *fragility of marine ecosystems* became obvious to me. Unfortunately, I had soon to acknowledge the fact that the oceans were rapidly deteriorating worldwide. Groupers were virtually eliminated from the Mediterranean; the Great Barrier Reef was slowly decaying; coral gardens of New Caledonia were choked and buried under millions of tons of waste from a huge nickel-mining complex; whales were decimated in the Antarctic; men were no exception: I have visited the vanishing Kawashkars in southern Patagonia, the last authentic "nomads of the sea," rugged people who had for millennia successfully lived naked on their boats, diving in ice-cold water for clams; there are fewer than twenty of these people left today. Fragility of the marine environment, fragility of man, precariousness of resources, are facts that we have to reckon with.

At sea level, caves are carved into the cliffs by waves and sand beaches are created by marine

erosion. Under the sea, such caves and beaches are found and are evidence that during glaciation periods the surface of the oceans was considerably lower than today. Conversely, if the ice caps were to melt, the surface could rise substantially. Seventeen thousand years ago, the Bering Strait was dry, Asian hordes of hunters walked from Siberia to Alaska and populated North America; the sites upon which we have built most of our modern harbors—New York, London, or Tokyo—were some 500 feet up in the hills! And the same harbors might very well be submerged 200 feet under the sea in a few centuries: The sea actually conditions all human activities, generating rain, floods, or droughts, bringing about constant changes, slow or abrupt, gentle or catastrophic. Our "liquid future" depends upon the foresight, the care, and the love with which we will manage our only water supply: the Oceans.

Meanwhile, the little pulse of life, thriving in the sea, still turns shining droplets of water into living jewels. The miracle of life defies the universal law of degradation, creates highly complex organic molecules, organizes chaotic matter into incredibly well programmed structures made of trillions of cells. It is the contemplation of life that inspired Father Teilhard de Chardin to meditate on the three infinities: in addition to the infinitely big and to the infinitely small, Teilhard told us there also was the *infinitely complex:* Life.

THE
ACT
OF
LIFE

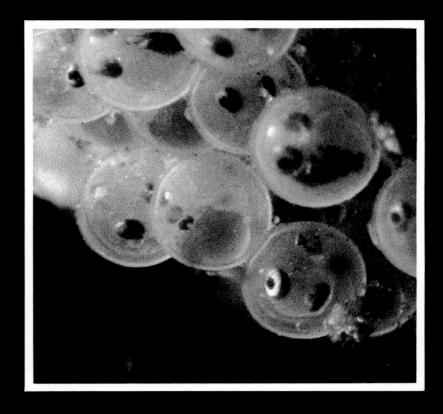

CHAPTER
1

DYING
FOR
SURVIVAL

Imagine an immortal animal. Evolution so fashioned his glands and organs that his parts replace themselves as soon as they cease to function. He is invulnerable to cancer and all forms of viral. microbial, bacterial infection. He exists in total harmony with his environment. As he has no reason to die he doesn't, but lives on and on through the ages—growing a bit bored, perhaps, but animals seem to agree that life is better than death on almost any terms.

Has evolution ever produced such a prepossessing creature? Theoretically it should be possible that physical obsolescence would simply be banished from an animal's life program. We know of certain plants—for example, lichens and the bristlecone pines of California's Inyo Forest—that live many thousands of years, near enough to immortality. But the oldest animal of which there is a record seems to be a tortoise that managed to struggle through 135 years or so—not all that much older than many old men.

Paradoxically, if immortality has ever been attained, it has quickly been eliminated, simply because immortality cannot survive. For earth is constantly changing, and animals must be ready to adapt to earth's changes. At least four times in the past 600 million years the reef communities around the world have been all but obliterated by upheavals in the environment still not completely understood. Fossils of palm trees have been discovered in Antarctica, now under hundreds of feet of ice. Faced with this dimension of drastic environmental transformation any immortal animal would be helpless. His ideal adjustment to the old environment spells certain extinction in the new. Locked into his "perfection," he cannot adjust. Immortal or not, he must die.

The mechanism by means of which each species responds to the challenges of a changing environment is the births and deaths of individual animals. In any large population there is one individual with a thicker hide, another with a more flexible snout, another with a bigger cerebrum, another with the tendency to bear twins, another with acuter hearing. And so on. In other words,

any successful species presents the environment not with an army of perfect individuals but with a smorgasbord of different characteristics dispersed through its membership. Then, when the environment challenges the species, the species has a chance to come up with the answer.

The process does not always work so simply. Immense as the dinosaur population was at the end of the Cretaceous period, some 65 million years ago, the smorgasbord of natural variations within the species was too limited for the challenges the environment posed it; they died out. But man is the great zoological eccentric. For man's brain gives him tools with which to participate directly, consciously, in his own evolution. To start with, he has grown almost as interested in individual survival as in species survival. Egged on by this concern for individual survival he has learned to screen himself from many of the "natural" agents of the selection process. Puerperal fever, tuberculosis, pneumonia, smallpox, diphtheria, plague—all these grim reapers that for thousands of years winnowed out the human species are mostly fears of the past.

If man is a unique exception as an animal it is a recent phenomenon. He is what he is because uncountable legions of animals have lived and perished since the first tiny cells stirred in the ancient oceans three billion or so years ago. From these lowly entities the natural-selection process has moved man steadily forward: past the jellyfish and the mollusc, past the turnoffs to equally successful evolutionary strategies like the insects', into the early experimental chordates, the mammals with their invaluable specialty of caring for their young, to the primates—to *himself!* Trillions of generations, trillions of deaths—each one a small link in the chain of evolution, each one a survival-ticket for man. Biological sciences are only a few generations short of being able to interfere consciously with genetics and to produce eternal youth. If we are reasonable enough to avoid a nuclear holocaust and to control population, immortality will no more be a utopian dream.

Female prawn with eggs

THE BALANCE OF NATURE

Were the crust to be leveled—with great mountain ranges like the Himalayas and ocean abysses like the Marianas Trench evened out—no land at all would show above the surface of the sea. Earth would be covered by a uniform sheet of water—more than 10,000 feet deep! The earth is truly a water planet and water is indeed a rare and precious gift.

Fecundity of the Sea

From the vast expanses of its surface waters to its beaches and marshes and tidelands and mangrove swamps, from its many thousands of miles of rocky shores to its deepest and darkest abyss, the sea produces life in fantastic abundance.

No wonder. The oceans are superior to land as an environment for life support. They provide directly the water fundamental to all forms of growth, laden with vital salts, dissolved gases, and minerals. The water temperature is more constant than air, reliably warmer in shallow and surface areas, reliably cooler in the deeps—freeing many species from the need to adapt, as most land animals must, to wide variations in temperature.

The surface waters, rich in oxygen and bathed in sunlight, support a teeming variety of plant life equal to that found on land. In the average cubic foot of this water as many as 20,000 microscopic plants will be found, together with hundreds of planktonic animals. (The word *plankton* is from the Greek. It literally means "that which is made to wander.") What slips through the finest net is even more impressive. The same cubic foot of water may hold well over 12 million unicellular plants, or diatoms. Encased in its self-made crystalline housing of silicon, each cell reproduces so rapidly that it may have more than one billion progeny in a month.

In this broth of life float billions and billions of eggs. For it is only by immense fertility that many species can survive, and even the barest survival is absolute success. Many marine creatures, unable to protect their eggs after spawning, have developed the capacity to produce enormous quantities of offspring. The blue crab will produce several thousand eggs at a time. The average mackerel will lay as many as 100,000. This is nothing. A hake will produce perhaps one million eggs at a time. A haddock will lay anywhere from 12 thousand to three million, and a cod from two to nine million. The purple sea hare will produce 20 million eggs, and the mola mola as many as 20 million. In a year, an oyster will lay 500 million eggs!

A skein of sea slug eggs is enclosed in a protective coating to help ensure their survival. Sea slugs (Navanax) mate in shallow waters, spending about an hour with their supple bodies wrapped together, exchanging sperm. Then they separate, both partners laying eggs alone.

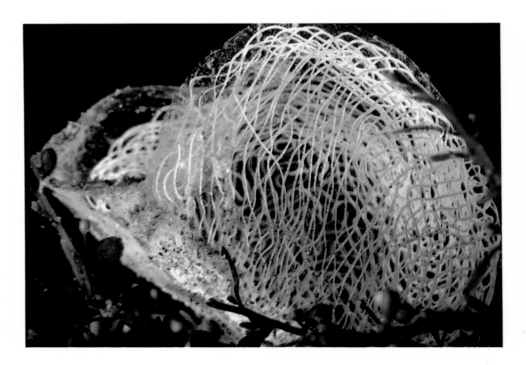

Controlling Factors

The success of a single species, if too great, can wreak havoc throughout the system and many species have a mortality rate of well over 99 percent. Many of the eggs and developing young, immobile and exposed in the rich surface waters, are ravaged by predators. The growth of each species is also controlled by a number of other important factors: water temperature, surface temperature, the strength and random flow of ocean currents, which may sweep vast quantities of helpless eggs and larvae away from their food sources or wash them ashore where the sunlight and the air will shrivel them dry.

While the most fecund species, like the cod, leave their eggs to fend for themselves, less prolific species choose other ways to ensure the survival of their kind. Some eggs are laid on the ocean floor, or in crevices of rock, and are heavy enough to remain on the bottom, though currents can still detach

Looking like a scarlet ribbon, this is the egg mass of a nudibranch, a shell-less mollusc. For protection, the eggs contain a substance that makes them distasteful to predators.

them so they become the prey of bottom feeders. Other eggs have a sticky surface, and affix themselves to rocks and weeds and shells. There are species like the grunion that bury their eggs on sandy beaches; others like the salmon travel long distances to freshwater spawning grounds comparatively free of predators. Still other eggs ride with their parents—some in the mouth or on the underside of the mother's body, others in a brooding pouch where they have added security. There are species that incubate their eggs within the female body, that nourish the embryos after they hatch. The sea mammals carry their offspring internally, suckle them for months, and help them to survive to adulthood.

SURVIVAL OF THE FITTEST

Four hundred million years ago fish had few fins and no jaws. They moved awkwardly along the ocean floor, sucking what food they could find from the mud.

Eventually some fish appeared with primitive jaws and with larger fins than others. This was accidental, but it gave them better steering and balance, more speed and the ability to catch other fish that had no jaws. They found more to eat. So a greater proportion of jawless fish with smaller fins, faring less successfully, died sooner, while a greater proportion of the new breeds survived longer. This latter group had the opportunity to reproduce oftener, contributing a greater number of offspring to each subsequent generation. Because, when they reproduced, they passed on their characteristics as genetic information, subsequent generations exhibited a larger proportion of faster fish.

Today, as a result of these processes, thousands of varieties of fish have developed, each species highly specialized for success in its environment. The parrotfish feeds on coral. It has developed a pair of jaws of solid hard bone, like a beak, plus a set of thick teeth fused internally. But there was a time when few parrotfish had beaks, few butterflyfish had spots on their tails; when flounder and sole swam vertically like other fish; when the viperfish did not have a stomach that stretched. All these special characteristics evolved by selection, in the course of which process the environment acted rather like a sieve to strain out individuals least suited to it, death in each generation reducing their number and increasing the majority of the fittest.

Survival Through Death

The Pacific ocean salmon pays for his success as a species with one of the most painful death agonies in the natural world. And only the superior individuals among the salmon could survive even as far as the death agony.

After four or five years in the open sea the salmon heads for home. Guiding himself by some not yet fully understood means, the mature fish presses on for the rivulet or pond where he began life. Swimming against the fierce flow of mountain streams, through rapids and over rocks and past an army of predators, terribly battered, fasting from the moment he embarks on this last trek, he forges on and on until he reaches the old spawning ground. Here in a final furious frenzy the female digs a hole in the stream bottom, deposits her eggs, and the male fertilizes them. A few days later the pair die.

In addition to the awesome demands nature makes on the salmon, scientists have been intrigued by the extremely rapid aging processes which are an aspect of this terminal phase of his life. In the final two weeks the salmon physically degenerates as much as a man would in 40 years: his arteries thicken, his liver gives out, his circulation weakens, he is subject to all kinds of infection and infestation. By the time he has spawned he is, in Dr. Andrew A. Benson's words, "a miserable shadow of the beautiful, silvery deep-ocean marine animal. His flesh has turned from orange-pink to pale tan. He has developed his hump and hooked jaw. His bones have become cartilaginous; his skin is peeling off. We even saw many with their tails falling off. His liver is a livid olive green because of the decomposition products of his hemoglobin. Only the heart of the salmon remains in good condition—and even this suffers from thickening of the coronary artery walls.'' Exhausted from malnutrition and the glandular ordeal, covered with fungus, the salmon has grown decrepit in a few days. Although the tale of the individual salmon is a sad one, it represents a triumph for the species. For in its death it provides for the survival of its kind. In the graveyard of the spawning grounds other fish will feed on the battered bones of this marine phoenix. The decomposing body will nourish its own fry when they emerge from these ''ashes.''

During the early part of their lives salmon live in fresh water. But they migrate to sea and spend their middle years in the ocean. When spawning time comes, the mature salmon brave many dangers and must often jump high rapids in order to reach the placid waters of the original spawning ground.

This lowly hydroid provides an insight into the evolution of its relatives—the true jellyfish and the corals. Most scientists concur that the hydroids are more primitive than their two relatives and sometime in the distant past gave rise to them.

THE CONTINUATION OF THE SPECIES

Reproduction is essential for the continuation of the species, but how is it ensured that young salmon hatch from their eggs as salmon instead of baby ducks or alligators? The answer lies in a genetic set of instructions that resides within the nucleus of every one of its cells. The alligator, the duck, and every other species of animal or plant have their own set.

Floating within the nucleus are a number of long strands called chromosomes, and arranged along the length of each chromosome are long molecules of a substance called deoxyribonucleic acid, or DNA, which contain the genes. The specific construction of these molecules and their position on the chromosome control every aspect of the lives of plants and animals, both their characteristics as species and as individual members of that species. They control how the body is constructed; they direct the fabrication of enzymes which control all chemical processes; they determine the coloration, size, and shape of organisms and their behavioral characteristics.

The Never-Ending Process

Animals and plants grow by reproducing their cells. This is accomplished by a division of the nucleus of each cell. The process is a continuous one, rapid in early stages of growth, enhanced by appropriate diet and ideal environmental conditions, somewhat hindered by aging and disease, but unending. It is called mitosis. Many one-celled planktonic plants and animals also reproduce by this method.

The several stages of the cell's development begin when the chromosomes, looped loosely in no particular pattern all over the nucleus, coil more tightly, taking the shape of thin rods. Each chromosome divides lengthwise into pairs, after which there is twice the number of chromosomes. The cell next squeezes itself into a spindly shape with half the number of chromosomes rushing to one end of the cell, half to the other. Then the spindle parts in the middle, making two cells, each with the same number of chromosomes as the original one. The chromosomes then uncoil.

Sexual reproduction, or meiosis, takes place in a way similar to body growth, but with an important difference. Genetic variability must be ensured so that the character of the strongest and most successful members of the species may have the maximum opportunity to express itself, giving the species optimum chances for survival. Cells identical in genetic character must be avoided as new generations are spawned. Thus, most plants and animals produce sex cells, or gametes—including the female egg and the male sperm. These are known as haploid cells: that is, cells with only half the required number of chromosomes for a normal (diploid) cell. The sex cells from different members of the same species unite, forming a diploid cell with a shared genetic heritage. This process of union is called fertilization, and when completed the new cell can grow by mitosis.

ASEXUAL REPRODUCTION

The asexual mode of reproduction does not allow for the great potential variations seen in sexual reproduction, but it also does not require another individual. Being, in a sense, simpler, a great number of offspring can be reproduced in a relatively short period of time. There is no energy spent in bringing the sexes together or in competitive battles for a mate. Most organisms that reproduce asexually simply divide into two individuals; each of these then grows, divides into a total of four, and so on, in a geometric progression. These include the unicellular phytoplankton species and some of the simplest forms of animal life, like the amoeba.

Common to some animal forms is another asexual form of reproduction known as budding. In budding, the same mitotic division of cells occurs as with the amoeba, with the difference that the cell divides itself into unequal parts. While both cells are diploid, the larger part is considered the parent, the smaller the bud.

Alternation of Generations

The great majority of plants have two alternating life cycles, one sexual and the other asexual. The asexual generation, having diploid cells, is usually the dominant life cycle and is called the sporophyte. Sporophytes produce thousands of spores. These unicellular diploid elements are capable of developing directly into adults.

Life Cycle of the Obelia

The obelia is an unusual animal. It is one of the hydroids, animals usually comprising two kinds of polyps, some specialized for feeding and some for reproduction.

The feeding polyps are long and thin with tentacles that catch food. They nourish the entire colony, from whose base grow the reproductive polyps, stubby and urn-shaped. Inside the urns little buds develop, and these are eventually released into the water. The process is asexual, the buds being equipped with the full number of chromosomes required in each cell, and developing by cell division.

But this is only one stage of the spectacular life cycle of the obelia. The little buds are jellyfish. Each of them grows rapidly. In its full-fledged size it is known as the medusa, so named after the Gorgon Medusa, the maiden of Greek mythology whose hair turned into a nest of snakes and who petrified anyone who looked at her. The name is apt, for the medusa's tentacles paralyze its prey most effectively. Bell-shaped, with a mouth for feeding and armed with stinging tentacles, it swims freely. There are both male and female medusae, releasing both sperm and eggs in the water. After fertilization, a larva forms, and it swims to the bottom where it develops into a new colony of polyps.

Parthenogenesis

Certain animals, including some fish and a number of invertebrates, will produce an egg which is not a gamete but a diploid cell. It develops without fertilization by a male, and in fact such a species has no male at all. Since all the individuals of the species produce eggs, this mode of reproduction, called parthenogenesis, has the advantage of producing a large number of new individuals in a relatively short time.

There are variations in this classic method of parthenogenesis. The *Poecilia mexicana,* for example, produces two kinds of females, one giving birth to the usual mixture of male and female offspring, the other giving birth only to females. Some of the male's characteristics are transmitted to the unisexed infants. The Amazon molly is a species consisting only of females all producing eggs with the full number of chromosomes. But these eggs will not develop without the attachment of a sperm to provide the stimulus for cell division. So the males are of another species, usually the common molly or sailfish molly. However, there is no fusion with the male nucleus, and the sperm is rejected.

SEXUAL REPRODUCTION

Early in the history of life on earth a method of reproducing developed which was in some ways better than the simple asexual method. This was a means of sharing genetic information so that valuable traits could be spread over a larger population and new genetic combinations be formed. In sexual reproduction, haploid cells from two different individuals combine to form a nucleus unlike any before it. Many creatures, either by physical forms of communication or by instincts we do not yet fully understand, come together at an exact time and place for the simultaneous shedding of sexual products. Some species are hermaphroditic, carrying both eggs and sperm in one animal. While they usually cross-fertilize one another, they often do not. The open sea, with its unpredictable shifts of currents and the constant threat

Squids are muscular, jet-propelled molluscs that reproduce sexually. This unusual photograph shows mating squid with the great gelatinous masses of eggs that the females have laid on the sea floor.

A grayish-green whelk lays a long chain of white encapsulated eggs.

of hungry predators, makes the task of development of an egg far more hazardous than it would be if the eggs were secured inside their mother's body and the sperm delivered there. Many of the more advanced creatures carry out internal fertilization, which provides the greatest chance for egg and sperm union as well as protection of the embryo, but females of most marine species are not equipped to receive sperm and fertilize their eggs internally.

Fertilization

Fertilization is the critical event in sexual reproduction. The male sperm, usually equipped with a head containing the genetic material and a tail for mobility, penetrates the egg, and the genetic material from both combine to form the fertilized egg, or zygote. As well as performing this function, the sperm also provides the stimulus which commences cell division and the growth of the embryo.

Most marine species simply shed their eggs and sperm in the ocean, relying on proximity and chance to bring them together. Even species that have pouches in which their eggs develop, such as the seahorse and pipefish, must fertilize the eggs externally. There are few exceptions to this rule. Mammals, some lower animals like the scorpionfish, crabs, shrimp, octopus, sharks, and certain rays, sea snakes, and the mosquitofish can transmit the sperm directly into the body of the female.

By what biological rhythms or subtle clues from the environment do millions of fish achieve the proper development of sperm and eggs and simultaneous spawning? We have not as yet unraveled all the mysteries of instinct. But we do know that fishes' reckoning is exact, their timing precise. Sometimes animals communicate with each other visually, sometimes by making sounds, and sometimes by releasing in the water chemicals that signal members of the opposite sex to join in the act of procreation.

Taking No Risks

The horseshoe crab is an ancient animal of the North American east coast that has remained unchanged for 175 million years. He is not a real crab at all; his closest living relations are the scorpion and the spider. The mature male is equipped with two strong hooks at the sides of his mouth, possessing an exclusively sexual function. By means of these hooks he grabs the female by the hind part of her shell and hitches a ride. Another male can clamp on to

Nudibranchs, like other gastropods, are hermaphrodites. When mating, each animal produces both eggs and sperm and each will bear young after exchanging its sperm with another nudibranch.

the first and so on—until a chain of as many as four or five has formed, holding on for days or weeks. The female drags her escort on to a beach at high tide, digs a hole for her eggs—and her males fertilize them.

Part of the Mating Game

The mating ritual of the grunt sculpin, a big-headed bottom-dweller, begins with a female chasing a coy male about until she forces him into a crevice or other chamber. Once inside, she blocks the entrance to make sure the male will be available to fertilize the 150 or so eggs she will deposit on the chamber walls. Only after this has been accomplished, does the female allow him to leave.

Hermaphroditism

Hermaphroditism is the condition in which an individual animal possesses the reproductive organs of both sexes. It can increase the reproductive potential of a species, since eggs are produced by all the individuals instead of only half, as is the case in species with separate sexes. Also, in solitary or rare species it guarantees that whenever two individuals meet, mating can take place, because each individual has both sperm and eggs.

Hermaphroditism is the normal mode of reproduction in many animals. For instance, the sea hare is a marine snail without a shell. At mating time one sea hare climbs on the back of another. Both face the same direction with the one behind fertilizing the eggs of the one in front. A third may mount the second so that its eggs are being fertilized while it is supplying sperm to the sea hare in front. In this manner long chains of mating individuals may form. Sometimes the first individual will swing around and mount the last so that the circle is closed. All the mating animals are thus simultaneously fertilizing eggs, plus being fertilized themselves. Some species don't rely on locating others of their species. They are capable of using their sperm to fertilize their own eggs.

PEACEFUL ROMANCE

Fish and other sea creatures in search of mates are equipped with many methods of attracting members of the opposite sex.

Love dances—the grace, beauty, symmetry of which can bring to mind a Viennese waltz—are practiced by many species. The male dolphin performs in any of dozens of ways for the female of his choice—often contacting and caressing her body to stimulate her interest, or displaying his skill and agility as a swimmer by swimming directly at her, artfully swerving at the last possible moment to avoid collision. While engaged in this behavior, they are very vocal.

The sense of smell also comes into play during the mating ritual and serves an important function in bringing the sexes of certain species together.

The ability of certain sea animals to phosphoresce is another method used to attract mates. When seeking a lover, these fireflies of the sea flash their lights at certain frequencies or in specific patterns. The result is clear: members of the same species are informed of the desire to mate, members of other species remain in ignorance.

Many animals in the sea are capable of changing color. In the pursuit of their normal lives, this ability gives them a better chance to escape a larger animal or, conversely, a better chance to capture an unwary prey. But in mating season the males of many species blossom forth in spectacular arrays of wedding displays. Normally drab creatures take on hues of brilliant intensity, making of themselves conspicuous advertisements. This gaudy appearance attracts females ready to mate.

The Dancing Cod

"Peaceful romance" doesn't always imply passivity. The male codfish is a dancer. Patrolling his territory vigorously, making sure that other males and unripe females give it a wide berth, he waits for a female who is ready for him—a lady of his species that moves into his water space with serene aplomb. Now the male goes into his act, raising his back fins, after a moment or two closing in on the female. Perhaps ten inches away from her he erects his median fins and contorts himself. The body undulates, the fins flap: this is the courtship display. If he has picked the right partner, she begins to "listen" to his enticements: his noises, dancing, displaying.

His point: to get her to swim to the surface where cod mate. At last, after much maneuvering and "dancing," he achieves his goal. Near the surface the male mounts the female. The female now spawns, her eggs are fertilized, and she swims out of her mate's territory.

By this courting behavior, the fluttering of fins and sinuous movements of his body, the male has lured his partner away from the near-freezing waters where adult cods spend most of their lives. Her eggs would stand no chance of hatching in the cold, hostile environment of the bottom. She has spawned at the warmer surface, and her lighter-than-water eggs can hatch there in about ten days. The little fish will stay in the rich planktonic layer, feeding and being fed upon for about two and a half months. The survivors will then sink to the bottom where they will live and grow to maturity.

Messenger Service

Pheromones are chemical messengers released for many purposes. The message may dispel fear of territorial threat or, on the other hand, produce it. In some species the chemical includes information on what species the intruder is, his social status, his sex, his age or size, his reproductive state or his intentions.

The chemical form of communication is especially helpful to creatures living in the unlighted deeps of the abyss and the murky coastal areas where vision is restricted or nonexistent. And consider its importance to the bottom-attached animals who, by the nature of their design, are unable to participate in mating rituals. Without some form of communication, mating

A female octopus rejects the male after they mate, and establishes a home where she waits alone for five or six weeks. The strands of eggs she hatches are usually in clusters—each cluster containing up to 4000 embryos. She shields the eggs, oxygenates them by squirts of water, and vacuum cleans them with her suction cups. The female usually stops eating at brooding time, is soon near starvation, and dies after the birth. Finally the babies burst from their egg cases but few survive the first few hours since fish may be waiting to gobble them up. Of the octopus' 200,000 or so offspring, only one or two will reach maturity and reproduce in turn.

29

This Indian lobster, living off the Grand Cayman islands, is more colorful than its northern Atlantic relative. The color does not assist the lobster in its mating ritual, however. It is believed that pheromones help these crustaceans locate and identify mates.

and reproduction would be haphazard at best. The oyster, for example, has an amazing method of reproducing based on sensitivity to chemicals. The male oyster can release its sperm on the call of many stimuli—among them the sperm of unrelated invertebrates. The female oyster doesn't release her eggs until she detects the presence of oyster sperm in the surrounding waters. The sperm-laden water is taken in and fertilizes the eggs inside her shell.

In certain species of crabs and crayfish the pheromone released by females is the same hormone that causes molting. As molting time approaches and the hormones are released, males are attracted to the premolt female. The aroused male will seize the female, holding and defending her against competitors and predators. When she has lost her shell, mating occurs. The male will continue to protect the now pregnant and extremely vulnerable female until her new shell begins to form, helping the species in two ways: the living female is not lost, the fertilized eggs she carries are sheltered. It is the response of the male to the premolt pheromone release which ensures that he will be on hand when molting occurs.

It has been shown in experiments that some species are able to remember for months the scent of a specific intruder—some catfish react to water from the intruder's tank as though it were there, presenting a threat as it had done once before.

Because the receptors for recognizing pheromones are extremely sensitive, man's pollution of the sea poses an enormous threat to animals depending on chemical communication in their own life processes.

Man may find uses for the pheromones—in marine farming, for example. They may be used to repel predators, to attract males with artificial baits for fishing, to inhibit aggression and cannibalism, and as growth stimulators.

Love in the Abyss

Finding a mate in the abyss, where no sunlight reaches, is resolved in a number of ways. The rattail, a relative of the common cod, vibrates its swim bladder to produce grunts, booms, and other sounds. This fish has a very sensitive lateral line which alerts it to nearby movement.

Although there is no visible sunlight that penetrates below 2000 or 3000 feet, a majority of deep-sea fish have the ability to produce light directly or indirectly. Bioluminescence is often produced by bacteria living symbiotically on the host. Other fish are able to make their own light with special glandular cells called photophores. The patterns of lights vary with the species. Some of the animals are able to turn the lights off and on to flash a variety of colors: white, green, orange, and red. In some species only the males have lights, leading us to believe they are as important as color changes are to shallow-water fish.

One of the most amazing and practical forms of deep-sea mating is performed by the anglerfish. The young male follows the scent of a female anglerfish avidly, for he has a lot of competition from other males of his species, and unpaired he is incapable of making a living in the abyss.

When he has found his female, a fish that can be as much as 25 times his size, he grips her body with specialized small teeth. His position on her body seems to be of little consequence. Once attached, the male changes in remarkable ways. His lips and mouth tissues fuse with the female's tissue, and his alimentary tract degenerates. In time their bloodstreams intermingle and he loses his now useless eyesight. Two small openings remain where the mouth was to allow water to enter for respiration. The male has become a parasite, but he apparently puts no strain upon the female, for some have been known to support as many as three males.

VIOLENCE FOR SEX

The cone snail does not need to dig nests or seek out secluded places. Its eggs have a distasteful outer covering that serves as protection, and it lays them in open, convenient places.

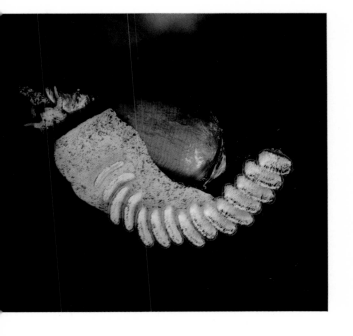

In the deceptively calm sea most species fight not only for the right to live but also for the right to mate. In the sea world, only the fittest and luckiest may reproduce.

There are several reasons why animals fight. Territory, for nest builders, is essential to success in reproduction. The choicest possible area must be found in order to attract a mate and to provide for the young. Nest-building fish, regardless of size, defend their territory with great boldness. Jawfish, for example, erect their fins, open their mouths incredibly wide (making them look larger to an intruder), and quiver their bodies. Should these methods fail to frighten the challenger into retreating, they may resort to butting, tail slapping, and to biting to ward off the assault on their territory and consequently on their right to mate.

The fierceness of the fight doesn't necessarily relate to the size of the combatants. Tiny tube worms engage in violent territorial battles—when an occupied tube becomes attractive to an interloper who tries to move in. If the rear entrance has been used to gain access, the intruder may first encounter the hind portion of the occupant, which he promptly bites with strong pincers. The defender may turn around while remaining inside the tube, thus exposing himself, undefended, to a flank attack. Or he may prefer to leave the tube and reenter, so that the two worms are face-to-face. They then may thrash violently around in combats lasting from 15 seconds to four minutes. They lock together, gripping each other with pincers, pushing mightily back and forth—until supremacy has been determined. Occasionally the fight ends with both worms occupying the same tube, facing opposite ends.

Cephalopods—with their many arms—tend to engage in ritualistic warfare for the right to mate. Before choosing mates, squids and cuttlefish ceremoniously rush at each other, making no contact. When mates have been chosen, the male will defend the female. But fights are often although not always avoided because the smaller combatant withdraws.

Battles for mates and territory may seem to jeopardize species, since males (and sometimes females) can die or be incapacitated in the course of them. But they are necessary. A harsh law in the sea is that only the most capable or fit should win the opportunity to breed.

Bloody marks on the neck of the walrus give evidence that he has just battled with an adversary over territory or a suitable mate. Like all such fighting among animals, though, the result is rarely death; the weaker invariably runs away or submits before serious injury occurs.

CHANCELINGS

Most sea animals reproduce by laying eggs. Some species conceal or expose them, deposit them in nests or in the open sea, guard them or leave them on their own to hatch. To the latter I give the name *chancelings*. Their survival depends entirely on chance. For once spawned and hatched, they have mainly luck and instinct to assist them.

Adults of chanceling species are provided with the ability to lay large numbers of eggs, by sheer number giving their species the best survival chance. The oyster, depending upon its surroundings, is able to produce from 500,000 to 500 million eggs.

Of course, we must realize that with these fantastic amounts of eggs deposited there is an equally large number of eggs which do not hatch, or which do not survive to reach adolescence and maturity.

Despite the fact that no parental protection can be provided to chancelings during incubation and after hatching, some parents do make attempts to ensure a somewhat increased rate of survival among the eggs and hatchlings by depositing them in relatively protected places. These practices, evolved over millions of years, are often ingenious. The grunion and the green sea turtle deposit their eggs in the sand, above the high-water mark, affording protection from nest robbers of the deep, but, on the other hand, exposing eggs and hatchlings to land predators like coconut crabs or frigate birds.

Nest Building

The stickleback presents us with one of the most dramatic examples of male aggression and domination of the mating ritual.

The male builds his tubular, roofed-over nest among weeds. Using a sticky, gluelike secretion from his kidneys, he binds the structure together, shaping the nest by rubbing his body against it much as a sculptor molds clay with his hands. Shortly the secretion hardens into a cementlike texture, resulting in a beautiful durable nest.

At the same time he dresses himself as handsomely as possible to attract the eyes of prospective females. The three-spined stickleback, normally blue or green with a silver belly, bursts into a flaming red below. The ten-spined stickleback turns brown; the 15-spined, blue. He now begins to approach potential brides. If when doing so other males contest him, a ferocious fight breaks out. He tries to coax a female to enter his nest and deposit two or three

eggs. He first tries gentle persuasion, including a courtship dance, but if that doesn't work, he tries to chase the female into his nest, hastening her along by nipping her tail.

After the female deposits her eggs and leaves, the male spews milt over them. Then he repeats the whole process several times with other mates until his nest is full of eggs. When this is done he guards the eggs for a month or so against intruders. Should the nest become damaged, he promptly repairs it. He aerates the eggs by swimming around the nest, fanning with his pectoral fins. When they hatch he destroys all but the foundation of the nest. This he leaves as a cradle for the fry, which he continues to guard until they are ready to care for themselves. When that time comes, the male simply swims away.

Mouthbreeders

At least two species of jawfish are known to be mouthbreeders. At breeding time, these fish scoop their fertilized eggs into their large mouths and carefully hold them there until hatching. Their mouths are not big enough to contain the egg masses when closed, so the fish keep their mouths open for the incubation. The mouthbreeding catfish and tilipia also protect their eggs in this way.

Specially Timed Spawning

A phenomenon occurs every spring and summer along the warm sandy beaches of southern California when they are invaded by seven-inch-long silvery fish coming ashore from the sea.

These smeltlike fish, called grunions, spawn only on the three or four nights following the new or full moon. Then the spring tides are receding from their highest, and the water tosses the little fish far up on the beaches, above the normal high-water mark where the sand is warmed by the sun and moistened, after having been touched by the sea. Finding herself in such a spot, the female grunion wriggles frantically on her tail, drilling a hole in the sand. When approximately two-thirds of her length is buried, with only head and pectoral fins exposed, from one to as many as ten males arch themselves around her body. The female then begins to deposit eggs in her sandy hole, the males to discharge milt that runs down her body reaching and fertilizing the eggs. When her eggs (she may lay from 1000 to 3000 every two weeks) have been safely deposited and fertilized, she leaves the hole and makes her way back to the sea. The outgoing tide deposits more sand on the beach, covering the eggs with eight to 16 inches of sand.

Approximately two weeks later, high tides wash the sand from atop the nest and stimulate the tiny grunion larvae to pop from their eggs and make their way seaward. The embryos develop sufficiently to hatch in from one week to ten days, but they wait for the high tide which will wash them out to sea. The grunion eggs are unusual in another way. Should the unhatched, mature eggs fail to be uncovered and agitated by the surf, they can remain

buried for an additional two weeks until the next high tide and hatch with no ill effects. The process repeats itself every two weeks or so for the several summer months, with the female making many trips out of the sea and up onto the beaches. She makes her run generally between nine and midnight under cover of darkness, fairly safe from hungry seabirds. By laying eggs in this manner, she offers her offspring a form of parental care.

Dolphins, whales, and humans are all viviparous—they give birth to living young. Viviparous females produce a few eggs, one or more of which is fertilized by the male while still inside the female. As the animal develops through its embryonic stages in the uterus, it is like a parasite of the female—attached by an umbilical cord through which food and oxygen travel from female to offspring and waste materials from offspring to female, the female sustaining the offspring by means of her own bloodstream.

Another experiment undertaken by nature occurs in the ovoviviparous animals. These species, including some sharks and manta rays, use a method of reproduction that may be considered midway between the egg-laying oviparous animals (such as birds) and viviparous animals like dolphins and man. The female of an ovoviviparous animal also releases an egg that is internally fertilized by the male, but it is retained in the oviduct. Lacking an actual uterus, the oviduct becomes a sort of womb. By being allowed to mature and hatch internally, eggs are protected from the hazards faced by eggs that are simply dumped in the sea.

The female seahorse plays only the briefest role in the prenatal care and delivery of the next generation. It is a strange reproductive procedure. The female produces the eggs and deposits them in a minute opening in the male's brood pouch located under his tail. As she swims away, never again to see her offspring, it becomes the male's responsibility to fertilize and incubate the eggs as well as to deliver the newborn. The incubation period lasts from eight to ten days. When the male is ready to give birth, his body begins to move back and forth, tensing at intervals, as he thrusts forward his bulging pouch. Slowly the pouch opens, and with continued rhythmic tensions and convulsive jerks he begins to eject his offspring, usually one at a time. Such "labor pains" continue for several hours after which several dozen to many hundreds of baby seahorses have been born, depending on the species. Having performed this birth function, the male seahorse leaves the babies to care for themselves, and even gulps a few. They at once begin swimming in search of suitable blades of grass around which they can twist their tails and begin growth.

INTERNAL PARENTAL PROTECTION

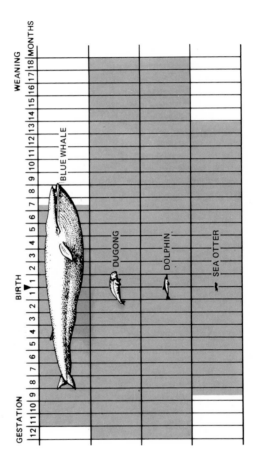

Despite enormous size differences, the gestation periods in mammals—the time it takes an offspring to mature from conception to birth—are remarkably similar. In the custody of the female's womb, the embryo is given full protection against the hazards that ravage the eggs spawned by lower forms. It takes all the nourishment it needs directly from the female's bloodstream. It need not grow at an accelerated rate, or undergo a sudden metamorphosis of form, in order to survive.

In this remarkable photograph of a dolphin birth, the tail of the newborn is just emerging. After about half of the dolphin's eleven- to twelve-month gestation period, the pregnant female isolates herself slightly from the herd. She may choose a single female to keep her company, to stand by and assist her. As the weeks go by, the pregnant female begins to flex her tail up and down, possibly to tone her muscles. Eventually the day arrives. The tail of the infant protrudes from the mother's body. Gradually more and more emerges. When the infant is fully expelled from her body, the mother whirls around, breaking the umbilical cord. She can now assist the young dolphin to reach the surface for its first breath, and join the herd.

Advanced Breeding

The mating practices of sharks and rays are among the most advanced of all the fish. The female carries eggs inside her body. The male is equipped with a pair of erectable rods called claspers. Through these claspers he releases his sperm into the female's body opening. Sometime after fertilization certain sharks and skates release the eggs directly into the sea. However, most open-sea sharks and rays are live-bearers, producing fully formed offspring at birth—pups who have experienced complete embryonic development inside the female's oviduct prior to birth and who during that period were nourished by the female and from the yolk of the large yolk sac surrounding them. As for rays, it is believed that the manta gives birth to a single pup. During the birth process (and often at other times), the manta flaps its fins much as a bird does its wings and leaps as high as five or six feet out of the sea.

PARENTAL CARE FOR THE YOUNG

We are often moved by the warmth, love, tenderness, and care usually given the human child. This devotion is unique.

The species which offer care after birth do not produce offspring in great numbers. Correlated with small fecundity is the relatively high survival rate of those few young. Parental care is extremely useful in improving the odds of the infant's survival during the early stages of life. The parent, or parents, protect the young from danger. In many cases the parent seeks out food for the young, sometimes partially digesting it. In other cases the offspring are led to the food, eliminating the perilous groping of the newly hatched young. Sometimes the female herself is the source of food. In mammalian species the offspring live on milk produced by the female until their digestive systems or teeth have developed. Mammalian offspring, therefore, are given the opportunity to develop other skills before having to learn to forage for food.

The young who are cared for are given opportunities to observe the parent and to learn efficient methods of locating food and self-defense. This capacity for teaching and learning is what probably sets animals that care for their offspring apart from those that do not. The former are able to make use of what has gone before them and to make refinements in life practices.

Harbor seals give birth to their single pups in late spring or early summer. The pups are born ashore, close to the high-tide mark, and can swim almost immediately and dive for 20 minutes without surfacing. They suckle underwater for about three weeks, and by then are strong enough to swim well and seek their own food, especially favoring small crustaceans.

QUEST FOR FOOD

CHAPTER
2

TO EAT,
TO LIVE

All kinds of prejudices encrust our thinking about the sea. Of no area, perhaps, is this truer than the quest for food that is so significant an activity in the animal world.

How many times, for example, have we read in books, in magazine articles, or heard in the conversations of swimmers and sailors that the shark is a "killer"? Quite a few highly successful films have been distributed that show closeups of large sharks in "feeding frenzies"—thudding into carcasses with robotlike persistence, fearsome teeth and jaws scooping out great hunks of meat as easily as I scoop up a spoonful of soft butter, the roiling sea filthy with blood and carrion. It is certainly an appalling sight—quite enough to send chills down the spine of anyone.

A man's carnivorous meal also includes an ugly scene that is simply concealed and happens in the slaughterhouse or on farms where pigs or calves are bled to death. But a shark has no way to pretend innocence. A primitive animal in many respects, he has evolved into an epitome of muscle power and streamlined design. It is by exploiting these advantages, plus his capacious mouth with its row upon row of razor-sharp teeth, that he makes his living today as he has been making it for millions of years. He has absolutely no choice in the matter.

Still, it is not as easy as it might seem to rid ourselves of the kind of projection that turns this limited fish into a "killer," because our mental responses to the outside world fall under the influence of thousands of silent cultural monitors every second of the waking day. Most of us think of the sea otter as an "adorable" creature. Like the koala and the panda, its facial features are arranged in a pattern slightly reminiscent of a child's and our hearts spontaneously go out to it. On the other hand, even as notable and experienced a naturalist as William Beebe has testified to an involuntary spasm of revulsion on seeing a large octopus slithering its way across a shallow tidepool. And the repellency with which the majority of mankind regards snakes may be something very ancient and deep in our genetic makeups—perhaps psychosexual in origin.

The senses too become involved in all this. There are occasions, like the height of the season at a seal rookery, when the life of the ocean produces a stench all but intolerable to our nostrils. At other times, facing into a bracing breeze on a calm morning in temperate waters, a man blesses the day he was born and the fate that made him a seaman. There is no way of avoiding such subjective impressions. But we must control them as much as we can, lest they disastrously interfere with our perception and understanding of the Ocean World. The sea is a huge and complex ecological machine in which every part interplays with every other, each death is the introduction to new life. To understand how it all works one must put aside everything in the way of sentimentality or personal bias. Then one becomes able to enjoy the beauty of any natural act.

What distinguishes a living creature from inanimate matter is the ability to appropriate materials from the environment and incorporate them into its tissues according to its own blueprint or, through metabolism, to break out from them the energies it needs for existence. The life-style of an organism largely determines what sort of food it requires. A one- or two-celled animal in the plankton layers has next to no energy expenditure, so it can subsist on a thin diet of microscopic plants. A larger animal will have a larger energy-budget; it may need its fuel intake in highly concentrated form and may be, as the shark is, a carnivore. The animals of the sea eat to survive. They don't kill for sport, they don't torment their prey before killing it, they don't exterminate whole species or render uninhabitable vast provinces of their living space. The shark is a most efficient carnivore, true. But he is no more a "killer" in the criminal sense of the word than the sedentary coral polyp that rapaciously ingests anything it can get its tentacles on, or the meathandler in a Chicago slaughterhouse—or, for that matter, the housewife who serves bacon at the family's breakfast table.

Parrotfish nibbling coral

FAMINE AND FEAST

Some years ago, while *Calypso* was making her way through the Indian Ocean, we encountered a large number of sperm whales traveling in small groups with young. We had a terrible accident. A young whale ran into one of our propellers and was hopelessly cut up, bleeding profusely. To cut short its suffering, we killed it, then secured it to the boat.

In warm seas, whales bleed profusely. The water about *Calypso* turned red. The spectacle began. Sharks appeared, more and more of them, attracted by the blood. For a time they nosed the baby whale, circling confidently, nudging and smelling. When the frenzy began it was sudden and swift. One shark lunged in and sliced away an enormous mouthful. The others joined. The animals would race along the length of the whale's body, taking chunk after chunk in quick succession like corn from a cob. There were more than thirty of these sharks. From their behavior, it appeared they had gone several days without a meal. But soon, as rapidly as it had begun, it stopped. The sharks had gorged themselves and could not finish all the food.

In less dramatic ways this pattern repeats itself everywhere in the sea. There is great deprivation, sudden abundance. In the lives of most marine creatures, feeding and the hunt for food occupy more time than any other activities. The competition for food is so fierce among most species that many animals must endure long periods of famine. When they are fortunate enough to find food, they feast.

The animal world depends on plants for its food, and the great abundance of microscopic plants in the sea provides the basis of a complex web of marine populations that support one another. Using the energy from the sun, plants absorb minerals to make basic foods. Grazing the plants, herbivores transform these simple materials into the proteins of their own bodies, comprising the food source for carnivores higher in the chain.

Generally food is scarce in winter because of fewer phytoplankton at the source of the food chain. In the summer months, when more solar energy is available and when more nutrients are brought to the surface due to shifts in currents, a greater volume of phytoplankton grows, providing a greater volume of food. This not only causes the individuals of a species to grow to greater size, but it also increases the supply of living organisms. Populations of the large predators high in the food web may also increase as a result of more phytoplankton supporting a greater supply of food.

When we begin to appreciate the apocalyptic scale of consumption that takes place in the sea, vast populations feeding one another, each creature a threat to another creature and/or threatened in its turn, we may picture the

The one-celled amoeba feeds on plants and animals smaller than itself: bacteria, yeasts, algae, other single-celled animals. Although one of the simplest forms of life, the amoeba has the power of distinguishing between grains of sand or other indigestibles—and food fit to eat. It does this by sensing the change of chemistry the food causes in the surrounding water. The amoeba eats in different ways. Should the prey be a tiny mote of green algae, the amorphous animal extends itself around, envelops it completely, and ingests it. If it seeks to take a smaller protozoan than itself, it will send out long flowing pseudopods. Once the prey is trapped, the amoeba draws it into its food vacuole.

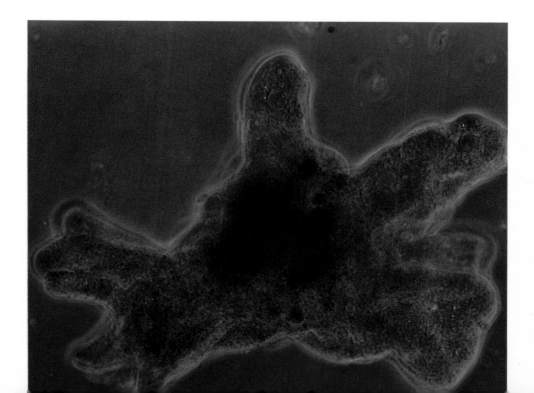

water environment as a kind of hellish bedlam of torn flesh and starving predators. It is not so. Except for certain hours when the sea boils with feeding activity, it is for the greater portion of each day calm and composed. Feeding is going on, but it is the quiet filter feeding of such animals as oysters, barnacles, feather-duster worms, and coral polyps, which, fixed in place, wait tirelessly for food to drift by.

As we have mentioned, when sunlight wanes in winter, the growth of phytoplankton slows. The resultant scarcity of this basic food source is felt throughout the marine chain. Many members of the sea's population move on to more fertile areas when the seasons change, others drastically alter their feeding habits. Some animals stop eating entirely for many months.

Dinoflagellates, tiny plant-animals in the plankton, take maximum advantage of the food supply by using remarkable versatility. For energy and growth, the same unicellular creature may carry on photosynthesis, like a plant, or utilize organic material, as an animal does, according to whichever is best suited to the food at hand.

Food Scarcities

An animal is a complex creature. Aside from its capacity to grow, it feeds, digests, respires, excretes, forms eggs or sperm, repairs its damaged tissue, and in most cases orchestrates a set of muscles in order to move in various ways at various speeds. All these functions most animals perform every day. They must do so or perish.

To keep the intricate array of its body machinery in constant running order, an animal uses a variable amount of the energy it derives from food. Some animals have more extravagant needs than others. The more simple its system the less energy a creature needs to keep itself alive. Seeds and spores, many of them unicellular structures, can remain dormant for years and still spring to life at the end of that time. Their energy demands are minimal.

The most expensive demand an animal can make upon its energy supply is a constant body temperature. All mammals have this costly feature, which is, however, a great asset. The animal is capable of sustained efforts and endurance several times greater than cold-blooded creatures of the same size. Marlins and sharks can beat a dolphin in a 100-yard race, but cannot compete at cruising speed.

The larger the mammal, the smaller its surface area is in proportion to its body mass. This means that large mammals, such as the polar bear, require less energy to heat their bodies than smaller mammals like the sea otter. This animal eats during most of its waking hours, consuming up to 20 pounds of abalone, sea urchins, and crabs a day.

Cold-blooded animals are another story. Fish and reptiles that live in the sea don't have a constant body temperature to maintain and most are capable of going without food for days, weeks, or even longer periods. Their body temperature varies with the water they swim and live in. Occasionally, it's substantially higher. Then a tremendous amount of food-supplied energy is needed.

In mammals, a complex hormonal mechanism stops the growing process when an individual reaches the adult state; the same does not necessarily hold with fish. Here growth is a continuing process throughout life. Nutrition plays a major part. Without proper amounts of the appropriate foods, fish will temporarily stop growing. But, unlike mammals, who suffer permanent and irreversible damage from malnutrition, fish may resume growth and overcome the effects of insufficient food. The sea anemone actually shrinks when food is not available and grows when there is renewed food intake.

Cold-water animals are usually larger than warm-water ones of the

FOOD IS NECESSARY FOR MAINTENANCE

same species. One explanation given by scientists is that in cold water animals need more time to reach sexual maturity and therefore have more time in which to grow. Oysters grown in the Cape Cod area of the Atlantic coast, for example, do not spawn until they are two years old. Those in the Gulf of Mexico may spawn at the end of their first summer—when they're three to four months old.

The starfish is an active feeder (in a slow sort of way) that preys on the bottom. It must extrude its stomach and digest the food outside its body.

To exist, most animals need three things: water, oxygen, and food. Once consumed, the three work together to maintain the body. But they must first be converted into chemicals the body can utilize.

Oxygen is converted as it acts on other materials—through oxidation—to produce energy. Water helps by carrying off wastes and by combining with the other chemicals. Food is the fuel the others help to convert. It breaks down into vitamins, minerals, fats, carbohydrates, and proteins. The oxidation of these chemical constituents is the metabolic process. Some metabolites build tissue for growth, some for replacement of old tissue or for healing sick or injured tissue in the animal body. The efficiency with which animals utilize food is a variable factor. Herbivores, for example, though they have longer digestive tracts, are less efficient than carnivores and even the meat eaters aren't all that efficient.

The means of consumption, what is being consumed, how it is consumed, and even the health of an individual animal are all factors in how effectively food is converted. Naturally, the nature and quality of the food is an important factor too. Some animals digest food outside their bodies and take in only what they can use. The starfish does this in consuming shellfish. Some take in not only food but extraneous matter, digesting the food and rejecting the rest. The goosefish may inhale rocks and sand along with an entire lobster.

CONSUMPTION AND CONVERSION

External Digestion

Not having teeth, starfish have evolved a unique method to digest food. Their diet consists mainly of shellfish and bottom detritus and debris. When a meal, perhaps an oyster, is found, the starfish engulfs it, attaches its many suction-cup feet to the sides of the shell, and secretes its digestive enzymes. Even though the shells are tightly shut, the mollusc will eventually feel the effects of the chemicals. As the muscles of the bivalve relax, the starfish opens the shells and everts its stomach to digest the meat inside. The enzymes reduce the meat to a fluid which can then be absorbed by the walls of the stomach. After digestion, the stomach is withdrawn. One species of starfish was observed eating 50 small clams in six days.

LOCATING A MEAL

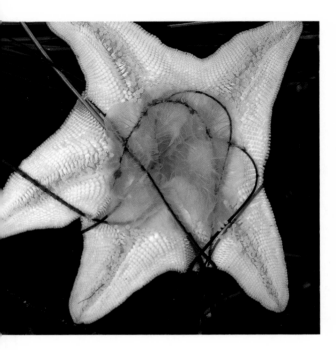

The difficulties involved in locating a meal in the ocean are partly determined by the differences between water and air. Because of the rapidly increasing pressure (14.7 pounds per square inch for every 33 feet) and the rapid falloff in sunlight (there is little left below 1000 feet), most sea animals cannot range freely in the vertical dimension. Still, there are animals specially adapted for "breaking the pressure barriers." The sperm whale can dive down about a mile in search of the giant squid, which for its part probably never voluntarily comes anywhere near the surface (and a living specimen of which has so far as we know never been caught or even seen by man). Other marine mammals—like the sea lion, seal, and dolphin—can dive perhaps a quarter as deep.

A significant problem for active feeders is the poor visibility underwater. In navigating through their three-dimensional water space in search of their meals, fish employ what we call their "sixth sense": the lateral line organ—a longitudinal system of canals and openings which starts over the head of most fish and runs under the skin along the body to the tail. This system almost certainly has two jobs: that of perceiving distant pressure disturbances as well as changes of flow patterns close to the fish. This additional sense helps a predator locate a meal, but at the same time warns the prey of the predator's approach.

Marine mammals have developed "echolocation"—a sonar apparatus like the bat's, capable in the dolphin and other species of elaborate sophistication. Animals can also taste and smell the water. Sharks, for instance, can home in on ounces of blood diffused in the sea.

Dolphin Sonar

The dolphin is an acoustic animal. On the "clearest day" in the sea the eye cannot reach much farther than 100 feet. To compensate, the internal and external ears of the dolphin are greatly modified in structure in order to receive and interpret a wide range of water vibrations. All this not only vastly improves the animal's hearing underwater but refines it for the purposes of echolocation, that sonarlike process with which the dolphin homes in on objects as distant as a mile away that he wants to inspect—perhaps to eat.

In practicing echolocation, the dolphin emits a fascinating spectrum of noises including clicks that last from a fraction of a millisecond to as long as 25 milliseconds: low-frequency "orientation" clicks to give the animal a general idea of his situation; high-frequency "discrimination" clicks to give him a precise picture of a particular object he is interested in.

The dolphin's larynx is an intricate structure, without vocal chords, but provided with a sphincter enabling the animal to make his sounds at the same time he is feeding underwater. Many experiments have shown us how finely tuned the apparatus is. At ten yards or more a dolphin can discriminate between two objects only a few centimeters in diameter.

Eyes

Among many fish and other marine creatures sight takes over where other senses leave off. This holds true for other animals of the sea beside fish. The octopus and squid, both molluscs, have highly complex eyes that compare favorably with our own in visual acuity.

Testing for Taste

Most of the nearly 4000 known species of catfish have special sensory barbels, or "whiskers," extending from their lower jaws.

The striped sea catfish, shown here, school densely and approach the

Striped sea catfish have barbels, or sensory "whiskers," under the mouth that give them the ability to taste food before eating it.

The eyes of the hammerhead shark are not unusual, but their placement is. They stand at the ends of the shark's broadened head, perhaps to give this predator the widest possible view of the sea around. The placement may also serve to separate the nostrils, heightening the ability to find the source of an odor.

ocean floor to grab around for food. The barbels serve a dual function. They locate food in the mud by touch as well as taste, to see if it is palatable. Through the use of these barbels, catfish feeding is not dependent upon sight, which is affected by water clarity.

A number of other families of fish have barbels serving the same function as the catfish's. Theirs too are movable extensions of the skin, rich in nerve endings for tactile and taste senses. The sturgeon cruises the ocean floor or river bottoms—keeping contact with four short barbels located inches ahead of the suctorial mouth on his ventral side. When the sturgeon tastes something good, he extends his protrusible mouth and sucks up the delicacy.

Another application of sensory organs similar to barbels is found in the mud hakes of the North Atlantic. These relatives of the cod have a single barbel; but their pelvic fins have also been modified into long, threadlike devices. The hake carries these highly specialized fins off to the sides. They have evolved into feel-and-taste organs.

The Nose Knows

Sharks can zero in on a wounded fish from a quarter mile away or more. Night-feeding fish can find food without light. Eyeless cavefish can too. How? Through a particularly acute sense of smell.

The shark swims in circles or random patterns, picking up odor gradients and, poor eyesight or not, pinpointing the source of an attractive smell. Most sharks have nostrils on the underside of the snout. Other fish have nostrils on the top or sides of their heads. In most, the nostrils are blind sacs not used for respiration but lined with odor-sensitive membranes.

Once, aboard *Calypso,* we tested the United States Navy's shark shield—the inflatable, plastic Johnson shark screen. Although the thin plastic was no protection from the powerful thrust and slashing teeth of a shark, it prevented man's scent from entering the water. The sharks attracted to the area passed the plastic bag with little more than cursory examination. But many times we have seen sharks give more than a passing glance at our unprotected divers in the water. Other experiments have shown catfish, sharks, and moray eels unable to locate food when their nostrils are plugged.

Limbs and Bristles

The spidery-looking banded shrimp uses its head in several ways to capture food. Besides four or five pairs of legs originating on its body, the shrimp has two or three pairs of appendages emanating from its head. These are antennae, antennules, and mandibles. They serve many functions including locomotion, feeling, setting up water currents, chewing, and seizing. The body legs also help crustaceans catch food. Typically a shrimp or other crustacean might first feel its planktonic prey, then move in more closely. The appendages provide information about taste. When close enough, it can set up water currents that will bring the tiny animals to its mandibles, with which it chews the food. In some lobsters and crabs, bristles on the body limbs also have a sense of feeling.

Busy Fingers

The big-headed, bottom-dwelling sea robin appears to walk along the ocean floor with "fingers" feeling for food everywhere it goes. These "fingers" are really the first two or three rays of the fish's pectoral fins. The rays are separate from each other and from the rest of the pectorals. They are busy almost constantly, sifting the sand or hard-packed mud, turning over bits of gravel or small stones, seeking marine worms, tiny crustaceans, and some of the smaller molluscs. While the fin rays drum along the bottom, the other flat, broad pectoral fin glides just off the bottom, sculling slightly to keep the fish moving over new ground in the food hunt.

Several species of the colorful sea robin inhabit the shallows and moderate depths of the Atlantic coast from New England to Venezuela.

SEDENTARY FEEDERS

Many animals in the sea fasten themselves in one place and spend the whole of their lives there. They have found evolutionary success by limiting movement and conserving energy. Some affix themselves to man-made structures like ships or piers, to the bottom, to shells of other animals, or even to swimmers, like whales. Some of them are able to propel themselves but do so rarely.

Some sedentary feeders move the water rather than move themselves through the water. Fixed animals are dependent upon their environment to

deliver food to them. They have developed remarkable methods of collecting it.

Filter-feeding worms like the beautiful feather-duster build stiff tubes around their bodies into which they retreat at the slightest disturbance. To feed, they extend feathery arms from the tube and strain out any food particles the water carries to them. In similar fashion barnacles feed, filtering the water with modified legs to catch whatever comes by. Some barnacles use the motive power of others by attaching themselves to animals from crabs to whales that will carry them about the oceans.

A few marine creatures have adapted to living in or eating the wood of ships and pilings. Some of these have an enzyme to break down cellulose and make wood into a digestible food.

The propulsive methods of sea urchins, starfish, snails, and nudibranchs do not give them much chance of capturing healthy, fast-moving fish. Instead, they rely mostly on fixed or floating food for sustenance: kelp, coral polyps, shellfish, detritus, plankton, and members of their own species.

Many animals and fish dig holes in the muddy bottom of the sea floor or take over a rocky crevice. Some just remain there waiting for something edible to approach, others occasionally venture outside.

Garden eels stretching from their holes look like cobras rising from a snake charmer's basket. The territorial eel digs a tunnel several feet deep 50 to 200 feet below the surface, claiming all the adjacent area. Each tunnel is about six feet distant from those around it. Garden eels live in colonies of 200 to 300 individuals. They are extremely cautious and withdraw into the safety of their homes at the first indication of movement around them.

Groupers wait in lairs for their meals. When a mullet, grunt, or crustacean comes within range, the grouper snaps his mouth open. The vacuum created by the open mouth literally sucks in the victim. If the prey is too large to fit the mouth, its tail will stick out and be held between small teeth and fleshy lips until the grouper's pharyngeal teeth have dealt with the victim's head.

Since grouperlike fish can grow up to 12 feet in length and inhale their meals, it is interesting to speculate whether a grouper may have been the

There are over a dozen varieties of mussels, most of which live near the shore. They attach themselves to pilings and rocks by means of a strong, threadlike material called byssus.

Sitting and Waiting

There were sponges on earth a half billion years ago. Sponges feed simply by sitting and eating. They remain always in one place, moving hairlike cilia imperceptibly to bring food-laden water into their bodies, then expelling it when they have filtered out the plankton and other nutrients. All varieties of sponges have collar cells with cilia.

These hydroids, looking dainty and feathery, are armed with stinging cells. They are facing the current to capture whatever floating debris or plankton touches their tentacles.

source of the biblical Jonah and the Whale story. What could have happened is this: Jonah fell or was thrown overboard, was inhaled by a grouper, temporarily lost consciousness, and was immediately exhaled by the fish, luckily close to land. Here is an experience no protagonist would cease talking about for the remainder of his life—and perhaps it is this wild ride that the world has never forgotten.

Stingers

The tentacles of many coelenterates are lined with small stinging cells called nematocysts. When an animal armed with these cells is stimulated by another living organism, the slightest brushing of its trigger causes tiny threads to extrude.

These threads are designed to act in one of three general ways. Some wrap themselves around the triggering object and thus entangle small prey. Others have a sticky coating that adheres to anything it touches. The coating holds the prey until it can be eaten. Another kind of thread has a hollow barbed tip at its end which works like a harpoon.

Generally the small darts can only penetrate the skin of tiny animals. Those of some jellyfish, for example the Portuguese man-of-war, can penetrate the skin of man—with severe and sometimes fatal results.

Another animal that uses poison to obtain a meal is the cone snail. Most snails scrape algae off rocks by means of a tooth-studded ribbonlike apparatus called a radula. The highly specialized radula of the cone snail is modified to form hollow teeth that are connected to a gland containing a very toxic poison. The snail strikes its victim with the radula, injecting the poison.

Deceivers

Many mechanisms assist both in defense and in food gathering. Camouflage is one of them. Camouflaged fish are able to stalk or wait for their prey.

Flatfish have both eyes on the same side of their head. They reside on the bottom and their coloring makes them difficult to distinguish from their "backgrounds." To add to the disguise, they cover their bodies with a light sprinkling of sand. All that is visible are their protruding eyes—looking for crustaceans, bottom-dwelling worms, and other fish. When something edible approaches, the fish springs to life and consumes the prey. If nothing passes by, it swims along for a short distance and settles again on the bottom and simply waits.

Many other fish make use of camouflage. Trumpetfish have elongated bodies that make them difficult to see as they hide among gorgonian corals. They stand on their heads and hang motionless in their hiding places until

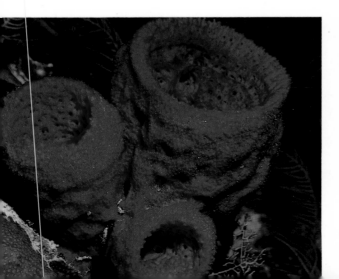

something nears. One blenny that lives around tropical reefs resembles a cleaner wrasse in the shape of its body as well as its coloration. It even imitates the dance the wrasse performs to attract customers to its cleaning station. When a customer (not expecting to be the victim of a fraud) approaches for a cleaning, the imposter gives him a series of bites.

Borers

The bane of wooden-ship admirals like Lord Nelson was a weird mollusc, the *Teredo navalis,* or shipworm, which, if not exactly an inactive animal, certainly may be considered a sedentary feeder.

The tiny larva of this animal swims through the sea until it finds a piece of wood lying under the surface. Here it settles down and develops a small shell—but a shell that is no more than a pair of one-half-inch plates, which it uses as augers even though the mature clam may grow to as long as two feet. With these miniature drills the young shipworm bores into the wood and is able to excavate a long cylindrical hole.

The shipworm can occur in great numbers, so crowded together that only filmlike partitions separate one burrow from the next. When this happens, the infested timber can bear no weight and crumbles. Curiously, the shipworm never enlarges the original opening to his hole, so in effect he has imprisoned himself in his own burrow—maintaining connection with the outside ocean by means of two minuscule siphons. It is through these siphons that his food comes. He gets no nourishment from the wood he is reducing to dust.

The peacock flounder, like all flatfish, is remarkable for its camouflage ability.

Nudibranchs, looking like beautiful, exotic flowers, live off prey few other animals can touch. The stinging nematocysts of the hydroid, for example, fail to bother the nudibranch. This shell-less snail, a slow feeder, moves about in hydroid colonies, leisurely consuming its victims.

ACTIVE FEEDERS

The relatively tranquil life-styles of the animals we have just considered don't suit all the sea's population. Many animals pursue their food vigorously—swimming long distances, following seasons and population fluctuations, resembling supermarket shoppers in picking up items here or there from the coral or the bottom. To these animals eating is serious business and they work hard at it.

The feeding habits of marine animals are reflected in their food-capturing mechanisms. Spikelike teeth can grab and hold a wriggling prey. The broad flat molars of manatees are suited to grind up plants. Sharks that take big chunks of large animals are equipped with sharp triangular teeth with serrated edges. The points sink into tough skins and the sides cut the flesh. Fish that scrape algae off rocks and coral have flat teeth with cutting edges similar to our own front teeth. Some animals don't have any teeth at all; they include the largest mammals on earth, the whales.

Most predators catch their prey tailfirst; then with subsequent bites and rejections the prey is manipulated into a headfirst position, where all the fin and body spines point harmlessly backward. It can then be swallowed.

Reef Feeders

Many fish that live on and around coral reefs differ from their open-ocean counterparts. Their bodies appear too deep for their length, and quite flattened, giving them a disclike appearance. However, this design is suited for maneuverability in the restricted environment they inhabit. Fish that feed on small plants and animals found in narrow cracks and small holes in the

51

coral have elongated snouts. The teeth of the reef dwellers are specialized, too, varying according to the fish's diet. Some use teeth to pluck animals from the reef, while others have piercing teeth which grasp and hold prey. Surgeonfish are herbivorous reef inhabitants. They usually feed on algae growing on rocks and coral. In most species the teeth have become sharp and flat-faced, which makes them well-suited for scraping up tiny plant life.

Half of the earth's surface is covered by more than 13,000 feet of water. It is the abyss—vast plains interrupted by volcanoes, rugged mountain ranges, and great scarring fractures. No man has walked here as he has on the moon. Only a few in bathyscaphes have visited it. Our knowledge and understanding of life in this hostile world is derived almost entirely from a limited number of automatic photographs or samples taken almost at random in trawl nets.

Most of the abyssal inhabitants are believed to take their nourishment from or on the mud or by filtering it from the water. The bottom is covered with plow marks of burrowing sea cucumbers and worms. Many deep-sea fish have large heads, tapering tails, and long anal fins. The body form gives the fish a downward thrust as it swims forward and the underslung mouth sifts through the bottom ooze. Even the mighty sperm whale sometimes feeds in this manner.

Many unusual anatomical systems have developed in deep-sea fish for food gathering in this impoverished zone. To many deep-sea fish, lights are important in the eternal darkness. Deep-sea anglerfish have luminescent organs located in or above their mouths to lure unwary prey. Many deep dwellers are known to possess sharp eyesight. Barbels under the mouths help fish to locate a meal in the mud. Some abyssal creatures, like the squid, rise toward the surface at night to feed in the rich upper waters.

Opportunities for predators occur infrequently. One group of deep-sea fish is equipped to handle almost anything that comes along. They are called the stomiatoids, scavengers of the deep, capable of engulfing dead creatures several times bigger than themselves. Most of these fish have barbels; one of them trails a chin whisker ten times its own length. The barbel in this fish is not used to detect food in mud, for these are not bottom feeders, so it would seem to be a sensory organ.

Feeding in the Abyss

The torpedo ray is among several species of flat cartilaginous fish with electric organs that generate significant amounts of electricity.

Stunning their Prey

The torpedo ray is a shocking animal. It has electric organs near its head in each wing capable of producing a charge of up to 200 volts, enough to knock down a man. When aroused, the animal sets off a series of electrical discharges of descending power. Once spent, the battery takes several days to recharge. Having electric organs helps the torpedo ray find and capture prey. Some members of this family are blind, all are sluggish swimmers. Other rays feed on fixed or fairly immobile animals. But the electric ray is known to eat active fish. It is possible that the ray emits weak voltage pulses that act as a form of detector, advising the animal that prey are near. When a codfish, for instance, wanders into its electrical field, the ray jumps from the bottom, enfolds the fish in his wings, and stuns it. Then he can eat at leisure.

Upheaval and Vacuuming

The gills of the bat ray are located on the underside of the flattened body. As the fish swims over the sea floor, a stream of water spurts from its gills, stirring up sediments and uncovering the animals hiding there. When the ray discovers an abundant food area, it stops its forward motion and settles gently to the bottom. The greatly enlarged pectoral fins begin to flap, whipping up a current. A cloud of loose sand and mud lifts from the bottom and in the cleared area large numbers of bottom dwellers are exposed. The ray eats them. Occasionally a shellfish, firmly attached to the bottom, refuses to yield to the hungry ray. To pry the animal loose, the bat ray presses its supple body over the creature, forming a vacuum between itself and the bottom. The thrust of the vacuum breaks the hold of the prey, which then is sucked into the ray's mouth.

Filter Feeder

In a series of languid loops—a three-dimensional ballet—the manta ray feeds. This large graceful animal, cousin of the sharks, is a filter feeder that passes many gallons of water through its huge mouth each minute. Two

feeding fins on either side of the mouth are directed forward and fan fish and plankton into it. In the throat a screen of small, spiny protuberances holds the food until it can be swallowed and keeps it from clogging up the animal's gills. In action the manta's feeding fins look like horns and have caused it to be given the name "devilfish" and a reputation it doesn't deserve. When we began diving, fishermen warned us that mantas killed divers by wrapping their wings around them and smothering them. This is not true. But their great size (one with a 25-foot wingspan has been measured) attests to the efficiency of their feeding methods.

Flying like a bird in its submarine world is the manta ray, a huge relative of the shark.

Lacking sharp cutting teeth to open stubborn shellfish, the sea otter would appear to be unsuited to a hard-shell diet. However, otters are almost unique in having learned how to use rocks, much as we use tools, to assist them in their efforts to get at the succulent meat of hard-shelled animals. When an otter dives to the bottom to collect an armful of mussels, crabs, snails, and sea urchins, he also picks up a large rock. Once safely back on the surface he rolls onto his back. Looking like an old man reclining on a beach float, he begins to dine. The rock is settled on his chest and used much as an anvil is. He places a shellfish in his hand and smashes it against the rock again and again until it breaks open. Then he pries the shells apart and picks out the meat, tossing the empties aside. Every 30 seconds or so he clutches the rock and uneaten food close to himself and rolls over, fastidiously washing any food debris from his gleaming coat. Sometimes the shellfish are firmly attached to the bottom, and the otter can't pick up his meal. Then the otter uses his rock as a hammer and pounds the shellfish until its viselike grip on the substrate breaks.

Utensils

Many whales, including the largest, have no teeth with which to chew their food. They are equipped with baleen, a feeding mechanism suited for ingestion of huge quantities of tiny fish and crustaceans, especially krill.

When filter-feeding fish extract plankton, they allow water to enter their mouths and exit through their gills. The food is strained out by gill rakers. Whales lack gills and gill rakers. Water entering their mouths must also exit from them. So the teeth of certain whales have been replaced by efficient strainers called baleen.

The baleen apparatus varies between species, but the general principle is common to all. There may be as few as 250 baleen plates or as many as 400 hanging from each side of the upper jaw. (The baleen is made of keratin, the same material our hair and fingernails are made of.) The plates are roughly triangular in shape, wider at the top than at the tip, and about one-fifth of an inch thick, separated by a space of one-half inch or less. The edge in contact with the inner surface of the lips is smooth; the edge facing the inside of the mouth is frayed into strands of coarse hairlike material. These hairs become entangled with one another, forming a thick, fibrous mat. This mat is the real strainer. The length of the baleen plates varies from about eight inches to about 14 feet.

The fast, sleek finback and blue whales have a throat that can be blown out when the mouth opens to huge proportions, taking in up to 15 tons of seawater. Any sizable fish or squid can be trapped as well as krill.

Among the most popular and useful animals in the sea are the little creatures that specialize in grooming others. The ''others'' are larger animals that present themselves to be cleaned of parasites and diseased flesh. Relationships of this kind are known as symbiotic.

There are three main categories of symbiosis: mutualism, commensalism, and parasitism. In mutualism, both of the participants benefit. Commensalism describes an association in which one partner gains an advantage and does not harm the host. A parasite lives at the expense of its host, in extreme cases killing it.

Symbiosis can either enhance or diminish the reproductive capabilities of marine animals. The female squid attracts a mate by flashing borrowed luminescence. She has a special nidamental gland in which she grows and nurtures the luminescent bacteria that help her get her man. On the other hand, the root-headed barnacle, *Sacculina*, destroys its host's reproductive ability. This animal invades crabs to take nourishment from their blood and incidentally alters their sex hormones so that males at their next molt assume female form.

The delicate green tracery seen on reef-building corals is a sign of the continuing relationship between the coral polyps and green plants which might have begun some 450 million years ago. Today corals and green algae form the basis of the most complex marine ecosystem. Corals are animals that feed on plankton and extract calcium from the water to build their rocklike skeletons. They do not engage in photosynthesis; still, they can exist only in sunlit waters. They are efficient at building their skeletons because of a symbiotic arrangement they have with unicellular green algae

FEEDING RELATIONSHIPS

The Algae–Coral Relationships

The moray eel remains perfectly still while the cleaner shrimp picks minute parasites from it. The eel gets cleaned, the shrimp get fed, and they live harmoniously while services are being rendered.

living within them. There is an exchange between these two organisms: the plant photosynthesizes compounds that nourish the coral and in the process gives off oxygen; in turn, the animal's waste products furnish elements needed by the plant for growth.

Cleaning Stations

Today we recognize more than two dozen species of cleanerfish, six species of cleaner shrimp, one species of cleaner crab. They turn up in both tropical and temperate waters, but those in the colder waters tend to be drab in color and fewer in number. They have not specialized to the degree of their tropics-dwelling counterparts. When no customers are around, tropical cleaners signal their occupation with elaborate dances, or they may rush back and forth across the area of their "stations" inviting customers to appear. They are characteristically bold. When they detect one, they take the proper position at the station so the customer won't be kept waiting. The cleaners take great liberties with their customers' bodies, enthusiastically performing their duties by reaching deep into mouths and even probing among delicate gills.

It appears that the cleaning service is a vital necessity for marine animals. When those capable of carrying out the chore were removed from an isolated reef all but the most territorial fish left the area and those remaining were in poor condition.

Commensalism

Small, vertically striped pilotfish enjoy close associations with sharks and other giant-sized animals, often swimming in large numbers around them, frequently directly in front of their snouts.

Sharks have poor eyesight, and legend has long held that the little fish guide their companions to prey. The story is unfounded. Actually, pilotfish station themselves in front of sharks and mantas to take advantage of pressure waves formed as the hosts move their large masses through the water. On them the little fish hitch a free ride by surfing on underwater waves, as humans and seals do on surface waves. Traveling this way, they conserve energy, not having to exert themselves to keep pace with their powerful companions. Pilotfish also conserve energy by not having to seek and kill for their meals. Sharks are messy eaters and many scraps of meat drift free when they tear great chunks of flesh from their prey. The little fish are always quick to scurry around for this free meal. As far as we know sharks receive no benefits from the pilotfish.

Streamlined remoras are a bit more forward in their associations. They actually attach themselves to larger animals with a large oval suction disc, a modified dorsal fin. Having control of the suction, they can release themselves whenever they desire. When a shark eats, the remoras drop off and gobble up food scraps from the host's meal. Some species (there are ten, ranging in size from several inches to three feet) may supplement these meals by nibbling on parasitic crustaceans attached to their hosts.

Predators that Follow Predators that Follow Predators

Tuna, one of the fastest and hungriest of open-sea fish, tour the world in unwearying pursuit of potential dinners. Many of the great tuna schools are accompanied by herds of dolphins—equally fast, equally strong, but far more intelligent—who are looking out for their own diet. In turn, many sharks follow the dolphins at a safe distance, waiting for a slow or ill one to fall behind the group. Man also hunts the tuna and takes hundreds of thousands of them a year in his huge nets.

Unfortunately, intelligent as dolphins are known to be, for some reason they have not learned to make themselves scarce when a fishing fleet bears down on a school of tuna. The terrible result is that about 200,000 dolphins accidentally perish each year when they are caught up in tuna nets.

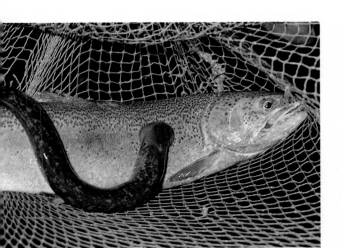

Lamprey eel clinging to its victim.

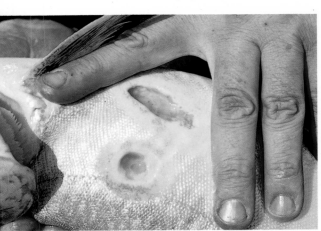

The damage inflicted.

A group of fast-moving yellowtail tunas.

WINDOW
IN
THE
SEA

CHAPTER
3

THE LESSONS OF LIGHT

Early before sunrise on a crisp winter morning, when the horizon reddens above a dark purple Mediterranean, I can see from my home in Monte Carlo the clearly delineated mountains of Corsica, more than 100 miles away. And at night I can see stars that are thousands of light-years away from me.

But when I dive in the clearest of sea waters the greatest distance at which I can identify an object is 100 feet. With or without any kind of artificial light a diver is imprisoned, as far as vision is concerned, in a bubble of perception only a few dozen feet in diameter. Most fish, most marine animals, are at best in the same situation, but they have developed other senses to extend tremendously their "spheres of perception," and those senses are mainly based on acoustics, pressure waves, or smell.

Underwater man has good hearing, but no echolocation system, no nerves sensitive to slight pressure waves, no smell at all. When diving, man suddenly realizes and appreciates how overwhelmingly dependent his race is upon its sense of vision.

Binocular vision gives man detailed information about shapes, colors, and distances. When I look at a robin singing, I know it is a small bird with a red breast sitting at some 70 feet from me. A blind cat would only know a bird was singing in such-and-such direction. A bat would know a lot more about the shape and the distance of the bird thanks to its echolocation system, but would have no information about its color. Dolphins have, like bats, an extraordinarily accurate echolocation device, enabling them to identify their prey hundreds of feet away at least. But when a victim enters their sphere of vision, they make good use of the precious additional information given them by their eyes.

The sea comprises layers of water, transparent or turbid, that are of a different temperature and salinity, or in which dead particles remain in suspension, or in which planktonic creatures struggle for their minuscule lives. The layers constantly move up and down, under the influence of the sun. Everywhere in the open ocean, at dawn, trillions of tons of creatures sink from the surface to the "twilight zone" hundreds of meters below, as if shy of light. After sunset all the multitude rushes back to the surface. This huge daily vertical migration, the "pulse of the oceans," is triggered by light alone, light mother of life through photosynthesis, light manufacturer of our oxygen, light architect of beauty.

Madeira is a Portuguese island 350 miles offshore from Morocco in the Atlantic Ocean. Its 400,000 inhabitants depend, for their food, upon a "deep sea monster," the espada (*Aphanopus carbo*). The espada is a ten-pound abyssal fish, the shape of a barracuda, with a bronze-black color and enormous reddish eyes. It is caught by local fishermen at night only, on hooks baited with squid, at 3300 feet if the sky is overcast, at 5000 feet by starlight, but as deep as 7000 feet if the moon shines. These abyssal fish behave according to almost insignificant changes in nocturnal light levels at the surface. How can the difference between starlight and moonlight be felt at 7000 feet of depth?

We know that light entering the sea is absorbed and scattered. Below 100 feet all is deep blue, shadows no longer exist. Below 2000 feet human eyes are unable to detect any light. Yet thousands of feet under the surface, in a world of eternal blackness, colors are revealed in the spotlights of a bathyscaphe. The animals of the abyss still possess pigmentation, mainly gorgeous reds and purples that could never have been seen by any fish since creation. Why? Probably pigments just happen to look red or yellow while accomplishing some other nonvisual function. The sense of beauty they convey to the brain of the human beholder is purely gratuitous. The universe far transcends what man can sense, what he can organize into thought, what he can assign a purpose to. Man receives into his brain only a few narrow bands of the gigantic spectrum of messages dispensed by the cosmos. Outside these narrow bands, only slightly widened by technology, man senses nothing and understands very little.

Sea fans

WHY IS THE SEA BLUE?

Why is the sea blue or green? Hold up a glass of seawater and it appears colorless. But stand on the deck of a ship and the sea will appear to have color. Why? To understand it, let us briefly outline what water does to light.

Visible white light is made up of a spectrum of all the colors—red, orange, yellow, green, blue, indigo, violet. When we look at an object and see it as blue, we are seeing the blue light of the spectrum reflected from the object. All other colors are absorbed and cannot be seen. In the case of the sea, red light is absorbed as soon as it breaks through the water's surface. And by a depth of about 25 feet virtually all the red light discernible to the human eye is gone; a bright red air tank on a diver, for example, would seem a dull dark brown. At a depth of 75 feet a yellow air tank looks more greenish blue, because the discernible yellow light has been absorbed by the water. The still shorter rays of light are almost all absorbed by 100 feet. All that remains are the shortest rays: blue, indigo, and violet. Below 100 feet or so, all light appears a monochromatic blue. So, when the sea is pure and clear, as often is the case in the open ocean, the least-absorbed shade of the spectrum, blue, is reflected to our eyes.

The sea isn't always blue, however. Some seas appear bluish-green or green or brown or even red. These colorations are partly due to the reflection of clouds, but are caused mainly by various particles, mineral or organic, that are in suspension in the water. In some areas, especially along coastlines and in shallower seas, organic matter decomposes and produces a yellow pigment that when mixed with blue light makes the sea appear bluish-green or green. Brown coloration may be caused by sediments, stirred up from the bottom, hanging in suspension in the water and reflecting its brownish coloration. In many coastal areas, at widespread intervals, blooms of a species of dinoflagellate that is red in color become so numerous they cause the waters to appear red. This is the famous ''red tide.''

When light passes from air to water—that is, from a relatively thin medium to another medium 800 times denser—its speed is reduced from about 186,000 to 140,000 miles per second. Crossing the surface, for the same basic reason, light is bent. This is called refraction. Each color of the spectrum has a different wavelength; blue has the longest and bends the most. The shortest, red and violet, are less refracted.

Penetrating the sea, light is not only refracted and absorbed. It is also scattered, slightly by the water molecules (some scattering occurs even in distilled water) but mainly by particles of sand, salt, and minerals in suspension. The light rays bounce from one to another until their energy is spent. The scattering process limits the distance one can see—even in the middle of the ocean visibility at 300 feet is exceptional—and reduces and subdues direct penetration of sunlight. In water 100 feet deep there are no shadows.

This series of photographs shows the effect of the filtration of light by water. Just under the surface the colors appear as in sunlight. At a depth of 15 feet the red has begun to fade. At 30 feet the red seems closer to a brown and the yellow is fading: green and blue predominate. At 60 feet the red has disappeared. The yellow is greener (compare with the first photo), even in the striping on the diver's suit and tank. The green is beginning to take on a bluish hue as compared with the first photo. The blues are more intense because other colors are not there to screen them. The white section, from left to right, also shows how the blue comes to predominate as the diver carries the chart deeper. This is because white is a combination of all visible light and as some of the visible light is screened out by the water, the white segment of the chart reflects the remaining color. The black section does not change.

MAN'S DIRECT AND INDIRECT VISION

Dive into the sea, open your naked eyes, look around. What do you see? Fuzzy shapes, completely out of focus. No detail. Washed-out colors, hardly anything recognizable. Out of the water, light passes through the air and into your eye—an eye containing a fluid similar in density to seawater. The difference in the density of air and this fluid bends or refracts light rays as they enter your eye. The refracted light focuses on the retina. Underwater, however, light passes from seawater into your eye bending very slightly because of the similar density of the two fluids, so everything appears out of focus. You become extremely farsighted.

If you interpose a pocket of air between water and your eye, your eye lens functions properly. To form this air pocket, you don a diver's face mask. Nevertheless, a mask is not a perfect instrument. The refraction of light through a flat surface of separation between water and air, like the plate glass of a mask or of an aquarium, has the effect of magnifying everything we see by 33 percent. A fish seen 12 feet away appears to be only eight feet away and looks the size you'd expect it to be if it were at that eight-foot distance. Additional problems in underwater vision include tunnel vision and distortion of peripheral vision.

In these cases, you have undistorted vision in only a narrow beam perpendicular to the plate glass. Lateral vision is heavily distorted, creating a

fuzzy image. Unconsciously a diver turns his head constantly from side to side, even more than a hammerhead shark does as it swims along; he keeps his face plate directly in line with the direction in which he is looking.

A mask not only reduces the field of sharp vision, but also drastically shrinks the field of peripheral vision, from more than 180° in air to less than 80° underwater.

Calypso's Observation Chamber

In seeking to inspect what lies beneath the sea's surface, man has devised all manner of inventions.

With the installation of a unique underwater observation chamber in front of the bow of *Calypso* we were able to seek shipwrecks, observe and photograph sea life, or navigate through treacherous shoal waters. The special steel chamber was bolted onto the wooden hull of our 140-foot, YMS-class minesweeper during a complete refit of the vessel at the shipyard in Antibes, France, after we had acquired her in Malta. Access to the mattress-lined chamber is through a 30-inch-diameter entry tube that runs vertically from forepeak to the chamber, eight feet below waterline.

Of the five circular ports for underwater viewing, two face forward, one looks down at a 45° angle, and one each faces toward port and starboard. The new bulbous bow added half a knot speed to *Calypso*; it resembles the Maierform bow of modern vessels.

Manned Undersea Stations

Divers outside and oceanauts inside Conshelf II's (Continental Shelf Station Number Two) Starfish House worked in close cooperation in an undersea village off the Sudan coast. Conshelf II also included a garage for the diving saucer, which was thus the first submarine to operate from an underwater base. A Deep Cabin at the 82-foot mark enabled divers to make prolonged observations in depths down to 330 feet without concern for time-consuming decompression.

To explore and exploit the sea's riches at greater depths we devised Continental Shelf Station III, a checkered sphere-shaped dwelling placed in

Man must place an air lens before his eyes to see clearly underwater. These divers demonstrate the versatility of design in face masks.

328 feet of water off France's Mediterranean coast. From Conshelf III's shelter six divers worked, ate, slept, and were monitored electronically from the surface.

Even in the dark waters beyond 300 feet, and with the aid of artificial light, vision was the necessary factor in showing man could, for example, set up an oil-well head. When they weren't outside working, oceanauts lived inside their checkered house. At work, they ranged outward some 200 feet and down to depths of 370 feet. Slightly more than three weeks later they surfaced, having proved man could not only exist but could do useful work at such depths—opening up new worlds to his enterprise.

Photographing the Deep

Not only man's direct, but man's indirect vision—through the eye of a camera—needs to be adapted to underwater existence. Since the early years still and motion-picture photography have come a long way both above and below the sea's surface. After 1943 the aqualung gave divers freedom and mobility and allowed them time to make true motion pictures underwater.

With the fast film and strong artificial lights it became possible to make them in color. In 1956 our first full-length underwater color film released commercially, *The Silent World*, introduced movie audiences to the startling beauty of the sea's inhabitants and landscapes. Today underwater photography is the hobby of thousands and an invaluable tool for marine scientists studying underwater life, geological formations, and other subaquatic phenomena.

One of the major steps toward this new and broadening interest was the invention in 1962 of the *Calypso* camera by Belgian engineer Jean de Wouters d'Oplinter, with some of the photographers of the *Calypso* team assisting. This is a still camera requiring no cumbersome watertight housing and consequently much easier to handle than earlier equipment.

THE BEAM OF LIFE

The source of all energy on earth is sunlight. It is the beam of life. Without sunlight—even the meager amounts of sunlight that penetrate our atmosphere and our seas—green plants could not exist. Without green plants there would be no life as we know it.

How does the sun support the rich and abundant populations of living

Underwater television has many applications. When the Calypso team was excavating the ancient Greek wreck at Grand Congloué Island near Marseilles, it used a television unit that gave an archaeologist at the surface an opportunity to literally participate in the underwater dig. To cut through the turbid water and make the image sharper on the television screen, technicians added a large cone of distilled water directly in front of the camera lens.

Deepstar's unique underwater observation chamber allows the photographer to observe and photograph sea life or shipwrecks that would otherwise be too dangerous to visit or impossible to reach.

Ten million square miles of the ocean floor are covered by 600 feet of water or less. The shallower of these depths have been invaded by inquiring goggled men, but, in practical terms, for any depths beyond 300 feet, submersibles such as Conshelf III are needed.

plants and animals? Sunlight that finds green plants—phytoplankton and seaweed, among others—initiates an amazingly complex and only partially understood life-sustaining process called photosynthesis. As a result of this process, organic matter (the basis of food) is manufactured from inorganic matter, something man has been able to imitate only with great difficulty.

Photosynthesis begins when visible light is absorbed by the principal green plant pigment, chlorophyll, and other accessory pigments. This energy is utilized to split water into oxygen and active hydrogen. The oxygen is liberated into the atmosphere where it may be subsequently respired by other organisms. The hydrogen is combined with carbon dioxide to form organic molecules, and more water. The organic molecules may be later rearranged, combined, modified, and coupled with other compounds to produce the broad variety of substances found in living organisms, that is, lipides, carbohydrates, and proteins. These fuel the growth of seaweeds, diatoms, and other plants, which in turn feed either directly or indirectly the ocean's animals. Ultimately these organic molecules are reoxidized through respiration in animals and plants to water and carbon dioxide with the release of energy. Then they are recycled in the biosphere. In this way, the process of life is ultimately driven by the sun's energy.

Thus, the plants provide not only a balance against our consumption of oxygen and our production of carbon dioxide, but provide us with nutrients and other materials we need to utilize our various plant and animal foods— sugars, starches, and enzymes to name a few. In the long run the animal matter we consume depends on plant life too.

Because plant life in the sea needs the sun's radiant energy to start this complex chain of events, and because the sun's rays penetrate in any

The Mid-Atlantic Ridge, a mountainous upthrusting from the ocean floor, emerges here and there to form islands like the Azores. This close-up photograph of a small section of the Ridge underwater looks like a miniature mountain range.

The sun radiates energy needed by sea plants and animals in several forms, including visible light. Those rays that come to earth traverse more than 90 million miles of space, penetrate our atmosphere, and strike randomly on land and sea.

significant amount only 200 or 300 feet, it is in those surface waters that the greatest amount of activity takes place. And since so many animals feed directly or indirectly on those algae, that is where most of the life in the sea is found. There are exceptions. For some algae, unexplainedly, are found in greater depths. Did they drift down? Were they carried there? Or do they occur naturally in those greater depths? No one knows for sure. But it seems evident that little or no photosynthesis can take place in those algae that live so deep the sun's rays do not touch them with its energy. There are animals living in those greater depths also. But these are not algae-eating fish and invertebrates. Rather they are animals that feed on organisms that feed on the algae of shallower water.

PULSE OF THE OCEAN

A heartbeat of majestic proportions is recorded by echo sounders on ships dispersed over the seas—a slow, daily universal vertical migration: the rise of the deep scattering layer (DSL) to the surface as sun's light fails, its return to the depths before dawn. Near the surface, in the upper 300 feet, the sea may be turbid with life. Here, triggered by sunlight, photosynthesis occurs; planktonic plants and animals abound. Beneath this relatively thin layer of living matter, in a twilight of diminishing sunlight, is a vast expanse of water sparsely populated with life. This desert stretches to a depth of about 1500 feet and contains clear water of great transparency, although natural vision is limited by the absence of light. Beneath the barren layer is another tier of living sea in which scientist Dr. William Beebe described animals "as thick as I have ever seen them." This is the DSL.

The layers of the ocean are not fixed to a single level, but the creatures comprising each cloud of life tend to remain in distinct groups. Those of the DSL rise to the surface and descend to the depths, responding, probably indirectly, to the light of sun and moon. As the sun drops below the horizon and the waters darken with the coming of night, deep dwellers respond to the decreasing sunlight and begin their long vertical migrations toward the surface. This is not to say that the entire DSL suddenly rises en masse, but that each species, in its own time, ascends. The animals at the upper side of the DSL may rise only a few hundred feet to reach the surface, while squids, spending their daylight hours at the lower limit, may climb thousands of feet.

In the darkness of the deep waters they inhabit during the day, the residents of the DSL are relatively safe from predators. At night they can rise, still relatively safe from visual detection, and feed in greater safety than if the waters were lighted. Also, in the process of photosynthesis phytoplankton may release substances toxic to them. By remaining below the photic zone at a depth where photosynthesis can't occur, DSL inhabitants avoid contact with these substances.

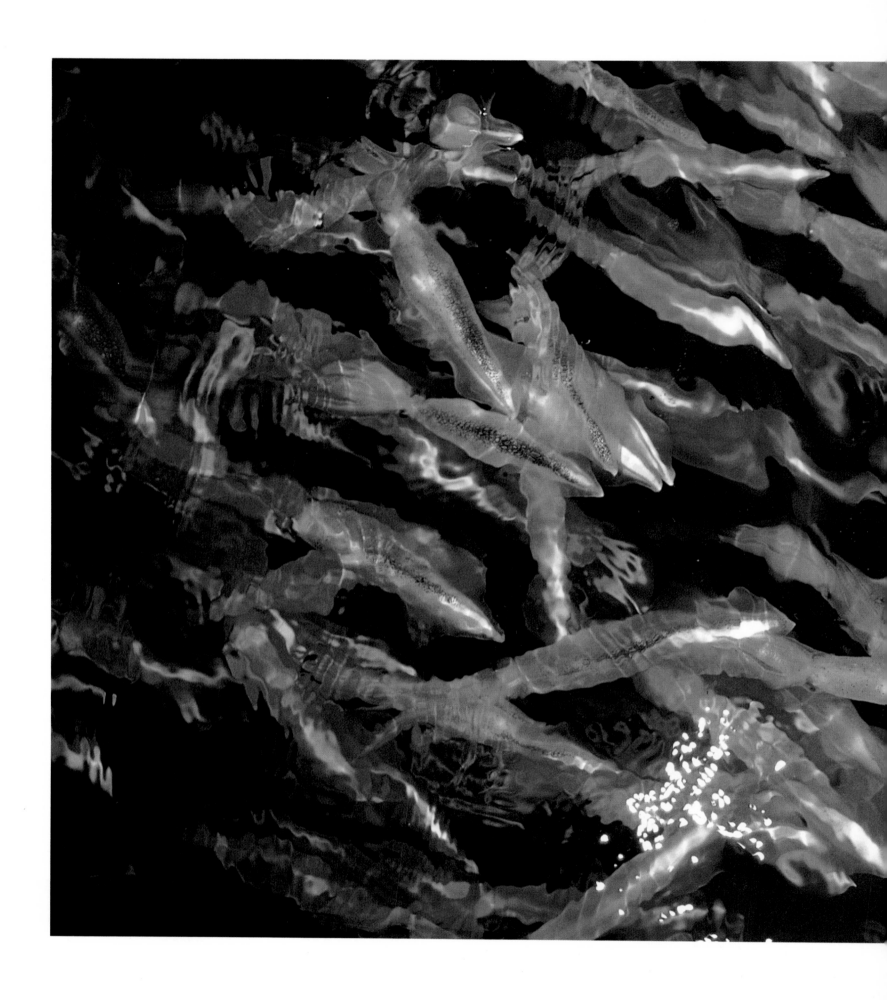

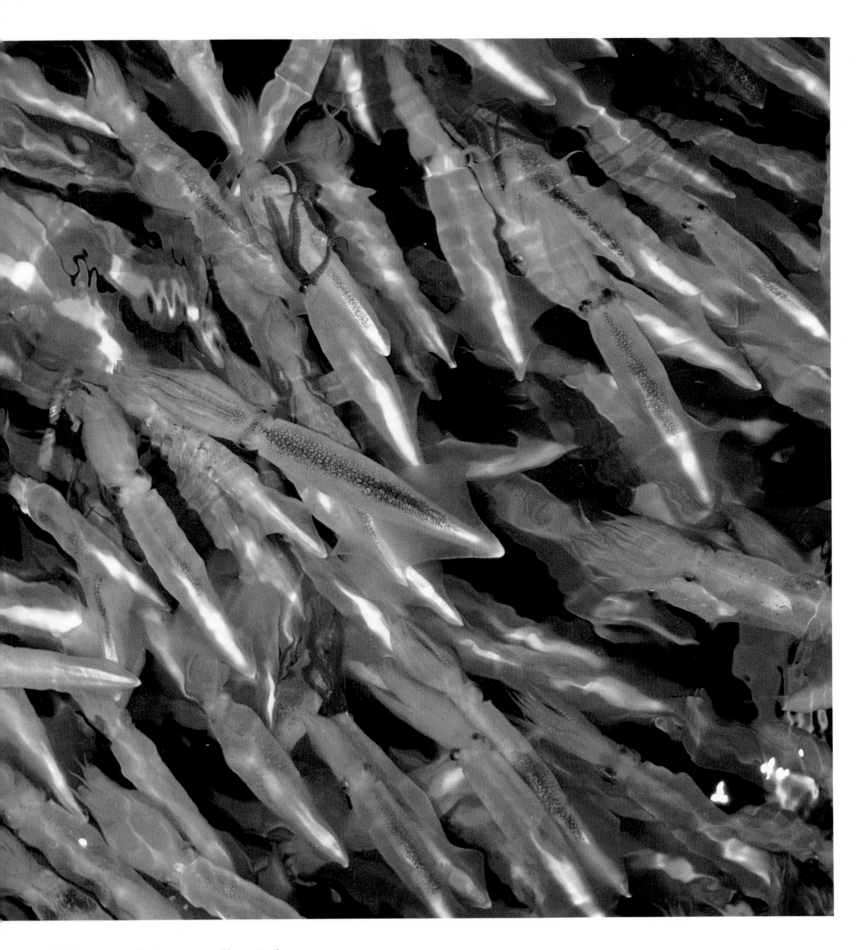

Squids are normally shy creatures. They retire by
day to the darkness of the depths and rise at night to
the surface. Some squids swarm around the lights
fishermen hang over the sides of their boats on dark
nights—perhaps they are attracted to the other tiny
creatures that gather around the lights.

LIGHT ALIVE

Before he mastered the art of sailing, man rowed his vessels to sea. As his boat glided through the sullen, black nighttime waters, the early seaman saw strange and mysterious stirrings in his wake. Wherever the water was disturbed by the passing of the boat or the stroke of an oar flashes of light occurred, flaring brightly for a brief moment.

Men wondered for ages about these ''underwater fireflies.'' But a number of other marine animals also bring light to the dark waters of the seas. Luminescent animals are found in nearly all ocean areas, from shallow coastal seas to the deep waters of the abyss. Little sunlight penetrates the waters beyond 1000 feet, and below 2000 or 3000 feet no visible sunlight can be detected. The luminescent creatures that live there provide the only visible light.

Many animals have unique cells called photophores containing specialized light-generating tissues. They vary in size, shape, color, and location on the body. One abyssal fish *(Malacosteus)* proudly displays reddish lights on portside and greenish lights on starboard. The light-producing cell may be isolated from other parts of the body by a cloak of black pigment; reflective tissue may line the photophore. In other cases living light is produced not by the animal that exhibits it but by bacteria in pockets on the surface of the host's body. These bacteria glow continuously. Some hosts have developed mechanical means to darken themselves—flaps of skin that can be pulled over the pocket of light like window shades. In some other fish the scale covering the pocket is thickened at the center and forms a lens that focuses the light and increases its effect. The position of the pockets the bacteria occupy varies, and males and females of a single species may even have different lighting patterns.

In a flash photograph tiny marine organisms seem to be a lifeless mass, but in the black of night they give off an eerie blue light. Luminescent creatures may be large or small, from shallow water or deep, but they have one thing in common: their light is nearly 100 percent efficient. Most of the energy used to produce it is converted to light; in contrast to our electric bulbs, little heat is given off. The light is usually blue, and is about the same wavelength as the blue in sunlight, penetrating farthest into the sea.

Bioluminescent Deep-Sea Fish

It has been estimated that as many as 75 percent of deep-dwelling species, and perhaps 80 percent of the total deep-sea population, are luminescent, including some worms and sea fans as well as fish. Luminescence differs from species to species, each having its own distinctive patterns and colors. Within a species the location of these organs is uniform, with all members of

the same sex resembling their relatives. Some have lights on their heads, others near their tails. Some have luminous fins, while in others the light organs line the sides of their bodies. In some species the luminous organs are located inside the mouth!

There are over 200 species of lanternfish, each a miniature constellation. The lights are generally round, looking like pearl buttons. Located along the sides of the fish, they are probably of little use in lighting up the water sufficiently to allow the fish to see and are probably used primarily as signaling devices. Some lanternfish have luminous organs on their tongues.

The first ray of the spiny dorsal fin of female deep-sea anglerfish, living in perpetual darkness at depths of 6000 feet, has modified and migrated to the front of the fish's head; from it hangs a fleshy bulb used as bait. The device is complex, usually with multiple lighted filaments to entice prey

closer. There is a great amount of variation in the characteristics of the luminous lure. Immature males, with tubular eyes and good vision, probably recognize the variations in the lights and identify mates by them. Since some males parasitize their mates, living off the female's body fluids, they have no need for fishing lures and do not have the apparatus.

The viperfish has a fishing lure, too, but his extends from the second ray of the dorsal fin. Equipped with a luminous tip, the elongated ray dangles in front of the mouth of a hunting viperfish. The viperfish has about 350 photophores on the roof of its mouth, and another battery lining the lower part of its body. These may attract the small fish and shrimp upon which the viperfish feeds. It can be said the viperfish feeds as it breathes, since by opening its mouth to admit water for respiration it reveals the photophores inside to the curious ones.

Under certain conditions otherwise drab marine organisms glow with splendor. An ordinary alga or piece of gray coral placed under an ultraviolet light becomes a shimmering gem.

Coral has the ability to fluoresce, to emit light of a color different from that of the external source; it is sensitive to a particular part of the spectrum of light. Other fluors (substances having the ability to fluoresce) are stimulated by pressure, heat, or X-rays. Fluorescence occurs when energy is absorbed by an atom or molecule, forcing electrons spinning around the nucleus into an orbit of higher energy; when these electrons return to their original orbit, energy is released—often in the form of light. When the stimulus is removed from the fluor, the light dims.

In the sea there are a great number of plants and animals with this characteristic. The ability to fluoresce varies between species, and even between individuals of a species. In some corals, for example, only the limestone castle protecting the living polyp glows under the influence of our lamps, while in other corals, just the polyp glows.

Some marine organisms fluoresce when placed under an ultraviolet light, others have the ability to do so in the presence of natural light. Here is one type of fluorescent algae.

Fluorescence

EYES

Eyes usually come in multiples of two, except in those cyclopean creatures with only one eye. Some flatworms have hundreds of clusters of two- or three-celled "eyes." Light is perceived by these animals, but vision is not possible.

In lobsters, crabs, and many other crustaceans, as in many insects, the eyes are compound. As a rule, however, animals that live in the sea have spherical lenses that provide them with sharp vision underwater. Among fish a variety of eyes is found, each an adaptation to the special circumstances of that species. *Anableps,* the four-eyed fish, is not really four-eyed. It has only two eyes that are adapted to see above and below the surface of the water simultaneously as it lies at the surface. Some fish that live their entire lives in lightless caves have no eyes, no way of perceiving light. Other cave-dwelling and some deep-sea fish have vestigial eyes that may be able to sense light only dimly. Still others of the deep oceans have large eyes with the

The giant Pacific scallop, known as a weathervane, has many eyes rimming the periphery of the mantle just inside the fluted edge of its beautiful shell.

capacity for opening pupils to a great size, enabling them to gather what minute amounts of light are available.

Frogs, aquatic reptiles like the turtles and crocodiles, and the aquatic birds that dive for their food have a transparent third eyelid that slides across the eye to protect it underwater. The aquatic mammals are equipped with the necessary spherical lens. Some of them, like the sea otter, that need vision in and out of the water have adjustable lenses that can change shape as needed.

The Compound System

A lobster's eyes are at the tip of movable stalks, able to scan the whole vicinity. In a close-up we see a grid of fine lines dividing the animal's compound eyes into a large number of separate facets. These are ancient eyes—the first examples of which are seen in fossil rocks dating back 500 million years to the Cambrian period. The now-extinct relatives of the crustaceans, the trilobites, had them. Today many crustaceans and insects have them, although some of those forms are eyeless or bear different kinds of multiple eyes.

Compound eyes differ greatly from the camera-type eyes of humans. They consist of many individual light sensors operating independently but functioning together as a single unit. A multifaceted cornea covers the entire eye. Each of the many facets is a lens for one receptor. The unit consisting of one receptor leading to a nerve fiber, its crystalline cone, and the single facet

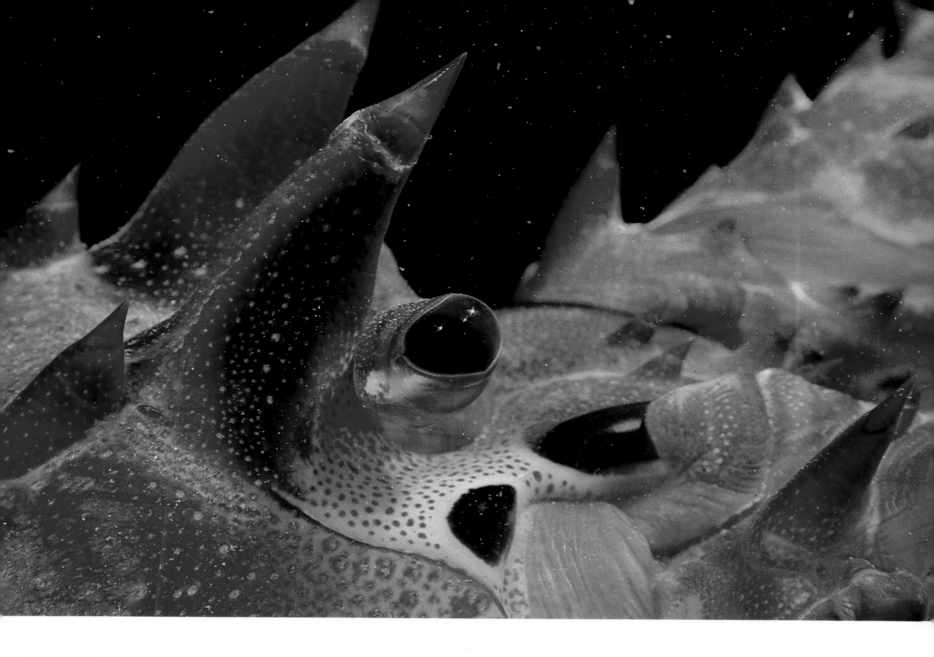

of the corneal lens is called an ommatidium. Sleeves of pigments surrounding each ommatidium prevent the escape of light. In poor lighting these sleeves retract, permitting maximum use of available light, since light passes freely from one ommatidium to another until reaching a nerve fiber. Light entering any facet merges with light entering other facets. The information then is somewhat disorganized, since what is "seen" by one ommatidium may be transmitted to the brain by another's nerve fiber. Vision is less precise, but poor vision is better than none. In deep-sea crustaceans, living in the virtually lightless aphotic zone, the eyes are heavily endowed with reflective pigments which enable them to utilize the tiny amounts of light more efficiently.

Compound eyes are well adapted for detecting movement. Each ommatidium points in a slightly different direction from those around it and may perceive an event that by itself could be meaningless. But when the brain integrates hundreds of signals, the sum of the images is believed to produce a mosaiclike image.

A fish's eye view? We have no way of knowing exactly what a fish perceives. Although it may seem, because of the curvature of their lenses, that fish are nearsighted, experiments show that in fact they are more farsighted when seeing laterally through their lenses than when peering forward. They appear to have sharper vision looking forward, when they need it most.

Since most fish, including the red snapper, have no muscles to modify the shape of their eye lenses, they move the entire lens backward to see

As a further aid to vision, the lobster's eyes are carried at the end of a movable stalk atop its head, like the periscope of a submarine.

The Eye of the Fish

74

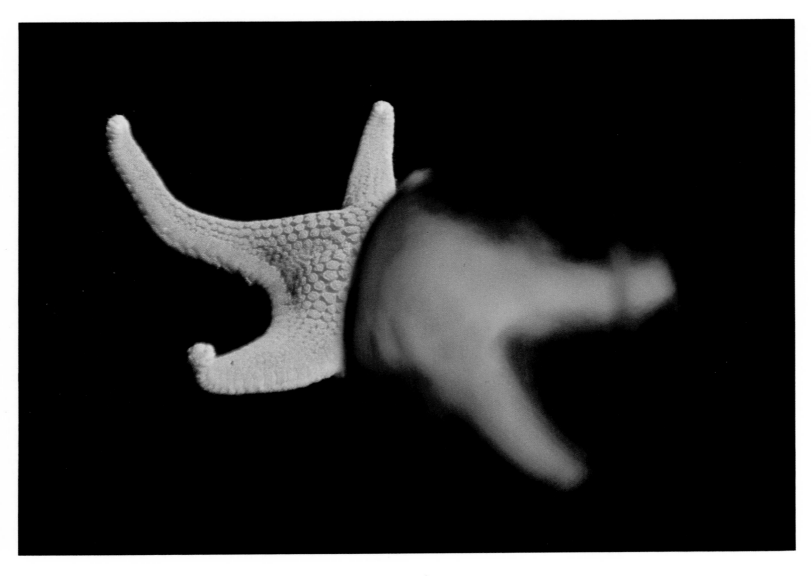

In an experiment to discover what a lobster sees, the photographer removed the lens from the eye of a lobster, placed it across part of his camera lens, and focused on a starfish.

A close-up of a lobster's compound eye reveals the netlike grid that divides the eye into many facets, multiplying the animal's visual acuity.

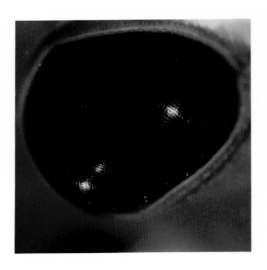

A flatfish that spends most of its time lying on the bottom, the turbot has both its eyes on the same side of its head. In the course of evolution, when the fish flopped over to live its life on one side, the eye on the underside migrated so it could function.

Like the lobster, the hermit crab's eyes stick up on stalks away from its body, which is often hidden or buried, allowing the animal to see without being seen.

things at a distance or forward to view near objects. This is exactly what we do when we focus a camera.

As most fish swim along, they move their neckless bodies from side to side, scanning as they go. Fish use their two eyes independently for such scanning which gives them monocular vision, each eye encompassing a wide range of 150°. But some fish can face an object directly by swiveling their eyes forward, aimed almost in the same direction. This is binocular vision.

Vision in Darkness

Most of us are familiar with the glow of a cat's eyes caught in the beam of a car's headlight. Many fish also have reflecting eyes. Sharks have particularly spectacular ones. These eyes have an iridescent layer called the *tapetum lucidum* (bright carpet), a series of precisely placed mirrors behind the retina of the animal's eye. They enable the animal to see in very dim light. If a weak ray of light passes through a shark's retina, it may stimulate only one visual cell. The mirror behind the rod cell picks up the light and sends it back to the retina at such an angle that it stimulates another rod in the same cell grouping.

Abyssal residents have adapted in many ways to the rigors of life in the darkness of the deep oceans. Their coloration differs from near-surface dwellers. Many are luminescent, bringing their own light (and the only light) to the deeps. Finally, their eyes have had to evolve in unique ways. Most abyssal residents have large eyes, orbits occupying more than half the length of the animal's skull. But the size of the eye means little unless the pupil is enlarged too, allowing more light to enter the eye.

Then, of course, large eyes are often associated with large retinas and with many more receptor cells in the retina, which in turn give greater visual acuity. But the critical factor is the area of open iris. So, to improve vision,

Because most fish have eyes that are set wide apart on their heads, they are said to have ''monocular'' vision. This means that each eye sees something different—the right eye sees things on the right side, the left eye sees things on the left. But since fish move their heads back and forth as they swim, the areas seen by each eye overlap a good deal and improve the fish's perception of objects. Although there is only a small area where both eyes ''see'' the same thing, fish have the ability to swivel their eyes forward in order to focus on something in front of them.

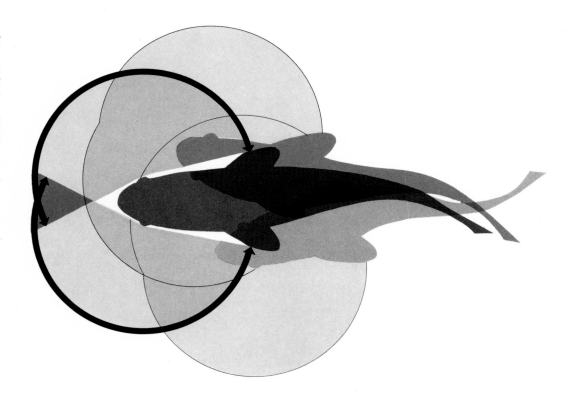

Behind the retina of a shark's eye is a group of mirrors that reflects any light entering the eye back to the retina. This has the effect of doubling the amount of light, dramatically enhancing the predator's vision.

the pupils of deep dwellers must be proportionately larger than those of shallow dwellers. This permits more light to enter the eye. Many deep-sea fish can multiply the light their eyes receive and see even in the pitchblack areas of the deep sea. They do this by reseeing the same light—that is, light reflects within their eyes and they sense that same light several times.

THE ART OF MOTION

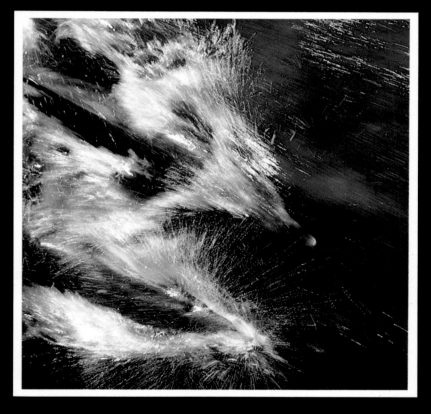

CHAPTER
4

LIFE MOVES

One of the clichés of the horror filmmaker is the mobile vegetable, especially if it is carnivorous! That a carrot should painfully pull itself up out of the ground and wriggle toward us: that is terrifying. It was, after all, the ambulatory wood of Birnam which persuaded Macbeth that all was lost. Plants should stay put, our instincts tell us. Movement is for animals!

In the ocean the general challenges of propulsion are complicated by the density of the medium through which the motion must be made. In the sea we find not one but three elaborate near-perfect animal machines—each employing a different mode of propulsion. First, the ultimate jet-propelled creature: the giant squid. Second, the ultimate streamlined cruiser: the tuna. Third, the ultimate high-efficiency thermodynamic machine: the dolphin.

How did propulsion fit into the evolutionary picture? We may find clues by imagining the lifestyle of an extremely primitive one-celled creature, looking much like our familiar contemporary, the amoeba. This infinitesimal proto-animal must eat. As he wraps his protoplasmic arms around an even smaller prey, or after digestion rejects the unusable excrement, tiny spasms ripple over his cellular membrane imparting an impulse to the little body. Soon he finds himself in another part of the sea with fresh food supplies. Again he feeds, again his cell walls convulse, again he forges on to new regions.

In other words, the act of nutrition itself may imply propulsion in animals. As the millennia and the millions of years pass by, those amoebalike animals that move faster find more extensive grazing and tend to survive. Later another theme joins the evolutionary tale. Those animals that learn to *sense* a food particle, and to move *purposively* toward it, not only survive but set foot on the path leading to nervous systems and the manifold differentiated forms of animal life we find in the fossil record and around us on earth today. Those animals that must count on chance to carry food to them tend to die out or to invest in alternate strategies of species survival: i.e., sheer fecundity.

All really efficient systems of propulsion are found in the ocean. In open water the two main mechanical ones are the jet system and the body/fin/undulation system—represented by the squid and the tuna. In addition, countless minor systems have been spun off to cope with other styles and niches of life: crawling, walking, jumping, burrowing, etc., etc. In the case of the third near-perfect animal machine—the dolphin—superb propulsion equipment is supplemented by related systems: a complex skin structure that dampens out the turbulence the dolphin's velocity creates and the mammalian warm-blood physiology that enables the dolphin to function with exceptionally high efficiency across a wide range of temperatures.

It is fascinating to trace the parallels between the naturally evolving modes of propulsion in the sea and man's artificial ones. Man swims, but with ridiculous clumsiness as compared to any fish. Yet the skindiver's flippers are surrogate fins. The wet suit is the analogue of the dolphin's insulating skin and blubber layer. The jet principle, of course, man reserves mainly for travel through the air. Our engineers have not achieved the skills necessary for constructing a vehicle that moves by undulation, any more than they have built a machine that truly flies as does a bird. Instead, to propel most surface ships and submarines we have coupled a rotating fin—the propeller—to highly inefficient forms of power like the internal combustion engine. In terms of shapes our borrowings from the sea have come easier. The same hydrodynamic principles that over the ages have polished the streamlined forms of the great fish are applied in designing hulls, rudders, keels, or planing surfaces.

One of these days man will voyage from solar system to solar system in vessels powered by gigantic sails that catch and direct the ions flowing out into space from all the active stars. Even this has its underwater inspiration. One of Albert Einstein's earliest scientific papers concerned Brownian motion—that random, never-ceasing dance of microscopic particles in a fluid as they are struck by the molecules of the fluid. Here again, at the outset of the career that transformed our world, the sea provided the clue.

Banded shrimp

THE DENSE MEDIUM

Land-dwelling animals, like ourselves, constantly fight gravity. When we climb a flight of stairs or walk across a room, we lift our entire weight with each step. Even standing still doesn't free us from gravity's bond—our legs must continue supporting our weight, and thousands of small muscles around our veins push the blood up toward the heart. While we must constantly exert ourselves to remain upright, we have no difficulty moving at low speeds through the air surrounding us. Seawater, on the contrary, is 800 times denser than air and greatly hinders movement through it, as you know if you have ever tried to walk through waist-deep water.

The water in the oceans is the same as that in lakes and streams, but ocean water has a greater quantity of materials dissolved in it. Most significant of these are the salts, which increase the density of seawater.

In spite of the handicap of moving through a dense and viscous medium, marine animals enjoy an advantage over land-dwellers. In water, a buoyant force pushes objects toward the surface, nearly balancing the force of gravity.

Buoyancy

In the third century B.C. the Greek scientist Archimedes carried out an experiment to find whether the gold crown of King Hiero II of Syracuse had been adulterated with silver. Archimedes took samples of pure silver and pure gold, each equal to the crown in weight. He immersed each in a container of water filled to the brim and measured the amount of water that overflowed. He discovered that the volume of water displaced by the gold sample was less than that displaced by the silver. When the crown itself was immersed, the volume of water that spilled from the container was less than the amount displaced by the silver, but more than that displaced by the gold. Archimedes concluded that the crown was a combination of gold and silver, and the attempted deception was uncovered.

Archimedes' experiment demonstrated for the first time the buoyant influence of water, and the principle underlying this phenomenon was named after him. Archimedes' principle states that an object immersed in a liquid is buoyed up by the force equal to the weight of the volume of liquid it displaces. An object floats when the buoyant force exceeds its weight. It sinks when the displaced volume of liquid weighs less than it does. If an object displaces a volume of liquid that weighs exactly the same as it does, it is said to be neutrally buoyant; it neither sinks nor floats.

Swim Bladder

Rock and reef fishes need not cover great distances; most of their lives are spent moving slowly about in a home territory. They are greatly helped by a built-in gas-filled chamber, or swim bladder: without it, they would slowly sink to the bottom, because their bodies are slightly heavier than seawater. The swim bladder increases their volume without increasing their weight and keeps them perfectly balanced.

Most of the fast open-ocean swimmers, fish, squids, sharks, or manta rays, have no swim bladder. They all sink slowly when they stop swimming. Some of these creatures spend a substantial part of their lives resting or hiding on the bottom. But negative buoyance forces tuna, squids, and most of the sharks into a mobile way of life: at the depths where their food can be found they swim forever without rest.

FORM AND DESIGN

An Apollo spacecraft hurtles through the vacuum of space, free from the force of gravity and from drag. But in the atmosphere, propulsion obeys the laws of aerodynamics; and the streamlining of modern aircraft is designed to reduce turbulence and offer the least possible resistance.

In water, the laws of hydrodynamics are generally similar but much more critical, because of the high density and viscosity of liquids. Aquatic

organisms experience very strong drag, which is a function of speed, of shape, of the surface of the body, and of the nature of the water flow over it. These four factors cannot be considered separately because they act heavily upon one another. The drag forces become very large at rapid speeds because they increase approximately as the square of the velocity. Two types of drag can reduce the speed of large or small marine animals in the water: pressure drag and friction drag. Pressure drag is a direct function of stream-lining and design. In friction drag, which mainly depends on the smoothness of the skin, some water is moving with the fish against water farther away which is not moving, thus creating eddies and turbulence. Fast swimmers are covered with a slimy mucus secretion.

Streamlining

Water can be *laminar* (smooth), *turbulent* (irregular), or *transitional*. Laminar flow creates the least drag, and turbulent flow the most. In practice, any fish or mammal or man-made hull produces transitional flow, as perfect laminar or totally turbulent flows are never encountered.

Turbulence begins to set in when the thin layer of water immediately next to the moving body (the "boundary layer") becomes unstable. If the turbulence in the boundary layer can be stabilized, then laminar flow can be maintained over the entire organism. Fish and marine mammals have such a stabilization mechanism: by constantly changing their shape to conform

Laterally flattened. The blue- and yellow-striped angelfish is more laterally flattened than most other fish. This shape allows them to retain their swimming speed and gives them the additional advantage of fitting into narrow crevices, whether seeking protection or food.

Combining shapes. This palometa, a member of the pompano family, has a somewhat flattened fusiform shape and reinforced narrow tail base that places it among the most powerful swimmers in the world.

Attenuated. The sharptail eel has an attenuated shape that enables it to slip in and out of openings in its habitat. Having a smaller diameter relative to its length, the sharptail is able to seek food in small holes.

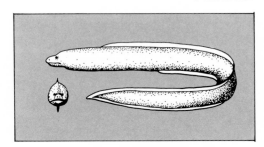

Attenuated and cylindrical. Trumpetfish are attenuated in shape and cylindrical in general form. They are not so flexible as the eels or morays, but their proportions are eel-like, enabling them to mimic certain species of coral in their normal habitat.

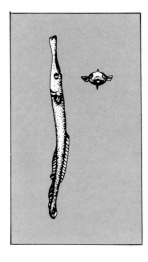

their body surfaces to the lines of flow, they are able to move at speeds that cannot be matched by exact, but rigid, replicas of their forms.

Fish Shapes   The shapes of fish's bodies fall into general categories: fusiform, like the shark, barracuda, and codfish; laterally (side to side) compressed, like the angelfish, spadefish, and filefish; dorsoventrally (top to bottom) compressed, like the skate and guitarfish; attenuated, like the conger or the American eel; and a number of infrequently encountered shapes (these include the seahorse, the triangular cowfish, and the globular porcupinefish). But whatever their shape, fish are similar in their bilateral symmetry.

Most species exhibit characteristics of more than one category. Few fish, for example, are precisely tubular, fitting the exact definition of a fusiform fish. Most that come close to this shape are usually somewhat flattened dorsoventrally or laterally. A few are long and drawn out as well as tubular and therefore combine fusiform and attenuated shapes.

Dorsoventrally compressed. The flattened form of this stingray enables it to move along the bottom almost surreptitiously. Flattened top and bottom, it is the epitome of the form. In adapting to life on the ocean floor, rays, and their relatives the skates, have become flattened and have developed a unique means of locomotion dependent on their body shape.

Fusiform. This coney, a member of the sea bass family, comes close to the torpedo shape of fusiform fish. When it must, the sea bass can move with exceeding swiftness to strike its prey.

Fusiform and laterally flattened. Squirrelfish are among the many species that combine fusiform and laterally flattened shapes. They dwell in the open areas of reefs. Other fish that share these characteristics are soldierfish, cardinalfish, several species of wrasses, and some of the small species of sea bass.

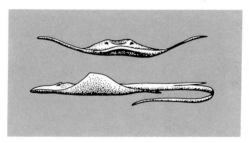

Cetacean Shapes

The cetaceans, which include the whales, dolphins, and porpoises, have adapted to a totally aquatic life since their ancestors returned to the sea nearly 70 million years ago. The most important changes were those having to do with the way the animals moved and breathed. They assumed the fusiform shape of early fish. The bones in their necks became shorter until there was no longer any narrowing between head and body. With water to support their weight, they became rounded or cylindrical in body shape, reducing the drag irregularities. Front limbs adapted by becoming broad, flat, paddlelike organs. The hind limbs disappeared, leaving only a trace internally. The tails developed into flukes. It is the flukes, combined with the powerful muscles of the trunk of these animals, that provide the propulsive power enabling them to swim and dive so efficiently. Unlike the fish, cetaceans' flukes are horizontal, moving up and down. In the fish, tail fins or caudals are vertical and move from side to side, providing an easy-to-recognize differentiation between fish and mammals in the sea.

Today's whales, dolphins, and porpoises are totally aquatic. Their streamlined forms and powerful flukes enable them to swim with ease. The big whales have grown to be the largest animals that ever lived, outweighing the biggest dinosaurs three-to-one.

Imitating Sea Animal Forms

Man has taken many lessons from animals that inhabit the sea. We seek to imitate the streamlined form of fish like the tuna, the mackerel, the shark, and other fusiform fish that are built for speed. Submarines, for example, have cylindrical bodies, tapering at the ends, and broad plane surfaces analogous to a fish's pectoral fins. Both have a vertical plane analogous to either the dorsal fin or the caudal fin of the fish. Our surface craft are handicapped by the fact that they move half through water and half through air; their design is thus a compromise between good streamlining and seaworthiness. Modern ships' propellers are really just an assemblage of several rotating fins.

The most common form of locomotion among aquatic animals is undulation. The body is thrown into a series of curves that begins at the head and passes along the length of the body as a *traveling wave*. Among most fish these body waves move in the horizontal plane, but in flatfish and in many marine mammals the body waves move in the vertical plane. One of the most typical examples of pure traveling-wave propulsion is an aquatic reptile, the sea snake, quite common in some equatorial seas.

The earliest vertebrates probably swam in a similar manner, with rhythmic muscle contractions flexing their bodies from side to side. Pushing against the water this way resulted in forward movement.

Nearly all fish employ one of three general swimming techniques stemming from these traveling-wave movements. The first is that employed by the sea snake and attenuated fish, like eels and ribbonfish.

Fish with rigid bodies, like the armor-plated cowfish, cannot flex their bodies. They use only their short tails, swishing from side to side in a short arc, to push them along.

Between these two extremes is a third method, used by most fish, combining characteristics of both. The wigwagging of the caudal fin is coordinated with subtle body undulations. This method yields smooth, rhythmic motion and improved efficiency. The fish's head swings in a small arc, and the body and caudal fin curve to form a complete transverse wave.

TRAVELING WAVE

The Living Box Moves

The inflexible, armored body of the trunkfish is fine protection against predators. But it renders the strangely shaped fish awkward and slow moving. Under normal circumstances, with no cause for alarm and no imminent danger, the trunkfish sculls about with its transparent, fan-shaped pectoral fins. Its equally small, delicate dorsal and anal fins contribute a wave-form motion in this normal mode of swimming. Its caudal fin barely moves. Frequently, the fish hangs in the water on the reef, barely moving any of its fins. When danger approaches, the awkward trunkfish and cowfish flail about to reach cover. Their caudal fins lash violently from side to side on the hinge at the base of their tails—the peduncular hinge marked with a sharp spine that offers additional protection against attack from behind. But the main thrust of swimming comes from a special way of moving that has developed over many thousands of generations.

This unique method of swimming, peculiar to the trunkfish, cowfish, puffers, and porcupinefish, is called ostraciiform movement.

The trunkfish's bony armor is best described as a living box of polygonal bony plates. Some of the plates have minute spines and tubercles giving their surface a rough feeling. One curious habit some divers have observed in trunkfish and cowfish is that of blowing jets of water at the fine sediments on the sea floor around the reefs. Each time they use these water jets, the sand in front of them billows up in clouds bringing up a few tiny organisms to eat. At the same time, the jet pushes the fish back an inch or two. The visual effect is a back-and-forth movement which is interrupted periodically when the fish darts forward a few inches to suck up the small animals it has found.

Effect of Muscle and Bone Action on Swimming

It was once thought that fish swim solely by using their tail fins, which they swept in an arc behind them. But motion-picture analysis of fish shows that fish swim by side-to-side undulations of their entire bodies. In fact, fish whose caudal fins have been amputated can still swim, some almost as well as intact fish of the same species. We know the tail fin facilitates swimming, but it is not the exclusive source of propulsion power for most fish.

The muscles along the sides of a fish are its strongest, while those associated with the fins are relatively weak. In swimming, a succession of contractions passes along each side of the fish. Some of the W-shaped

Startled into action by a diver, the sea snake moves in a perfect example of the traveling wave. Distinct S-curves travel from the animal's head and along its flattened body, and, in increasing breadth, finally reach the tail.

muscle segments, or myomeres, contract on one side, and those opposite them relax and stretch. This bends the fish's body, and the fish pushes against the water first with one side and then with the other.

The flexible frame of the fish is a good foundation for the muscles. The backbone extends from the head to the tail and is made up of many interlocking vertebrae. The vertebrae are jointed to allow side-to-side movement and are strong enough to withstand the great strain placed on them by the flexing muscles.

Orca's Traveling Wave

In an easy undulating motion, the orca (killer whale) breaks the water's surface with the top of his head. A puff of vapor issues from his blowhole as he exhales. As he quickly inhales, his back with its tall dorsal fin breaks the surface too. He bows his head and points downward even as his great, broad flukes flash momentarily out of the water. And he's gone—submerged beneath the sea's surface as smoothly as if he hadn't passed that way with his multi-ton hulk.

Those great broad flashing flukes on orcas, and other whales and dolphins, are planes that normally are parallel to the surface of the water. They are moved up and down driven by powerful muscles in the body ahead of the tail. The muscles are connected to the tail and flukes by a series of tendons. When the orca bows its head downward to sound, its back arches and a traveling wave passes along the length of the animal's body with increasing amplitude. As it reaches the part of the body that houses the muscles that power the tail and flukes, the great animal "snaps the whip" and the broad flukes drive the animal forward.

THE ROLE OF FINS

Fins are to fish what arms and legs are to men. And even a little more. Most fish have two general types of fins—median, or vertical, fins, which originate along the midline of the animal, and paired fins on their sides. The median fins include the dorsals on the back; the caudal, or tail, fins; and the anal fins on the belly just behind the vent or anus. Some fish have as many as three dorsal fins. The paired pectorals at each side near the head are

A smooth trunkfish sculls about a coral reef in search of food. Its rigid body makes it a clumsy swimmer, but in an emergency it can usually escape approaching danger fast enough.

analogous to our arms and to birds' wings, and most fish have them. The paired pelvic, or ventral, fins are located below and usually behind the pectorals.

Fins have been adapted for many purposes, but they are mainly used for propulsion, stability, steering, and braking. In some cases they have been modified for other functions. As fish evolved toward faster speeds, their fins, like feathers on an arrow, provided stability and made it possible for them to propel themselves where they wanted to go. As its head moves from side to side when a fish swims, the animal has a tendency to veer off from its forward path. To resist this condition, known as yaw, the fish erects its dorsal fin. The tendency is further reduced in fish with long, slender bodies and by the long, trailing dorsal and anal fins of some reef fish with deep, short bodies. These deep bodies keep the fish from rolling, much in the way a sailboat's keel keeps it steady. Fish of other body structures extend their paired fins to avoid rolling. To change direction vertically, fish bring their paired fins into play. The pectoral fins act as hydroplanes to raise the nose of the fish, while ventral fins bring the rest of the body into horizontal plane.

The Important Propulsive Fin

Whatever the shape of the caudal fins, their function is essential for aquatic animals. The caudal fin, or tail, is used in coordination with the massive muscle segments of the body to propel the fish through water. Most fish tails originate at the end of the vertebral column. In heterocercal (uneven) tails, the spinal column extends into the larger upper lobe of the caudal. The heterocercal tail of sharks gives these heavy fish without swim bladders an upward thrust. In homocercal (even) fins the vertebral column ends at the fin base and supports a symmetrical "tail" which produces only a forward thrust. Some homocercal tails are forked, some are square, some are rounded—but each serves a specific function. For example, the lunate (crescent-shaped) tails of mackerels, tunas, and jacks indicate fast swimmers. The broad tail of a grouper gives him the ability to accelerate very quickly.

To Start, Hover, Turn, Stop

The basic simplicity of design of a typical fish or of a marine mammal when compared to that of a lobster, of a giraffe, or of a man, is obviously due to neutral buoyancy and to the absence of strong, large, complex limbs. Thus the muscular structure can be concentrated in one solid pack. We have seen that fins other than the tail are used for fine maneuvering which enables the fish to master its liquid environment.

Dependence on Fins

The pufferfish moves principally by sculling with its pectoral fins. Its small dorsal fin and its caudal fin come into play to some extent, wiggling rather feebly from side to side. When a puffer inflates itself with water, its body becomes even more stiff than it is normally. In such a hopeless situation, it cannot depend on even a slight body motion to help it move, but must rely entirely on its fins.

Because the puffer is so slow moving, it depends on its ability to inflate itself for protection. Two additional defenses puffers have are hardly visible. One is their ability to bury themselves in sandy ocean bottoms by squirming. The other protection is the toxicity of the pufferfish for those eating them. The poison is concentrated in the liver, the viscera, and the gonads, especially around spawning time. Because some species of puffers (the "fugu") are considered a delicacy on the Japanese table despite their poisonous qualities, cooks preparing them for human consumption often must show proof of graduation from a special school that trains them in detoxifying puffers. Seven thousand tons of "fugu" are eaten in Japan each year, but the Emperor is not allowed to enjoy that dangerous delicacy.

Undulating Fins

The seahorse's fins are nearly invisible, but close observation shows that the animal has control of each individual ray. High-speed photography has revealed that each ray is capable of moving at a rate of 70 times a second in an

Pufferfish move about in a unique way. Their bodies remain practically rigid and they move by waggling dorsal and anal fins synchronously to the same side, which gives them a readily recognizable appearance. When undisturbed, they leisurely beat these fins left and right like living metronomes.

91

The scallop moves by clamping the two halves of its shell together, which causes a stream of water to shoot out and propel it forward.

action similar to slats falling in a sagging picket fence. As an undulation passes from one end of the dorsal to another, the seahorse moves forward or backward, or up or down, in its own peculiar, very slow but versatile version of a traveling wave.

Being a poor swimmer, the seahorse usually avoids areas where strong currents occur. It has sharp eyes and sits with its tail wrapped around seaweeds or gorgonians. When it does choose to swim, it unwraps its tail, straightens itself out, and begins to flutter its dorsal fin. Even so, these slowpokes may take one-and-a-half minutes to cross a one-foot area.

THE JET SET

When a child blows up a rubber balloon and lets it go, it will fly about erratically as long as the pressure of the air inside the balloon is higher than that of the surrounding environment. The air streams out through the narrow neck of the balloon with a force proportional to the difference in pressure.

In the sea there are animals that move about in much the same way as

the balloon, making use of jet propulsion. The simplest are "salpa" and jellyfish. Some of the bivalves, or two-shelled molluscs, accomplish jetting by very sudden contractions of their muscles: when the shells clamp shut, water is forced out and the animal travels in the opposite direction of the stream.

Jet propulsion is also used by cephalopod molluscs like the octopus and squid. These animals energetically contract their mantle, forcing out a narrow stream of high-pressure water that can be directed to steer the creature with precision.

Fish that normally swim by ordinary methods sometimes use what is called "a jet-assisted takeoff," accomplished by the forceful closing of their gill covers. As the covers snap shut, water is squeezed out and the fish gets an additional thrust forward.

The diving saucer from the research ship *Calypso* uses the same type of propulsion as a number of animals. Like the siphon of the octopus, the saucer's nozzles can be pointed in any direction. Jets provide not only propulsion but unique maneuverability and safety, for jet nozzles do not get so easily fouled as propellers.

Along the ocean floor this batfish is awkwardly walking on its elbows, which are actually part of the fish's pectoral fins.

WALKERS AND CRAWLERS

Some marine mammals—sea lions, walruses, and fur seals—use their fins to walk when on land. The octopus usually employs jet propulsion, but it can also creep along on its tentacles.

The tiniest of animals, single-celled protozoans, can move by any of three methods. Some have minute cilia, hairlike projections that they use to set up currents that push them along. Others have whiplike appendages with which they flail their way. And some move by extending false feet ahead of them and pulling their bodies after.

Many bivalves, the group that includes clams, oysters, scallops, and mussels, extend a muscular foot and by alternate contractions and expansions creep along.

Cautious Stalking

Out of the water in a manner rare for a fish, the mudskipper raises itself up on its forelimbs and levers itself forward. Then it moves its pelvic fins up and pushes off.

Lobsters, crabs, crayfish, and shrimp use four of their five pairs of jointed legs to walk about on the ocean floor. But their armored suits make walking an awkward process, for they limit the distance and direction their legs can spread. Because many are scavengers, however, this method of propulsion is adequate: they move across the bottom picking up animal and vegetable detritus. If endangered, shrimp and lobsters can swim very quickly backward, by flipping their tails beneath them.

The batfish is a fish that "walks" the ocean floor. It has no swim bladder and therefore tends to stay at the bottom of the sea, in shallows and in deeper water, where it elbows its way along on jointed and heavily built pectoral fins. It uses pelvic fins to a lesser extent. A batfish swims with an ungainly awkward style.

The batfish number about 60 species found almost exclusively in tropical oceans of the world. They grow to a maximum size of about 14 inches and they eat a wide variety of animals, including crabs, smaller fish, worms, and molluscs, which they capture after lying in wait, hidden by their cryptic coloration and sometimes by covering themselves with sand. A batfish, approached by a diver, will "freeze" or perhaps cover itself with sand, then lie motionless until the diver leaves.

Sea stars, often called starfish, have a complex, but not very efficient, system of propulsion. To understand it, we must look at their anatomy.

The central disc and arms are covered with a skeleton of shell-like plates or rods that are loosely meshed together to allow the animal great flexibility. On the underside of the sea star's body, along the center of each arm, is a V-shaped furrow, called the ambulacral groove. This groove holds nerves, blood vessels, and a water canal, all radiating from the central disc. Outlining the ambulacral groove are rows of tiny muscular tube feet, which end in suction cups or points.

It is on these thousands of tubes that the sea star moves. Unless the star is climbing on a smooth surface, it does not use its suction cups to pull itself forward. Instead, it depends upon a pushing action of its tube feet, and it uses hydraulic pressure for this thrust. Water channeled from the central disc flows into little bulbs above each individual tube foot. By muscular contraction of the bulb, water is forced into the tube through a nonreturn valve, and extends the tube foot. The tube is shortened by expelling the water.

When a sea star wishes to move, it extends some of its tube feet in the

Walking on Fins

How Starfish Move

Equally at home on land or sea the marine iguana of the Galápagos Islands is a mediocre swimmer, using its body and long tail to undulate through the water. It rarely goes more than 100 feet offshore, however, since it finds the seaweed on which it feeds closer to land.

desired direction, places them firmly, and uses them as levers to push its body along. All of the sea star's thousands of feet (as many as 40,000) must be coordinated in order to move the animal effectively. It is not surprising that most sea stars are unable to move rapidly. Their average speed is about six inches a minute.

OTHER WAYS TO GO

Dinoflagellates, tiny plant-animals, have two whiplike organs they beat against the water to propel themselves, pushing the water behind them. Some dinoflagellates travel 150 feet each day in vertical migrations. This sounds like a very short distance to us, but it is almost two million times the length of the dinoflagellate. The equivalent for man would be a daily 2000-mile underwater journey, clearly beyond our most advanced technical capabilities.

The turtle's use of its flippers as paddles for swimming is not unlike the underwater swimmer's use of his arms to do the breaststroke. The most efficient undersea paddlers are the giant sea turtles. Clumsy, almost helpless when they have to creep ashore to lay their eggs, they are fast migrators in the sea, can reach peak speeds of ten knots, and are very agile. Nor is the penguin's use of its wings for underwater flying much different.

Man has copied the webbed feet of sea birds and the flippers of seals for swimming. Flippers give us more surface to push against the water, but our legs are peculiarly unsuited for an aquatic life. Many serious swimmers, interested in improving their technique, have attempted to emulate the dolphin's swimming movements. To do this, they keep their legs together in one unit similar to the dolphin's flukes and flex them from the hips with only a limited knee action.

The Waggler

The marine iguana tucks its four legs close to its sides and uses its body to wend its way through the sea. While most lizards are terrestrial, the marine iguana of the Galápagos Islands in the Pacific lives in and at the edge of the ocean. It lies quietly in the sun along the rocky coast or swims offshore and dives to the beds of algae on which it feeds. To move through the water, it

The little hydrozoan velella, known as "by the wind sailor," floats on the surface with its tiny iridescent sail standing high to catch the breeze that will push it across the water.

uses the traveling wave; through evolution, the tail of the marine iguana has flattened laterally like the tail of the sea snake.

The Blubbery Athlete

The huge, ungainly, blubbery walrus, which often wallows on the ice floes of the Arctic Ocean, becomes a graceful, smooth, efficient swimmer when it slips quickly into the sea. The transformation is astonishing because the contrast is so great. On land the walrus moves ponderously. But in the sea the enormous beast becomes agile, tucking its forelimbs out of the way as it oozes into the water and using its hindlegs—broad, flattened, paddlelike limbs—to propel itself.

Webbed-Foot Swimmers

There are fifteen species of penguins, found in various parts of the southern hemisphere, and as far north as the Galápagos Islands. Seven of these species live almost exclusively in the frigid waters of the Antarctic, which are rich in their basic food, shrimp called "krill." Penguins spend most of their lives in the sea; they "fly" very swiftly underwater using powerful strokes of short, smooth wings, and also jump out of the water, moving very like dolphins. Their webbed feet are used mainly as rudders. Some species are capable of cruising great distances at speeds up to seven knots and to dive as deep as 900 feet. Cormorants also "fly" underwater with their big wings and can cover at least half a mile without surfacing.

Underwater Fliers

The manta ray is another underwater flier. There is a certain similarity in the underwater swimming of birds like penguins and cormorants to that of fish like mantas, skates, and other rays. In each case the forelimbs, analogous with the arms of humans, are used for underwater flight: While birds use their wings, mantas use their pectoral fins which are developed into enormous, flexible, triangular, winglike structures. And in flapping they can change the pitch of the wing to get the utmost efficiency during upward as well as downward beats.

Catching the Wind

By twisting its sail, a tiny velella is able to steer itself slightly, but is at the mercy of the winds and is often stranded on beaches. Great flotillas of these little animals are frequently seen in tropical waters, dotting the ocean for miles in all directions. Beneath its float of air cells, the velella has short tentacles with which it stings and entraps its prey. Fortunately, its sting is harmless to man.

Dolphins Riding Bow Waves

As a ship plows its way through the ocean it creates waves. Dolphins occasionally come to play and ride on these waves. Sailors have often seen this. From all indications, however, the dolphins are neither seeking food nor being lazy. These playful mammals probably simply enjoy the game of surfriding.

The theory behind this phenomenon is that as a ship moves along, water piles up in front of its bow and is forced forward. The waves that are thus formed travel with the ship. Dolphins typically scout the edges of this bow wave, "feeling out the pressure field." If they move into it, they position themselves on its forward slope and gravity keeps them falling downward and forward just like a surfboard rider.

Dolphins often leap out of the water seemingly for the sport of it—and to the delight of anyone watching.

One of the most spectacular methods of escape for sea creatures in danger is to get completely clear of the water even if for just a few seconds.

But survival isn't the only motive for getting out of the water. Some animals quite literally jump for joy. Other creatures leap out of anger, fear, or hunger. And there are a myriad of mysterious reasons for jumping, reasons that scientists still seek. Whatever it is, the ability to get clear requires special mechanisms as well as special motives. To break free of the sea, a marine animal, whatever its weight, must reach a certain vertical speed. For example, to leap two yards above the surface, a creature must break the surface with a vertical speed of 12 knots, whether it is a sardine or a blue whale. To reach eight yards, the speed must be 24 knots! Consequently, only fast swimmers, large or small, have the ability to jump into the atmosphere.

GETTING THE HULL OUT OF WATER

Fish that Fly

The adaptation that has enabled the flying fish to survive through millions of years of predation is its ability to get up and out of the water for up to 20 seconds—long enough to escape predators. As this adaptation continues to develop, it may someday enable flying fish to soar through the air above the water for minutes at a time. One flying fish was seen to reach a height of 36 feet. And 20-foot-high flights are not uncommon. Soaring along at speeds up to 35 miles an hour, flying fish may travel 250 yards or more in a single glide. Shorter flights of 60 to 70 yards, lasting six seconds, are more typical, but they can be repeated several times in a row.

Typically, a flying fish launches itself through the surface first by gathering speed swimming upward toward the surface. This speed helps it leave the water. With the enlarged lower lobe of its caudal fin trailing the water it moves rapidly from side to side and extends its broad winglike pectoral fins. The forward thrust given by the caudal fin plus the lift provided by air acting on the pectorals helps the fish become airborne. When it drops back to the surface, it is the whirring action of the caudal fin in the water that helps the fish repeat its flight. Often the predator follows the flying fish, seeing its silhouette through the surface, waiting for it to splash back. In this case, the flying fish tries to fool its chaser by making a right-angle turn in flight. Some fish, notably the dorado, have been seen to chase flying fish right out of the water, sometimes capturing them in their mouths in midair.

Jumping Billfish

Swordfish, marlin, and sailfish, collectively known as billfish, are highly prized by anglers. When it has been hooked, a billfish leaps out of the water and attempts to shake loose the barb that pierces its mouth. The angler must keep a taut line to prevent the fish from escaping.

Sailfish, with their high dorsal fins, and marlin, which share a lunate tail with their fellow billfish, are fast and powerful, ranging the open seas of the world. The main source of thrust for the phenomenal leaps made by these great fish comes from the segments of powerful muscles in the rear third of their bodies.

Leaping for Play

Sea lions often leap from the water. Their streamlined shape and muscular hindlimbs and flippers enable them to do this. Sometimes they arc through the air with a headfirst reentry. Other times they may do backflips. Typically, sea lions will pick up speed underwater and then break through the surface, arcing upward and forward through the air before reentering the water. They may leap clear of the water when they chase fish as well as when they are courting or amusing themselves.

Whales Clearing the Surface

Whales come to the surface in a variety of ways and for at least two reasons. Being air breathers they must come to the surface. If they are lazing along, they may rise vertically, quickly exhaling, and then take in a new breath, close their blowholes, and submerge quickly. If they have been moving along underwater at a good rate of speed, they surface at an angle—about 30°—exhale, close off, and dive back in, all in the space of a few seconds. There are other times, though, when whales seem to enjoy leaping high into the air, and then flopping back into the sea with a resounding smack.

Humpback whales especially seem to enjoy this sport, slapping the water with their flukes. Sperm whales also jump clear of the water and fall back in, raising gigantic columns of water. Such spectacular jumps are probably done after very deep dives, when the whales, out of breath, speed up for air at more than 20 knots and leap.

The gray whales that migrate along the California coast each year engage in something called spyhopping. Observers believe the gray whales spyhop to check their location along the coastline.

Jumping for Air

Dolphins are the high jumpers of the sea. Scientists are still investigating how and why they jump.

When barracuda move in schools, they typically remain almost motionless for a while and then dart quickly off without any apparent motive.

To catapult themselves up and out of the water, they use the mighty thrust provided by their flukes, which are powered by masses of lumbar muscles and connected with their tails by a series of tendons. Rigid vertebral columns and these powerful muscles give dolphins the strength and speed to launch themselves.

Sometimes dolphins leap high into the air, reentering the water head-first or belly-flopping back. Sometimes they make flat jumps, arcing out of the water and splashing back in again.

There are two main reasons that dolphins behave in this manner. First, they do it for the fun of it, jumping as a form of play. Second, they leap to get air which they need because they are mammals.

TOWARD HIGHER SPEEDS

Using Bursts of Speed

Bursts of speed are required by many different kinds of marine animals—by slow or sedentary creatures like the triggerfish or the octopus; by trans-oceanic cruisers like pilot whales; and by swift swimmers like the barracuda. To perform such "rushes," a creature must have the ability to generate a strong acceleration of its body, an instant tremendous thrust forward. The necessary energy must be readily available. It has to be stored, chemically, *inside* the muscles. The translation of this energy into a large, short-duration force is best illustrated by the jet squirts of squids—they use such pulses constantly to prey on flying fish at night or to escape porpoises, and they remain almost motionless between darts. Another short-impulse transducer is the deep, long, soft tail of the grouper. When a grouper springs to attack, the first stroke of its tail is so powerful that it creates a strong sonic boom. Such organic explosions are familiar to divers. Open-ocean fish—mackerel, tuna, and barracuda—are well streamlined and powerfully equipped to cruise at fairly high speeds, but are also capable of accelerating suddenly for short bursts of speed. Such speed is needed mainly to catch prey, to escape predators, or for the strenuous love dances performed by such fish as ocean jacks in the spawning season. To achieve these bursts of speed, cruising fish call upon special sets of tail muscles, which—curiously—are light rather than dark in color, and have built-in stores of energy.

Anatomy of a Tuna

The bluefin tuna is one of the fastest fish because of its design, the high "aspect ratio" of its tail fin, the reinforcement of its caudal peduncle, its powerful muscles, and the chemistry that supplies energy to fuel its fast motion. Laboratory measurements prove that the limiting factor is the amount of oxygen extracted from the water by the gills: per volume, water contains 25 times less oxygen than air. The mouth of the tuna is relatively small to create the least possible drag; the quantity of water it can pump in while swimming is proportional to speed. This explains why the tuna's metabolism is in equilibrium only at cruising speed. The tuna is sentenced to be a perpetual traveler. It has two different types of muscle tissue that provide a high gear and a low gear into which the tuna can shift.

Slime for Speed

The scales of most fish are covered with slime. Secreted through mucus glands in the fish's skin, this slime serves two functions. It acts as a lubricant to ease the fish's way through water with the least possible amount of friction. But probably its most important function is that of sealing the fish and making it watertight. Like all animals, the skin of a fish is semipermeable, and without a slime coating, the salt concentration inside the fish would rise through osmosis until it equaled that of the sea, thus poisoning the fish.

Do Dolphins Defy the Laws of Hydrodynamics?

Performance of aquatic animals is usually estimated indirectly by calculating the operative drag forces and then figuring the muscle power necessary to overcome such drag. From computations of this type emerged Gray's Paradox, named for the famous scientist of animal locomotion. To explain the dolphin's ability to exceed speeds of 30 knots, the Paradox insists either that the dolphin's muscles can produce ten times the power of terrestrial muscles (man can develop about .024 horsepower per pound of muscle), or that their bodies generate complete laminar flow. But pure laminar flow has

Designed for fast, efficient movement through the sea, the submarine bears a strong resemblance to the ocean's mightiest creatures—whales.

never been observed in animals as large and fast as dolphins.

The drag calculations, therefore, had to be inaccurate. Further research demonstrated that dolphins possess a flexible and pressure-sensitive skin that dampens boundary-layer turbulence. The dolphin's skin comprises three layers: a thin, flexible outer layer; a thicker middle layer containing channels filled with viscous substance; and a stiff, thick inner layer. When the boundary layer of the water tends to thicken into eddies, the dolphin's skin is depressed; similarly, it is pushed out if that layer thins out. This passive adjustment system requires no energy from the dolphin and dramatically increases its performance.

Auxiliary Traveling Wave

Squids rival fish in their design for speed during brief runs, but they cannot compare with bonitos for long-range cruising. There is one field, however, in which squids are unmatched: maneuverability. Capable of swinging their siphon nozzles almost 360°, they can aim the propulsive thrust of this jet propeller and dart in any direction. For traveling at low speeds, these members of the jet set have also adopted the traveling-wave system as an auxiliary power transducer. On each side of their bodies is a horizontal muscular fin that can be controlled to undulate in both directions, so that they can operate in low or high gear as well as backward and forward!

The Fastest Underwater Vehicle: Nuclear Submarine

With its periscope, snorkel, radar, and radio antennae retracted and all hatches battened down, the nuclear-powered submarine moves through the sea. The hull is a deliberate and careful imitation of the body streamlining of swift-cruising, large sea animals—mainly whales. Buoyancy of the submarine can be adjusted by filling or emptying ballast tanks, which more or less replace the mammal's lungs or the fish's swim bladder. Powered by nuclear sources, with unlimited range underwater and capable of producing oxygen from the sea, submarines have become nearly as natural inhabitants of the sea as fish.

ATTACK
AND
DEFENSE

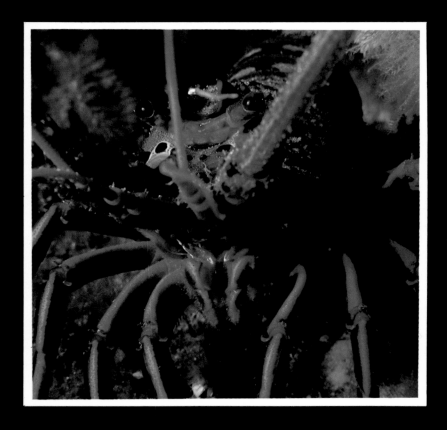

CHAPTER
5

IT'S A HARD LIFE

"Nature red in tooth and claw"?

Civilization has the fundamental ambition of introducing a degree of order into the primeval struggle for existence. But to this day it has only succeeded in drawing a curtain between life's harsher manifestations and certain of our human sensibilities. Most of us have never visited a slaughterhouse or a fish cannery or a 24-hour-a-day illuminated chicken-and-egg farm. When we fall in love we rarely have to dispose of our rivals by means of a billyclub. When a neighbor turns nasty and tosses beer cans into our backyard, we call the police. When the police collude with the neighbor, we call the state or federal authorities. When we play rough games, we institute rules and time periods and an umpire.

But the classic, bloodstained struggle simmers right along in easy view if we care to look for it. We need protein just as urgently as the shark needs his mouthful of red snapper even though we are fed by the farms and stockyards and trawlers of this world. The newspaper reminds us daily that lust and territorial dispute and injured vanity and greed turn into acts of savagery and murder in the best-regulated communities. Man has lots of shark in himself still, as almost all of us find out at least once or twice in our lifetime.

The creatures of the sea do not write poetry or paint pictures. Their lives are more obviously determined than ours by the basic quests of life: for survival, for food, for a mate, for a territory—for play. And in pursuit of these quests they have developed over the evolutionary eons offensive and defensive weapons in nearly every conceivable direction. Man can find a precursor of almost all his primitive or refined armament in the sea. There are animals that use the analogues of swords, spears, bows and arrows, nets, electric cattleprods, camouflage, armor-plate, speed, poison—often two or more instrumentalities in dazzling combination. Moreover, advanced animals utilize tactics and strategies. (A lioness will stampede a herd of zebra toward her invisibly crouching mate; the stratagem is roughly that employed by Napoleon at Jena.) Barracuda "herd" schools of smaller fish. Dolphins hunt in packs. With the notable exception of the explosives man concocts in his laboratories (culminating in the biggest bang of them all—the atom bomb) man can learn everything he needs to know about the principles of attack and defense from some creature in the sea.

Still, in the sea, as elsewhere in nature where it has not been contaminated by civilized man's ecological irresponsibility, there are balances struck between attack and defense. For example, predators survive only if prey also survive. If too many sea otters devour too many urchins and abalone, the otter population soon suffers from the effects of starvation. Theoretically a too-successful predator will ensure his own extinction. The thought that this might be man's destiny is intolerable, but we will need a great amount of vigilance, imagination, and sacrifice to avoid such a fate. In the world we see and live in the efficiencies of attack and defense systems have become attuned to each other over the two or three billion years that life has evolved on our planet.

It is these dynamic balances between prey and predator, aggression and withdrawal, that Western civilization is disrupting—with the consequence that the job of restoring an order to nature is now man's alone. Man is an animal, and the "nature red in tooth and claw" side of his animal nature is never far from the surface. Yet with his brain, and his languages, and his prehensile hands, man has liberated himself from most of the laws that rigorously limit the possibilities available to the rest of the animal world. In the field of weaponry he has borrowed a tiny chunk of the sun's own fire for his thermonuclear devices; armament cannot go much farther than that. If he wants to, he can reduce earth to a nightmarish desert, or blow it up altogether. But why should he do these things? How much more in harmony with the other side of his animal being—his instincts for survival, for mating, childbearing, playing—he is when he turns the marvelous tools of intelligence and analysis which produced the ultimate weapon of attack to the fabrication of the ultimate defense: peace—a human species living at peace with its world and with the other inhabitants of its world.

Sea otter eating a rock crab

To restate an old saw—when a man bites a fish, that's good, but when a fish bites a man, that's bad. This is one way of saying it's all right if man kills an animal, but if an animal attacks man, the act is reprehensible. The animal is labeled "killer," something to be feared, hated, shunned, punished, even killed by man.

How dangerous are those sea animals with bad reputations? A few actually kill. A few maim. Some are poisonous when eaten by man. Most sting, stab, or poison and cause mild to severe discomfort to man. Yet man is one of the larger beings that sea creatures encounter, and these poisons usually can't kill him. Very often these poisons are used defensively against predators and offensively in food gathering.

There are a few animals that have won themselves a bad reputation even though they have little or no effect on man. They have won their rating through man's interpretation of their attitude toward lower animals. These animals have been seen feeding in what appears to be a savage manner. But this behavior may perhaps be comparable to a man tearing the flesh off a chicken leg with his teeth.

The word "shark" strikes fear into the hearts of many men. In some cases this fear is justified. In most it is not. Of the 250 species of sharks currently recognized, only about 35 to 40 are actually dangerous to humans. The more than 200 species of sharks not considered a peril to humans are ill equipped, physically or temperamentally, for such activity. Some have flat-topped teeth, others are too sluggish, a few are too small, and many just aren't interested in man as food or as a threat. So when they do encounter man, they go the other way—and so does man. However, those species that will attack

BAD REPUTATIONS

Toothy Threats

man are unpredictable and if you're uncertain about the species swimming close by it's better to get out of the water.

Sometimes the sharks may swim casually about a diver for hours without showing any interest in him, and on other occasions they may behave erratically, the ambient field becoming electric as soon as a diver enters the water. Sometimes sharks flee from an unarmed, unprotected swimmer, and at other times they may deliberately crash into the steel bars of anti-shark cages and bite furiously at them in unprovoked attack. The blue is usually a solitary hunter, but when the scent of blood is in the sea, many will appear as if from nowhere, like vague shadows suddenly come to life. They circle cautiously sometimes for hours until they are sure that there is no danger. Then one of the circling pack rushes the intended prey, brushing or bumping it. If the object seems edible and harmless, the boldest of the cautious group approaches for the first bite. Then a feeding frenzy may begin, when the behavior of those active sharks stimulates all the sharks in the vicinity to become many times more aggressive than normal.

The Shark's Eversible Jaw

In spite of what some scientists from Aristotle onward have thought sharks *can* bite a very large object, such as the side of a whale, with their dorsal sides up. They do so by using the lower jaw and snubbing up their upper jaw in a grotesque manner. This snubbing action opens their mouths so wide that their jaws are nearly vertical, and the huge cavity of the fish's mouth is revealed, as are its rapierlike teeth. In addition, strong muscles allow the upper jaw to be thrust outward to grasp the flesh and then rotated downward in a cutting action.

When a shark takes a bite of a large animal, like a whale or dolphin, it clamps onto the animal with the great jaws and sinks its teeth into the flesh. Then it seems to go into convulsions, violently wriggling its body from head to tail. Its razor-sharp serrated teeth are twisted from side to side, and they scoop easily through the captive's flesh. This awesome spectacle is over in an instant. The bite of a shark leaves a cavity in its prey's body.

Poisoners

Jellyfish have a bad reputation, which is deserved by some species but not by others. *Pelagia* have tentacles that carry poisonous stinging cells named

With its rows of sharp curved teeth, the sand tiger shark may menace man, although not in North American waters. Sand tigers generally eat fish, crustaceans, and squid. They are also among the few predators that attack the vicious bluefish.

The mottled appearance of a scorpionfish makes it nearly invisible against the sea floor.

nematocysts. So do many other jellyfish. In man their sting causes burning and itching of the skin and at worst, swelling, redness, and breathing difficulty. A few, like the white moon jellies that range the oceans of the world, are harmless.

The Portuguese man-of-war secretes poison powerful enough to do serious harm to men. Under a pink-blue gas-filled float hang many filaments reaching down 40 to 60 feet and including nematocysts, or stinging organs, capable of injecting poison. The man-of-war stings indiscriminately to feed and thus to survive. Most predators dare not touch it and only a few seabirds are known to eat it.

Armless and Armed

Some of the most venomous creatures in the sea are the sea snakes. Their venom paralyzes the nervous system, and the victim, unable to activate muscles for breathing, soon dies of suffocation. It is often said that sea snakes have a small mouth and can only bite man's tender skin at the base of the thumb. This is not true. These snakes can bite anywhere, but they only do if seriously disturbed. In the Persian Gulf many pearl divers who did not wear goggles have been killed by sea snakes because they could not see the snakes and accidentally grabbed them.

The blue-ringed octopus is a unique type of octopod. It is rarely larger than four inches long, yet in spite of its size its bite is often fatal. Beachcombers of Australia are therefore warned that these ''cute'' animals can be deadly, if sufficiently provoked.

Scorpionfish

Scorpionfish are most deserving of their bad reputation. In some species their 13 dorsal spines carry the deadliest poison of any fish—poison lethal

108

enough to kill a swimmer or beachcomber in two hours. This group includes the lionfish and the stonefish. Scorpionfish so closely resemble stones that, unfortunately, they are almost unnoticeable, and a person walking in shallow water can easily fall victim to one of them.

Manta Rays

Mantas are often called ''devil rays'' or ''devilfish,'' and Cuban fishermen have superstitions about them, ranging from their hypnotic powers to their habit of jumping out of the water onto fishing boats. The largest observed specimen of this animal had a wingspan of 22 feet and weighed almost two tons. Fishermen who harpoon one of these giants soon discover its strength. A manta can demolish an ordinary fishing boat in a matter of minutes, but this is a normal fight for survival. Far from dragging sailors to watery graves, rays are content being left alone, peacefully eating huge quantities of small fish and plankton, occasionally jumping clear of the water as many as three times in a row.

Stingrays

The stingray has a fearsome, whiplike tail longer than its body, and near the base of this tail are one, two, or three flattened spines with small, sharp teeth—coated with venomous slime which can bring serious injury or even death to man. But our misconceptions center on the manner in which the stingray uses this formidable weapon.

A stingray leads a quiet life on the ocean's floor and never attacks man. If approached, it will flee. The stinger is used only as a defensive weapon, not as an offensive one. Its position on the tail enables the ray to sting an enemy above it. When threatened, the stingray whips its tail around until it finds its attacker. The stinger is not even used to obtain food for the ray; the ray feeds by sucking molluscs and crustaceans into its mouth. If a diver or swimmer steps on a stingray's poisonous spine, who is to blame?

Barracuda

The barracuda's razor-sharp teeth and powerful jaws coupled with its ability to strike its prey with lightning speed have given it its reputation as a man killer.

Actually this sleek powerful fish, which may grow to a length of six feet and a weight of 113 pounds, has only been involved in a few dozen cases of attacks although, in most instances, the reports are unreliable. .The swimmer was in turbid water, and the attack was probably accidental; the identity of the attacking fish was generally not ascertained. When a diver, even un-

Though it is fierce looking, the moray eel would rather turn away than fight. Once morays feel threatened enough to fight, however, they can do great harm.

Blending with its coral background, this octopus is barely visible even to a discerning eye.

It is almost impossible to see the Corambe *(right), a little nudibranch, nibbling on a lacy colony of bryozoans (left). The reason for the* Corambe's *resemblance to the bryozoans remains a mystery. The* Corambe *does not need to surprise the bryozoans, because these animals, like corals, live in rigid houses from which they cannot move. And since it has few enemies, the* Corambe *should not require camouflage for protection.*

armed, swims toward a barracuda, the latter dashes away—but not very far—and soon it is back behind the diver, close to his swim fins. As the impression of being followed by such an enigmatic predator is somewhat disagreeable, the wary diver turns around and threatens the fish away—for a few seconds. This game of intimidation can last for hours with no apparent lassitude in the barracuda.

Bluefish Bluefish are known as one of the sea's most bloodthirsty fish. These fast-moving fish do not just kill to eat—they often kill for no apparent reason. Long after their hunger is gone, they continue to slaughter, leaving shredded, half-eaten fish in their path.

In addition to being able to attack other fish (for whatever reason) with remarkable effectiveness, the bluefish, once it has been hooked by an angler, fights furiously and jumps violently at the end of a line. When finally caught and in the boat, it has been known to grab for and bite off a fisherman's finger. Conversely, the bluefish can be made into a good meal itself once it has been caught, scaled, gutted, and broiled.

Moray Eels

Moray eels have long had a reputation for being attackers. It seems to have begun in Roman times—historians tell of Nero throwing slaves into water-filled pits of moray eels just to amuse the bored aristocrats whose pleasure came from seeing people being eaten alive. In point of fact, moray eels are retiring and would rather hide than fight, and so if this tale is true, Nero must have either given the moray eels no place to escape when the slaves were cast in, or starved them until they were desperate. These eel-like fish, which are usually four to five feet long but sometimes reach ten feet, attack only when threatened. Once they bite, however, they can do considerable injury with their strong jaws and many sharp teeth.

COLOR ME INVISIBLE

Camouflage creates deceptive appearances. Animals deceive for two main purposes: to avoid predators and to obtain nourishment. With some animals camouflaging is a full-time occupation, with others a part-time one.

Probably the most widely used form of camouflage among fish is countershading, or obliterative shading. It renders a fish almost indistinguishable to a potential predator or prey as the animal's color blends into the surrounding water and matches its reflection of light. Even in crystal-clear waters sharks appear suddenly without warning. A look around may give a diver no hint of a shark's presence, yet only an instant later one may be curiously circling only a few yards away. How can these large animals approach unnoticed when a diver has carefully scanned the surrounding waters? The answer is that sometimes we cannot see them even though they are nearby. Their sleek, muscular bodies are countershaded, dark on top and light underneath. We can better distinguish countershaded animals swimming at the same level as we are, since we can see the contrast of the light underside with the dark dorsal side. But even then it is not easy.

Disruptive coloration is especially common among fish of coral reefs. In this form of camouflage, broad, bright bands break up the readily recognizable outline of the fish. Directive and deflective markings are another form of deception practiced by the animals of the sea. These markings usually draw attention to the least vulnerable part of the fish; or they may draw attention away from a particularly vulnerable area.

In mimicry, an animal may seek to look like something else so it won't be recognized for what it really is. In some cases the animal mimics something inedible. In other species an animal which is not poisonous might imitate one which is. Yet in others, a predator resembles something harmless.

Some fish possess bizarre body extensions and protrusions to assist in camouflage. The scorpionfish family is a master at this; some produce fleshy extensions on the fins and head with hairlike projections all along the lateral line. The advantages derived are most likely a breaking up of the fish's outline and an irregular texture not unlike the algae-covered bottom on which it rests.

Another camouflage technique is that of body posture. The animal may assume a position which makes it less apt to stand out in contrast to its background. Pipefish are probably the best example. Their elongated shape, in a vertical position, closely resembles the marine grass in which they live. On the other hand, a horizontally swimming fish stands out dramatically against the background. Predatory trumpetfish often position themselves parallel to a slender gorgonian in an attempt to remain unnoticed.

Long and narrow, the trumpetfish often hides by attaching itself next to a whip sponge.

Find the Fish

The ability to change color helps some fish avoid detection. They change color to match their backgrounds, blending so completely that even a person knowing of their presence may have trouble finding them. This concealing kind of pigmentation is called cryptic coloration.

It is remarkable, considering the variety of colors fish can assume, that

there actually are only four variables within their color repertory. They possess only three pigments—black, white, and orange—and special reflective cells. It is these cells that permit fish to become an iridescent green or blue. Somehow they separate the spectral colors selectively, reflecting some and absorbing others. Many fish have nervous control over the cells and can quickly alter the colors reflected. One of the most dramatic examples of this is the multicolored flashes that progress over the body of a spawning dolphinfish or mahimahi.

In the well-lit levels of the sea an adjustable color system is essential. Bottom texture and colors vary from one place to another as does the color of the surrounding water. But as one goes deeper the colors of fish are less important and as a result become more limited. The most logical reason is that deeper in the sea colors cannot be seen and therefore are of little use. This is because a particular colored pigment only reflects its own color of light. If orange light is not present in the first place, orange pigment will have nothing to reflect, appearing dark gray or black. Under these conditions a fish finding it advantageous to appear obscure or dark could be pigmented brown, black, or even red, as many deep-sea fish are, since red light does not penetrate deep.

Quick-Change Artist

The cephalopod cuttlefish, a mollusc relative of the squid and octopus, is a master of disguise. It can change color in an instant or more gradually. Unlike many other animals that can alter color only slowly through hormonal action, the cephalopods have nervous control over their skin pigment. Colored pigment is contained in cells called chromatophores. Under stimulation many tiny muscles attached to the cells pull on the edges causing the cell to form a large flat plate making the pigment more apparent. Relaxation of the muscles causes the cell, and thus the pigment, to become concentrated into a dot and thus essentially invisible. The varied colors that these remarkable animals can achieve result from a blending of pink, brown, blue, purple, and black pigments. Coloration reflects the mood of the animal—white for fear, red for anger, multicolor for sexual display.

Chromatophores also can exist in fish but differ in the mechanism of pigment dispersion. For example, the melanophores, dark pigment cells, have many branching extensions into which pigment can be forced. The pigment granules move while the cell remains basically the same shape. By a similar controlling mechanism, orange and white pigments can greatly increase a fish's potential for coloration.

Deceptive Eyes

The jet-black pupil in the eye of a fish, even though it may be a target for attacking predators, cannot be altered much to avoid its conspicuousness. Because this black spot on the head is inevitable in all fish, many have evolved with color patterns that obscure the eye or detract from its prominence. For instance, a black bar on the fish that extends through the eye region makes the eye very difficult to distinguish. Another even more creative color pattern is the eye spot near the posterior part of the body seen on many butterflyfish, flatfish, wrasses, damselfish, gobies, blennies, and even on some rays. One example of the effectiveness of such an eyespot is in the misguided attacks of an Indo-Pacific blenny that preys on other larger fish by tearing off pieces of skin. It often goes for the eyes, but when confronting a butterflyfish, it is directed to the eyespot at the rear of the body, which does much less harm to its victim.

LIVING IN ARMOR

In the undersea world of predator and prey one of the best defenses is armor. And animals display a wide variety of armor, some of which man has been able to copy very effectively in his own weaponry.

But animals that adapt to life in armor face a major problem—they are cumbersome and slow-moving.

A false eye near the tail of this spotfin butterflyfish may draw the attack of a predator away from the fish's head—its most vulnerable part. A vertical black line through its real eye breaks up the animal's silhouette for further deception.

Animal armor comes in several styles. There are exoskeletons—skeletons that are external instead of internal as is man's. There are tubes, shells, and cases that are part of the animal. Exoskeletons are widespread even in the plant world. Diatoms, basic ingredients in the ocean's intricate food webs, have silicate shells that enclose the cellular components of these plants. Other microscopic marine plants—the silicoflagellates—have shell-like plates protecting them. Many seaweeds secrete and deposit on their exteriors a coating of lime that armors them.

In the animal world, the examples of exoskeletons are legion. Some of the one-celled protozoans, sponges, and corals have exoskeletons. The armor of crustaceans, including lobsters, crabs, shrimps, and barnacles, is familiar to many. And molluscs, like clams, scallops, mussels, and oysters, are also well protected.

Then there are the creatures that are born without armor but eventually live within cases, tubes, and tests of their own making. Some armors start out as soft mucous secretions; these secretions combine with lime solutions and develop into tough outer coatings, which can ward off physical or chemical attack. Some other creatures that are born without armor inhabit the abandoned shells of other animals. Or these creatures make a safe haven in the substrates. They find protection as they gradually envelop themselves in the substrate.

In a class by themselves are certain fish and reptiles that are clad in a different kind of armor—scales, which can be thick and tough. Some fish, however, have evolved tough, scaleless skins.

Heavy Armor

Molluscs owe much of their success to the heavy armor they carry about. Over 500 million years ago their ancestors probably possessed a simple horny covering for protection. They subsequently gained the ability to impregnate this covering with calcium carbonate. This shell, along with a strong muscular foot, may have allowed snails, the largest and most success-

From the safety of its shell, the triton leisurely consumes a sea urchin.

ful group of molluscs, to exploit habitats too inhospitable for other animals. Fossil evidence indicates that much of molluscan evolution took place in the shore zone where an abundance of food and a variety of habitats existed. An impervious shell would have provided protection from both drying at low tide and abrasion, and a strong muscular foot would have enabled them to hang on tightly to wave-swept rocks. One mollusc that successfully endures such a habitat is the limpet, a small animal which possesses a pyramid-shaped shell. The pointed shell probably reduces the unsettling effects of waves, allowing the animal to graze on algae in the intertidal zone where the greatest wave force is exerted.

In addition to the protective coiled shell seen on most gastropods, a horny "trapdoor" or operculum further isolates the animal from outside. This durable shield may be any shape depending on the aperture of the shell. The queen conch has a curved narrow operculum that fits way back inside the shell while the turbo snail has a beautiful, circular calcareous operculum often called the "cat's-eye."

A hard shell is an effective protection, but these molluscs, however, are not invulnerable. A number of predators are successful at circumventing this deterrent. Bat rays possess platelike grinding teeth able to crunch the heaviest shelled clams; some starfish extrude their stomachs to digest the protected snail; others ingest the whole snail.

Some fish have sharp protective spines. These spines may occur almost anywhere on the body and in some cases occur everywhere on the body.

Animals not built for speed rely on other protections. The pufferfish's body is densely covered with short, sharp spines, which are actually modified scales. When threatened, the puffer inflates itself with water, an action that makes it more imposing and also erects its piercing spines. The puffer has another defense: its poisonous flesh contains a toxin that affects the victim's nervous system.

Spines that Protect

Spines are either modified scales or spiny rays of the fins or bones that project out from the fish's body. A classic example of such modified scales is found all over the spiny puffer. Surgeonfish and tangs also possess scales modified as razor-sharp scalpels which evolved from bony ridge scales on the body near the tail. These lancelets are attached at the posterior end, projecting forward, and can be erected or depressed at will. Another example is the dagger on the tail of the stingray.

Spiny rays of the fins are of great importance for the defense of many fish, acting as instruments for the injection of poison, making the owner difficult to swallow, or merely providing an unpleasant stingy surface to deter enemies. The development of elaborate spinous fins are a characteristic of the scorpionfish family, and spines are notorious in the lionfish and turkeyfish and in many other bottom-dwelling fish. The dorsal spines of the weeverfish are actually used in offense—it has been said to attack divers with its venomous weapons. Spiny rays can be found on the dorsal fin, the pectorals, the pelvics, and the ventral fin—or on all of these. In some cases the mere erection of these spines as a threat may deter a predator. Some reef fish, like the triggerfish or the filefish, are able to erect their spines in such a way as to make them impossible to pull from the hole in which they are hiding.

STRATEGIC WITHDRAWALS

The eighteenth-century British author Oliver Goldsmith wrote: "For he who fights and runs away/May live to fight another day." This statement has truth for life in the sea as well as on land. An animal outmatched in a fight is wise to withdraw if it can. Most simply, some escape by turning and fleeing, outdistancing or outmaneuvering their opponent. When we think of one animal outrunning another, we usually think they must have great speed and so it sometimes is. But many animals we consider to be incredibly slow-moving can move just fast enough to outrun an animal seeking to make a meal out of them. On land, a fugitive tries to escape in two dimensions. In the sea, an unpredictable three-dimensional sharp turn gives the advantage to the pursued over the pursuer, even if he is substantially slower. Some burst out of water in flight, returning every few seconds to scull with their tails. If the deck of a ship is not too far above the surface, a variety of sea animals is sometimes found there in the morning. Surprisingly, some creatures not even known to be "fliers" show up, thus giving us new insight to their ways and habits. Until the voyage of the *Kon Tiki,* authorities generally ignored reports that squids jetted themselves right out of the ocean. But with the evidence gained on this voyage, they began taking a closer look at these remarkable molluscs.

Like a naval ship hiding in a fogbank some animals like octopus, squid, or sea hares release an ink cloud that is equally effective. And since they have control over its liberation, they can decide when it should be brought into play.

When threatened, some marine creatures that live in the substrate or among plants may duck into pockmarks in coral reefs, cavities in rocks, or other holes. Or they may bury themselves in the sand.

Protection by Boring

Some animals spend nearly all their lives in hiding. Boring molluscs find safety by tunneling through mud, wood, and even rock.

Most clams burrow into sand or mud using only their soft foot as a digging implement. In Puget Sound, along the coast of Washington, one species of large clams, the geoduck, may weigh twelve pounds and has a body long enough for its neck to reach the surface from a burrow as deep as four feet.

The largest and most efficient rock borers are the pholad clams. Near the end of their larval stage, the clams fasten themselves to a surface of heavy

To escape danger, an octopus may change color to camouflage itself, retreat into a hole its pursuer cannot enter, or release a cloud of dark ink to confuse and mislead a predator. Or the octopus may simply jet away as it is doing here.

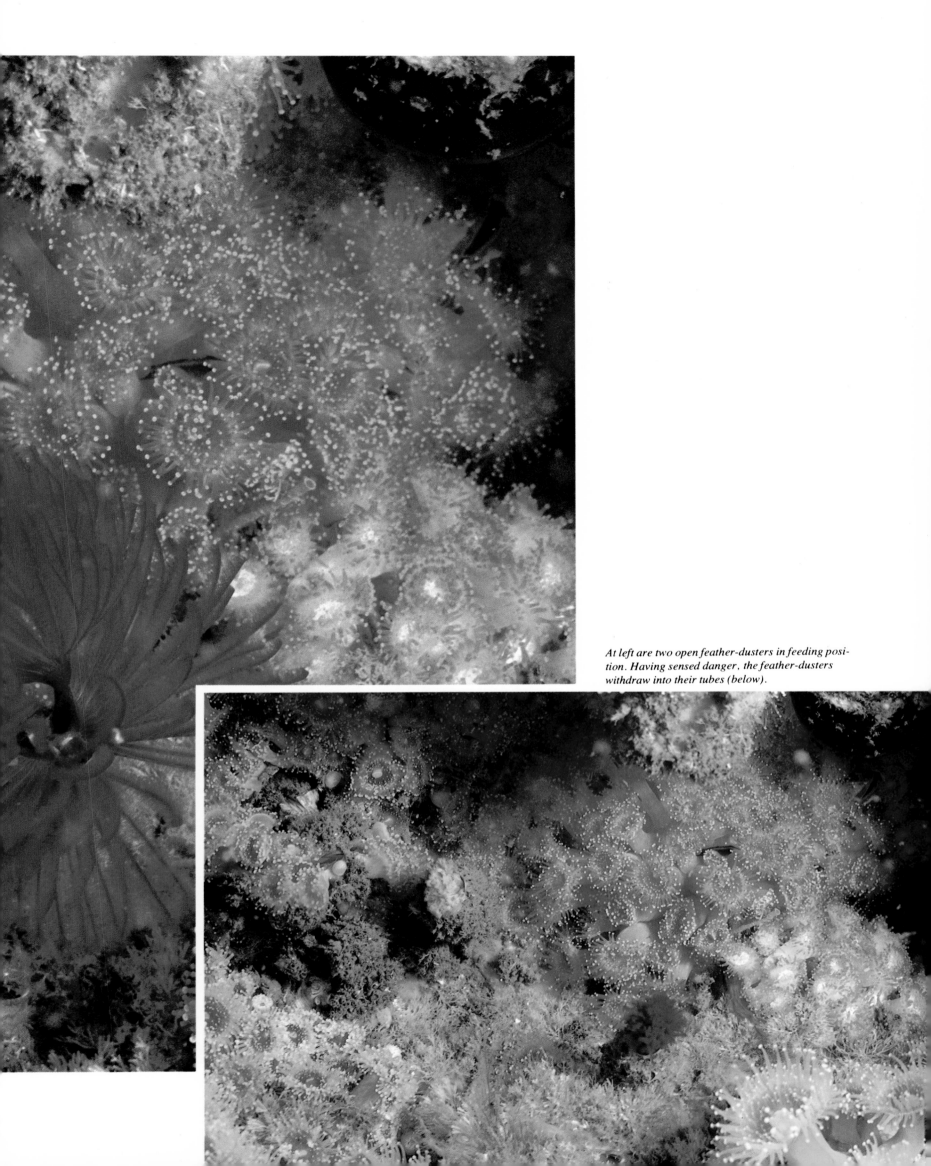

At left are two open feather-dusters in feeding position. Having sensed danger, the feather-dusters withdraw into their tubes (below).

clay, sandstone, or limestone. They begin to make their burrow when their shells begin to form and harden. They rasp the burrow face with the shells until a hole is gouged out. The movement and grinding continue, and soon the hole becomes a tunnel, which the clam increases in length and diameter to accommodate its own growth. In fact the mollusc, like the shipworm, becomes completely trapped within its protective chamber. The orientation of the shell is such that it can burrow forward but cannot reverse the direction. Thus, its only connections to the outside world are two siphons to the opening of the tunnel—one to bring in water and food and the other to expel water, body wastes, and gametes.

Closing Up

The feather-duster worm and the sea anemone are sessile animals that live fastened to the substrate. When they are threatened or disturbed, they cannot move away easily, so they have devised other systems of "escape." The worm has very fragile gills, which it extends for feeding and respiration. If they are damaged, the worm might die. Fortunately the feather-duster secretes a rigid, fairly strong, tubular structure around itself. So, when the worm senses danger, it quickly withdraws into its tube.

The anemone does not construct a tube or other protective device, but it can withdraw into its own body cavity. When danger threatens, the anemone folds its tentacles toward its oral disc and then rolls them inside, until the sensitive tentacles are covered.

Bag of Tricks

As they inch their way along the seabed, sea cucumbers may seem extremely vulnerable. In truth their defenses are formidable. Predators are discouraged by the poisonous skin of some sea cucumbers. A cucumber that is disturbed reacts by expelling water from its body and contracting itself. A truly desperate sea cucumber resorts to a remarkable defense: it turns itself inside out, spewing out its insides—respiratory and reproductive organs and even its intestines—which entangle its hapless attacker while the cucumber escapes. In about six weeks the eviscerated organs are regenerated.

Mucous Shields

The brightly colored sea slug (*Navanax*) is relatively safe from predation. It seems that only others of its own kind find it palatable. Perhaps one reason why predators leave it alone is the yellow fluid it gives off when disturbed. Another reason may be its odor, which is unpleasant to us and may also offend fish and other animals. The parent *Navanax* imbeds its eggs in a

One of the sea cucumber's effective defenses against attackers is to eject skeins of sticky white mucus.

mucous coating to discourage predation. Adults also form mucous cocoons for their own protection.

Parrotfish are brightly colored residents of the coral reefs found in warm tropical seas. By day these fish graze on the reefs, biting off chunks of coral which they eat for the algae on and in them. By night some species sleep in a mucous envelope. This covering, which may take the parrotfish as long as 30 minutes to secrete, completely surrounds the fish and may help protect it from its enemies. Some scientists think that the cocoon may act as a barrier to prevent the fish's odor from attracting a predator.

Special glands in the parrotfish's skin secrete a delicate membrane that looks like cellophane and creates a protective mucous cocoon.

Some animals take the initiative in their own protection. Poison is one of the defenses developed by a number of sea creatures. It may be administered by teeth, spines, beaks, or barbs. A few animals have electrical properties they use to stun prey or predator. Many crustaceans and some other animals have pinching claws to capture, crush, and rip food or to defend themselves against each other or other predators.

Whether poison is or is not involved, stabbing alone with sharp spines is often enough to discourage many predators from pursuit. Biting with teeth is one of the commonest defenses, especially among vertebrates. Some bite and hold, some slash as they bite, and some bite repeatedly. Larger animals may use brute strength to club or ram an opponent. Sounds are sometimes used to frighten attackers away as much as to paralyze with fear a potential meal, while some sea creatures change their shape to appear too large for the predator.

Modern man's devices for attack and defense seem more elaborate than

OFFENSIVE DEFENSES

animal systems. In reality, however, they are all inspired by nature, with the erratic exception of the nuclear bomb. One wonders if with this invention man has not dissociated himself from nature forever.

Multiple Defenses

The delicate flowerlike sea anemones and their relatives, the hydroids, have stinging cells in their tentacles to stun their prey or protect them against predators. Some predators, however, like some of the 5000 varieties of delicate, soft-bodied nudibranchs are unaffected by stinging cells. In fact they eat them without destroying them and use them for protection. These beautiful molluscs have the amazing ability to prevent the discharge of the stinging cells as they are consumed, as they pass into the digestive tract into and up a special canal, and are finally incorporated into the cerata or gills on their backs. When predators attack nudibranchs, the ill effects of the stinging cells are passed on.

In addition, the little animals secrete a mucus, which smells unpleasant to man and perhaps makes them unappetizing to fish and other predators. One species has a specialized acid gland from which it releases a slimy sour secretion containing sulfuric acid.

Some of the nudibranchs enjoy even further advantage. They are able to swim. They propel themselves by bending their bodies from side to side with head and tail almost touching. They can also beat the water with their cerata for additional speed. By moving up into the water, they get away from any possible danger from an obstinate foe. Others are able to cast off parts of their bodies when they are under attack and get away. Later, these parts are regenerated.

Nudibranchs are among the most vividly colored animals in the sea, possessing vibrant orange, blue, purple, yellow, and red pigments. No predator could mistake any of them for a conventional prey. This is precisely the object of this eye-catching publicity, for any animal that recognizes them will not want to eat such a stinging snail. In contrast to camouflage, in other words, they are brightly colored, and advertise themselves as unpalatable.

Nudibranchs are not immune to all predators, however. Some of them fall victim to parasitic worms and copepods and are eaten by some starfish. These few predators aside, it seems that nudibranchs are quite safe in their gaudy dress and therefore make very little contribution to the ocean's food chain.

Poison-Fang Blenny

The Pacific Ocean blenny has poisonous fangs, canine teeth in its lower jaw that deter most predators. Groupers have been seen spitting out a poison-fang blenny immediately after ingesting it. The blenny had undoubtedly bitten and poisoned the grouper. The blenny's bright coloration probably serves as a warning to would-be predators that the blenny is a bad risk. As with many poisonous animals, the blenny has no fear of others larger than itself and even acts aggressively toward them. It threatens any intruder into its territory, including divers, with a series of short, jumplike strokes in the intruder's direction. There are, incidentally, two species of nonpoisonous blennies, which look very much like the poison-fang blenny, and they probably benefit greatly from this mimicry.

FIGHTING FOR TERRITORY AND SEX

When combat takes place between members of the same species, it is almost always the result of competition for territory or a mate. In both cases, possession is the motivation.

Animals generally do not fight among themselves the same way they fight an enemy. In fighting members of their own species, they meet in ritualized combat, involving the use of threat displays and other noninjurious means to defeat a challenger. The weapons they use against predators and prey (such as teeth, claws, and poisons) are rarely or partly used against each other. These weapons are intended for killing, and killing is not in the interest of the species. So they resort to violent, but usually harmless, jousting. Though there are notorious exceptions, such as deadly fights of octopus for a shelter, a growl, grunt, or squeak may be enough to drive an opponent away. Some fish beat at each other with their tail fins. The pressure wave they set up is indicative of the strength of the fish. Often, though, more subtle means are employed to show superiority. Sometimes opening the mouth wide is enough of a threat to frighten off a challenger. Color changes, ritual movements, or flaring of the fins may do the job too.

Garden Eels   Colonies of garden eels live in individual burrows dug in the ocean floor. So timid of exposure is the garden eel that none has ever been seen wholly extended by uncamouflaged divers. There appears to be a rigid social order in garden eel colonies, with dominant males, harems, and firm property lines. A challenger from an adjoining territory is met with a ritual display. Stretching from the burrow, the eel performs snakelike undulations, ripples its dorsal fin, and turns its profile and then its back to the attacker. If the

Sea anemones and hydroids (left) stun almost everything that comes in contact with them.

123

challenger persists, the eels will square off, snout to snout, and strike at each other.

Lobsters in Combat

Atlantic lobsters in combat use their large pincers and, unlike many other creatures in the sea, actually do bodily harm to each other. In this fight for territory, one lobster may succeed in literally disarming the other. The loser could be killed and eaten by the cannibalistic opponent. Or it could limp off to regenerate a new claw to replace the old one. Regeneration to full size may take several molts over a two-year period. Atlantic lobsters usually have their hard exoskeleton to protect them against other lobsters or other marauders, and if necessary, they can shed their claws to escape.

Lobsters are most vulnerable immediately after they have molted. Males and females alike are subject to harassment and dismemberment by their own kind and others when their new shells, composed of soft, proteinaceous material, are no stronger than wet paper. During this time they have their greatest need for calcium carbonate, which provides the hard substance of the shell, and it is not uncommon for a lobster to consume its cast-off exoskeleton immediately after shedding it.

Fighting for Social Status

Battles between elephant seal bulls often occur during mating season and are usually intended to determine the social order within the herd. Before body contact is made, the bulls threaten each other with an inflated snout, a raised stance, abrupt aggressive movements, and alarmingly boisterous bellows. Only one in 60 of these confrontations gets beyond the threat stage. But if the males are comparable in size, weight, and aggressiveness, threats are ignored, the animals square off, and the fighting begins.

The two behemoths stand chest to chest, feinting and faking, waiting for an opening to fight. Finally, with a fast and powerful blow, one strikes at the neck of its opponent. The attacker's head slashes downward and its sharp teeth rake the opponent's flesh.

When one has had enough, it backs away, conceding defeat. The loser has not necessarily received the worst of the battle, but for some reason it chooses not to hold its ground. The bleeding is profuse and the wounds are deep, but they heal quickly. Fortunately the neck and chest of the elephant seal can stand up to this rugged treatment and have a horny layer of tissue to help protect the animals from serious injury.

ANCIENT ANIMOSITIES

Relationships among people and other animals take many forms. Some species get along with each other. Other creatures have so little to do with one another's lives that theirs must be termed a nonrelationship. Some animals even help each other. But there are a few species which seem to have a natural animosity toward other particular species. Such are cats and dogs, cobras and mongooses, and, in the ocean, sperm whale and giant squid.

Why and how do these traditional animosities arise between species? When two species don't get along, their aggressive behavior could be emotional or calculated. Perhaps hundreds of thousands of generations ago the members of two species competed for the same habitat, territory, food, or ecological niche. In competing, they may have resorted to combat. And perhaps this combativeness has continued as one species faces the other even after many generations and the two are no longer competitive.

Octopus-lobster, moray-octopus, and lobster-moray are three couples with such built-in animosities. Mediterranean fishermen tell stories of traps that they pull back to their boats, sometimes containing an octopus, a lobster, and a moray eel: the three retreat to the three corners of the trap as far from the others as possible, because they know the first one to attack will be immediately killed by the third party.

Sharks vs. Dolphins

In the deadly but careful game of shark vs. dolphin, the mammals usually eventually win. But not always. The greater intelligence and vitality of the dolphin give him the advantage. In great numbers, sharks always trail packs of dolphins, waiting for a dropout—an ill animal or a young one—to fall behind the rest of its group. When feeling threatened by the shark, the dolphin turns on the speed and slams beak-first into the gill area of the soft lower abdomen of the sharks, their most vulnerable spot. The dolphins' beaks—their pointed jaws—are efficient weapons for such blows.

Several marine laboratories are studying the shark-dolphin relationship in the hope of making the dolphin's behavior useful to man. Experiments on lemon sharks and bottle-nosed dolphins show that if given the choice, sharks will avoid dolphins. The researchers have also been training dolphins to be used for shark control. One dolphin has been taught to ward off sharks in captivity on command from a sonic device. The dolphin, on cue, will chase and hit the shark. Soon the scientists will conduct these experiments in the open sea, hoping to employ dolphins to defend divers from sharks. Someday such trained dolphins may help oceanauts by acting as watchdogs around undersea habitats, or they may police coastal beaches, warding off sharks and protecting swimmers.

INVISIBLE
MESSAGES

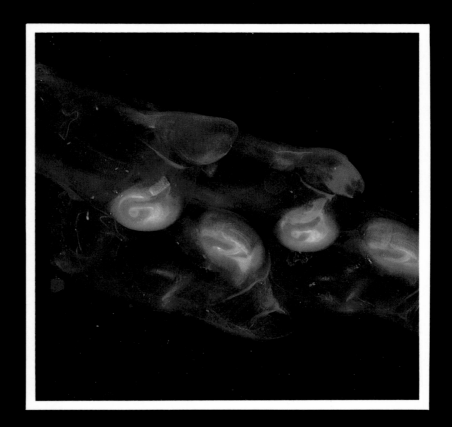

CHAPTER
6

SEEING
WITHOUT
EYES

Soon after the laws of gravity were formulated, astronomers painstakingly calculated orbits of celestial bodies. They had time to spare and no deadlines. This is no longer the case for astronauts who are launched on rockets for interplanetary voyages. When a capsule makes a soft landing on a foreign planet, its rockets have to be fired within milliseconds, the time lapse being a function of various data (speed, acceleration, gravity, distance, etc.), which change very rapidly. No human brain could match the requirements, and high-speed computers had to be produced and programmed for such vital reckonings.

In the animal world, on land and sea, it is the central nervous system that acts like a high-speed computer, receiving and interpreting the approach of enemies, or of food, or of a mate—and directing the musculature to respond appropriately. The switchboard of the system, the brain, is naturally only as efficient as the senses that inform it. And the priority of senses is not the same on land and in the sea. In the sea the paramount land sense, sight, is rarely prevailing. A marine animal relying mainly upon sight is imprisoned in a bubble of perception only 200 feet across. In the ocean, animals have developed nonvisual senses, which are able to receive "invisible messages."

Many of the "invisible messages" produced by undersea creatures cannot be read or understood by man. Many years ago we took *Calypso* into the equatorial area of the Indian Ocean three different times for the entire month of April. The two first years, for many days from dawn to dusk, the ship was escorted by dozens, sometimes hundreds, of sperm whales, spouting, breaching, and raising their great flukes in the air to sound. Herds of dolphins, nothing apparently on their minds but the joy of exercising their command of the liquid world, rode our bow waves and leapt out of the water in show-off acrobatics. The third year, in exactly the same region of the sea, all was a desert before us. We saw a few dolphins, but they nervously fled from us as though we harbored a mortal disease. What could have happened? It took us two weeks to find out. Then, miles away from the area, we encountered several groups of killer whales.

We humans had no knowledge that formidable carnivores were closing in on the playground of the previous springs. But the whales and dolphins knew it and cleared out.

Sound is the key information medium of the great sea mammals. The sound-producing and sensing organs of the whales and dolphins perform a function as vital for them as that of eyes for the eagles. By varying the frequencies of the vibrations they emit and correctly interpreting the reverberated sound waves as they bounce back, the big animals gain an extremely detailed picture of their surroundings in any condition of murk or darkness. With zero visibility they can judge distances, distinguish between the sizes (and probably the species) of fish in the neighborhood, find their way through the mazes and obstacle courses of jagged defiles in the submerged mountain ranges.

Sound is not the only important sense underwater. Fish have evolved an organ which has no parallel on land: the lateral line—sensors that pick up pressure disturbances in the waters around the animals, even if they originated very far away. Smell, touch, taste, and special senses to detect gravity as well as magnetic or electrical fields play their part. A majority of radiations remain unnoticed—even if they have an influence on our behavior.

Man is only beginning to comprehend the range of "invisible messages" the universe produces, much less read them all. Radiation and vibrations permeate all living things; each creature has its own limited scope of perception, and man is extending his own senses with the help of instruments. The sea is just different enough from land to be loaded with helpful hints. Replacement senses are inspired by the sea. For example, British scientists have built acoustic goggles for the blind, transmitting and receiving ultrasonic signals, very much in the way dolphins do. They enable a sightless man to "see with his ears." But essentially, as man develops his capability of tuning in more and more to the myriad "invisible messages" in our universe, he expands his intellectual horizons, his sources of inspiration, and his artistic and philosophical creativity.

Garibaldi evicting a sea cucumber

THE RECIPIENTS OF INFORMATION

A primitive animal, the flatworm has an organ analogous to an eye made up of tiny pigment spots. But the worm's ability to "see" is limited to the simple reception of light.

Every undersea animal receives a constant flow of information from its environment and from other animals. These messages may indicate danger or desire for mating; they may demarcate territories or announce the presence of food. Information is transmitted by a variety of means, including light, sound, pressure waves, chemicals, touching, and electricity. Even tidal rhythms, gravity, and the characteristics of water itself impart knowledge. Whatever the means, messages are constantly permeating the sea, and animals able to receive and use them gain great advantage over those that do not have this capability.

Some basic awareness is found in all creatures, but the complexity of the information an organism receives depends largely upon the complexity of the organism itself. In the simplest animals, individual cells respond to many stimuli, while specialized sensors are found in higher forms of life. Sponges, for example, are a collection of loosely organized, unspecialized cells. Each responds independently to stimuli. If a sponge is pricked with a sharp object, it responds by contracting, but this response is slow and may be localized. These nonspecific cells will react, as will any living organism, if they are bombarded with sufficient stimuli—whether it is sound, electricity, or pressure—but the amount and quality of the information they are capable of receiving is extremely limited.

Flatworms

Flatworms have primitive eyespots containing pigment, which enable them to distinguish between light and dark and, because of two receptors, determine the direction of the light. They also have two other types of sensors. One is sensitive to chemical stimuli, perceiving far-off substances by smell and sensing them on contact by taste. Another is stimulated by the passage of water over the flatworm's body surface. This is perceived by special rheotactic sensors.

The flatworm's brain is merely a swollen mass of nerves connecting two nerve cords, but it can effectively interpret information and even has the capacity to learn.

Sea Anemones

Like all animals, the sea anemone is sensitive in some degree to chemical changes in its environment. When juices of food are placed close, it reacts by expanding its body and waving its tentacles. When the food is removed, the reaction stops. And the sea anemone can apparently discriminate among the foods placed near it. It reacts most to meat and other forms of protein. Starches and sugars evoke only a slight response. There is virtually no response to inert objects. When food is sensed, activity begins; the tentacles reach out, grasp the food, and move it to the mouth.

This deep-sea shrimp has internal pockets lined with sensory tissue and containing sand that enable it to gauge its depth.

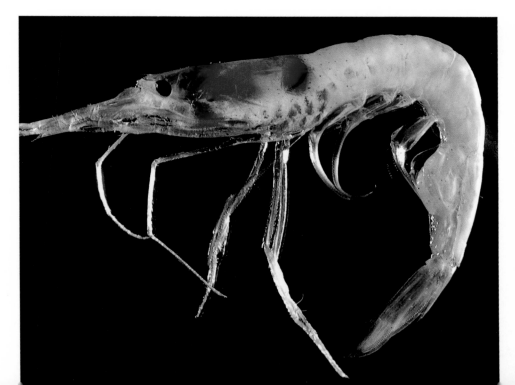

130

Salps

Barrel-shaped salps are classified in the same broad group of animals that also include men—phylum Chordata, animals with spinal cords. Yet the salps are relatively primitive creatures.

In their larval stages they have a spinal cord as well as a concentration of nerve cells that could pass for a brain. In the course of their development into the adult stage, the brain and spinal cord degenerate and disappear. What little sensory equipment they have consists of a number of tactile and chemoreceptor cells that taste and smell on the inner and outer surfaces of their siphons, the organs that carry water in and out of the salps for respiration and nourishment. Because their requirements are so elementary, whatever elaborate sensory system the salps may once have had has degenerated, until today it is very rudimentary.

Sea Cucumbers

In and under the skin of the sluggish, slow-moving sea cucumber is a network of nerves, stemming from a nerve ring located near the base of the tentacles. Interspersed with that network are a large number of wartlike projections on the body surface, which house nerves connected with the network. All of these nerve fibers are linked to sensors that receive messages through the water—chemical ones, which the sea cucumber may smell or taste, or tactile ones, which the animal feels. The network in and under the skin is like a mesh of fibers that covers the sea cucumber's body and is most sensitive at the two ends. The sea cucumber's tentacles are also supplied with nerves from the circum-oral nerve ring. These apparently are for taste, smell, and controlling tentacle movements.

Octopods

The octopus has a well-developed nerve ganglion that acts as a brain and a complex nervous system. Since its eye resembles our own, its visual sense is

When the sedentary sea anemone senses the presence of a meal, its mouth opens and each tentacle responds, gathering the food.

The moray eel, a fairly common reef resident, hides in caves by day and hunts by night, tracking its prey largely by scent.

thought to be very good, and the senses of touch and taste appear to be well developed too.

An acute sense of touch is especially important to the females. They lay their fragile eggs in long strings and attach them to the walls or ceilings of their secluded lairs. The mother manipulates the strings almost constantly for two or more months, caressing them with her suckers and tentacles to keep dirt and fungus growths from accumulating on the eggs. Apparently none is damaged, a most remarkable feat considering the amount of time she spends handling them. Were her sense of touch less delicate, this would probably be impossible for her to do.

How do octopods sense by touch? The octopus' eight arms, or tentacles, radiate from its bulbous head and body. Each arm is lined with two rows of cuplike suckers, which are used to grasp objects. The sense of touch is particularly acute in these suckers, and the rims of the cups are especially sensitive. When an octopus encounters an object that arouses its curiosity, it examines it with the suckers, pressing them over its surface. In experiments, blinded octopods, using only their sense of touch, were able to differentiate between objects of various shapes and sizes as well as normally sighted octopods were able to do.

SENSING THE PULL OF GRAVITY

Most animals are sensitive to the pull of gravity—even those that live in the sea, where this pull is nearly balanced by the buoyant force of the water.

In the higher animals the equilibrium organ is in the ear, but some lower animals, such as shrimp, lobsters, jellyfish, and sea cucumbers have special equilibrium organs called statocysts. Impulses are sent to the central nervous system from the statocyst, enabling the animal to tell up from down. An animal may have one, two, or several statocysts. Shrimp and lobsters have two, one at the base of each antennule (the sensors that are adjacent to, and look like a second pair of, antennas).

The statocyst is a chamber filled with fluid and lined with sensitive tissue. Hairs grow from this tissue, projecting into the fluid. Tiny particles, called statoliths, may be calcium carbonate, sand, or pebbles; they rest on the bed of hairs within the statocyst.

When a shrimp or lobster is standing on the ocean floor (its normal position), the statoliths, pulled down by gravity, press on the bottom of the statocysts. The sensory hairs on which they rest are bent by their weight and trigger nerve impulses to the nerve center. When the animal changes position, the statoliths move, bending other hairs and changing the nerve impulses that inform the animal of gravity.

KNOWLEDGE THROUGH SCENT

Blood gushes from a wounded fish and water carries it down current. Animals in the path of the blood sense it in different ways. In the distance some fish pick up traces of the blood with their olfactory receptors. They are said to smell it. Those close up may move in and mouth the wounded, bleeding fish and thereby sense directly the flood of the blood molecules. These animals are said to taste it.

The senses of taste and smell, although separate, rely on the same basic principle—the reception of various molecules. Matter, whether liquid or solid, is made up of molecules bound together chemically. When these molecules enter the water, they may separate from each other because of the chemical action of the water, a powerful solvent. The molecules are diffused through the water, carried on ocean currents, or even moved by electric currents.

All animals are sensitive in varying degrees to changes in the chemistry of their environment. A change may be caused by the release of blood from

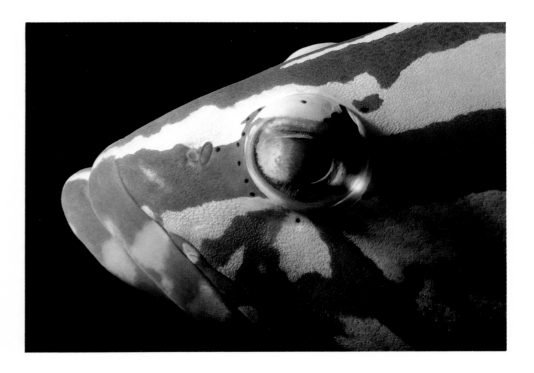

The Nassau grouper moves about very little. Its double nasal opening gives it a highly efficient olfactory sense, important in providing long-range information about its environment.

the wounded fish, by release of "alarm substance" from a frightened animal, by man's wastes dumped into the sea, by changing salinity, and by other factors. As already mentioned, some animals secrete powerful substances called pheromones that convey information to others and often evoke specific responses.

The senses of taste and smell are developed to different degrees among invertebrate animals. In the most primitive animals, the protozoans and the sponges, these senses exist, but not in the same manner we usually define them. In animals a little higher up on the evolutionary scale, the senses of taste and smell exist and function to some degree. Molluscs like the deadly cone snail have sensory patches (osphradia) located close to their gills, which react to chemical changes—that is, they are able to smell. Some also have taste buds. Some crustaceans have chemosensitive cells over their whole bodies; these cells are located directly beneath the animal's shell and are connected to the surface by minute pores and ducts. In fish the sense of smell lies in the nasal pits.

Man and other mammals cannot smell anything underwater, but fish can. Why? Our olfactory sensors are located in our noses. When we submerge, air is trapped in our noses; and if water is allowed to penetrate our nostrils, the burning pain due to the difference of salinity between water and human tissues obliterates any sensitivity to odors. Some marine mammals seal off their nostrils with flaps of skin to ensure that no water enters them. In contrast, fish do not breathe through their nostrils, so water may pass through them without affecting their respiratory process. Most fish have one olfactory receptor on each side of the upper portion of the snout. Each receptor is a pit, which is lined with sensitive tissue, folded into a series of ridges and valleys. The folds increase the amount of tissue exposed to the water, without increasing the size of the receptor. The nasal pit is covered by a roof, protecting the delicate tissue inside. One or two holes in the roof admit water.

There are several ways in which water can circulate through the nasal pit. If there are two openings in the roof and the fish is a fairly active swimmer, its normal swimming motions may force the water in one nostril and out the other. The water enters the forward opening, where it swirls over the sensory tissues and then exits through the rear opening. In some fish a small ridge at the back side of the front nostril acts as a funnel, directing the flow into the opening. The shortcoming of this type of circulation is that it only works when the fish is swimming or facing a current.

A second way in which water is brought into the nasal pit is by a pumping action. The movement of a fish's jaws, forcing water through its gills for breathing, is linked to the nostrils, and the motion draws water into and forces it out of the nasal pit. Fish that are not too active benefit from circulation by pumping, since they can smell even when they are not moving.

Finally, water may be circulated by the cilia, hairlike projections growing inside the nostrils. The beating motion of the cilia is only effective in driving the water through narrow, enclosed spaces. The tubular extension of the forward nostril of eels is ideally suited to this end. Circulation of water by cilia is not fast, but the water in the nasal pit is constantly changing.

Murex Snails

Although many snails are scavengers, the murex snail prefers the meat of a live clam or oyster to a dead one. And it has developed an elaborate apparatus for getting at the meat of its victims.

How can the snail tell the difference between live and dead clams? Apparently the answer is chemical. Experiments have shown that a small amount of a chemical substance emitted by live oysters is enough to attract the murex and other related snails to their prey. This sense of "smell" is very well developed, and although the exact distance at which it operates is not known, it is believed to be considerable.

When the snail discovers this scent in the vicinity, it goes in search of the meal, finds it, and attaches itself to its prey. It then bores a hole in the shell by means of a chemical and a filelike radula, then rasps out the victim's tissue.

Atlantic Lobsters

The Atlantic lobster has been found to have a chemical language. The chemicals used in this language are pheromones. For each message to be sent, there is apparently a different pheromone. There is one identifying the sex of an individual lobster. Another may tell of aggressive intent or intent to demand territory. Others tell of a desire to mate and can be detected at a distance. They are apparently strong and persistent.

When researchers placed a sex pheromone from a just-molted female lobster in with an aggressive male lobster, the male began a mating dance although there was no other lobster in his tank.

ELECTRICITY AND COMMUNICATION

All animals, terrestrial as well as aquatic, within their specialized muscle, glandular, and nervous tissue produce minute electric charges, once called "animal electricity." A few families of fish have learned how to receive and interpret these charges.

Some fish can also produce and deliver a considerable electric charge using organs made of highly modified and specialized muscle tissue no longer used for movement.

The ability is not equally developed in all current-producing fish. Those with the most powerful shock use it as an offensive and defensive weapon, discharging when they detect prey or when they are disturbed. Other animals, with less powerful abilities, use their low power discharges to supplement information received by their other senses. The electric field they generate aids them in navigating and finding food and may help them locate mates.

The tissue that makes up the batteries of electric fish looks something like a stack of coins, each coin being a single cell called an electroplaque. An electroplaque resembles a cream-filled cookie with a different type of wafer on either side of the filling. The two wafers represent the different membranes on each side of the electroplaque's jellylike center. Nerve fibers connect to only one of the membranes.

Each electroplaque is a tiny battery, producing a small charge between the two membranes. All the electroplaques in a single column face in the same direction, so they are connected in "series." Batteries connected in series add their voltage outputs to one another, so the column of electroplaques creates a difference of potential that is the sum of the many small voltage differences of the individual electroplaques. The charges are released on electrical impulses from the nerve fibers.

In some fish, such as the electric eel, there are only a few very long columns of electroplaques. These run lengthwise, parallel to the fish's backbone. In others, like the torpedo ray, there are many more columns, but these are not nearly so long, and they are situated perpendicular to the backbone. The longer columns produce a high voltage charge, and more numerous columns produce a charge with high amperage.

When a coil of wire is rotated within the field of attraction of a magnet, an electric current and resultant field is generated. Electrical charges are generated in the oceans of the world in much the same manner. Great rivers, the ocean's currents, flow across the face of our planet, and in doing so pass through the magnetic field of the earth. They become the equivalent of the coil of wire responding to the magnet. Electrical fields are generated. We

Olivella snails are small burrowing animals that usually travel beneath the sand, leaving long trails behind them. Tests on similar species of mud snails show that their movements can be influenced by the presence and intensity of a magnetic field.

know that sharks have special organs that can sense electrical stimuli, including that generated by the ocean currents. But do sharks use the electrical field in the oceans to navigate?

Many of these fish live in the open ocean (rather than in semienclosed bays and harbors) and some range thousands of miles about the sea. It may be that they have no need to navigate and that it doesn't matter where they are. Yet some studies have shown that a few sharks regularly inhabit a certain area or a particular type of habitat and navigational ability would help them locate the kind of area they prefer.

Electric Skates and Rays

Both the torpedo and narcine rays have a pair of electric organs on each side of their heads. The larger of these paired organs can generate enough power to stun a would-be attacker or the fish's prey. One species of torpedo ray has an additional pair of electric organs. It is speculated that these are used as a navigational aid. As they move about their environment, they may carry their electrical fields with them. When that field is disrupted by an obstacle, the ray may sense the disruption or distortion of the field.

The thornback skate's electric organs are more of a mystery. It has a pair of electricity-producing organs that are formed of modified muscles, one on each side of the base of its tail. No one knows how or why the skate uses its ability to generate electricity. The skate is reluctant to discharge current and does so only after considerable prodding. The skate's electrical charge is of low output, only about four volts, while the torpedo ray may discharge 40 volts or more. One torpedo was measured as discharging 220 volts and a fairly high amperage, but this is the highest voltage observed in the torpedo.

The Ampullae of Lorenzini

Sharks, rays, skates, and chimeras, known collectively as elasmobranchs,

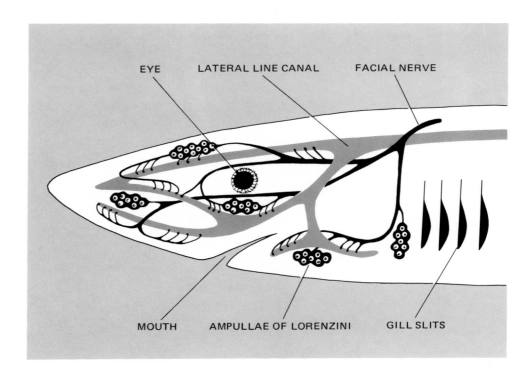

EYE LATERAL LINE CANAL FACIAL NERVE

MOUTH AMPULLAE OF LORENZINI GILL SLITS

A torpedo ray rests on a sandy sea floor ready to ward off intruders with an electric charge contained in its whiplike tail.

are among the most primitive of fish, with skeletons of cartilage instead of bone. But they have some advantages over the more advanced and more fully developed bony fish. One of these is the ability to sense electricity with the aid of special sensors called the ampullae of Lorenzini. These ampullae are scattered over the top and sides of the heads of these animals—evidenced outwardly only by small pores in the skin.

The ampullae of Lorenzini are small sacs filled with a jellylike substance. These sacs are connected to the surface skin by ducts leading to pores. The linings of the sacs are sensitive tissue with many folds, bumps, and other irregularities. The ampullae are able to sense electrical fields from various sources in the water. Careful measurements showed that the ampullae can sense as little as one-tenth of one microvolt (one one-millionth of a volt) of electricity.

In one experiment, a marine biologist trained a group of sharks and rays to eat in an area directly over a pair of electrodes buried in the sand bottom. When he fed the fish, the current was turned on and emitted four-tenths of a microvolt. Then he stopped putting food out, but when he turned on the tiny electrical charge, the sharks and rays swarmed about the electrodes, even uncovering them and snapping at them. Thus, he showed they could sense minute charges of electricity and could also trace it to its source.

The ampullae have been thought to have other functions than sensing electric fields. When tissue from the lining of the ampullae was taken from a fish and studied, it was found to emit rhythmic pulses of electrical energy in minute amounts. These rhythmic pulses varied with the stimulation given the tissue. When the temperature was lowered, pulses speeded. When temperature was raised, pulses slowed. The tissue also reacted to slight changes in salinity, leading some researchers to believe the ampullae can sense variations in salinity.

VOICES AND DRUMS

The sea was once thought to be a silent world. Today we know it can be a noisy place, with a variety of sounds. Scientists have found that no fish, crustacean, or other invertebrate animal has vocal cords to help it produce sounds in the same manner we do. But they use a number of other methods to do so. The mammals, too, produce several types of sounds. They may click, wheeze, cluck, or rumble.

139

This juvenile jackknife cartagena (Equetus punctatus) *makes sounds by vibrating strong muscles in its swim bladder.*

Besides the variations in the types of sounds fish produce—the scrapes, bumps, rumbles, clicks, and staccatos—the sounds themselves may be variable in length and may be given at different intervals. All these variables give the fish a sizable number of messages it could send, if, in fact, that is what the sounds are produced for. Evidence does indicate that fish use specific sounds for specific purposes—to signal aggressive intent or, in courtship, for navigation, or for schooling. Their messages must, however, be of only a very general nature and cannot be considered a language in our sense of the term.

Grunts, Croaks, and Drums

Fish of the grunt family get their name from the gruntlike sounds they emit when grinding their upper and lower pharyngeal teeth, far down in their throats. Their swim bladders act as sounding boxes, resonating and amplifying these sounds. The sounds can be heard distinctly by man when the fish are caught and removed from the water.

Croakers and drums are among the best known of the sound-producing fish. Their swim bladders serve as resonating chambers for the sounds produced by the vibrations of strong muscles attached to the bladder walls—in the same way as the sounding box of a guitar amplifies the vibrations of the plucked string. These muscles vibrate about 24 times per second. Drums can produce sounds at will, giving off noises of varying pitch—from deep, drumlike thumps to higher-pitched sounds.

Toadfish

The toadfish is a very vocal fish capable of producing two types of sound—grunts and boat whistles. The boat whistle (also called the foghorn) sound is a low-frequency burst of tone which lasts about one-half second. These sounds are emitted more frequently during the reproductive season. Highly competitive males are more vocal when a female toadfish passes by the male's nest.

Mussels and Clams

The primary background crackling sound produced outside tropical and near-tropical temperate waters is the sound of mussels and clams snapping their anchoring threads as they move. These tough byssal threads start as a viscous material secreted from between the clam's or mussel's valves. Exposed to seawater, the viscous protein sets to a tough, flexible consistency, strong enough to withstand the force of breaking waves.

The Poppin' Pistol

Pistol shrimp have been called the gunmen of the reefs and tide pools. The

140

pistollike sound of these one- to two-inch-long shrimp is made by clapping two parts of the large claw together. The small part of the claw is held perpendicular to the large part of it. When the small part, like a thumb on a mitten, is closed onto the large part, or palm of the mitten, it snaps over a ridge. This produces the sharp report that can stun a small animal nearby by the concussion it generates, and is loud enough to trip sonic submarine detectors. Pistol shrimp use their ''guns'' to defend the burrows they live in and to stun their prey. If they lose the large claw, the small one grows up to replace it while the missing one is regenerated.

Vibrating Antennas

Divers who have hand caught spiny lobsters in the tropical waters of the Caribbean have felt the vibrations produced by their stridulatory or noise-making apparatus. These noise-making devices are located at the base of each of the lobster's two large antennas. The antennas themselves pick up sounds as they vibrate through the water. They are covered with tiny sharp spines and also serve as defensive weapons. When the diver first grabs the lobster, the animal's stridulatory mechanism starts vibrating and can be felt even through the diver's gloves. The diver may also be able to hear the sound, which is audible to most aquatic animals. The sound may be meant to frighten predators.

Fiddler Crabs

Fiddler crabs spend much of their time near the entrances to their burrows on tropical beaches. They are able to hear vibrations through the ground they stand on. Males produce either a rapping sound or a honking sound as courtship signals. The rapping sounds are produced by drumming on the ground with the large claw. Leg movements are thought to produce the honking sound. The females can hear the low-frequency sounds the males produce from more than 30 feet away.

Extremely vocal, the toadfish belongs to a family with a wide repertoire of sounds.

Pinnipeds

Seals, sea lions, and walruses have well-developed vocal cords and make use of them to bark or growl for a variety of reasons. In addition, they make clicking sounds underwater, which also originate in the larynx or just behind it. Walruses click their teeth but also produce bell sounds using air bags they have in their throats.

A major reason sounds are produced is for recognition, a very important function between a mother sea lion and her pup. Pups produce a sound described as "aa, aa, aa." This provides a means of recognition from a distance for the mother sea lion. (Scent probably plays the most important role in recognition up close.) Clicking sounds may be part of an echolocating system. If one sea lion challenges the dominant animal of its group, vocalizations are one of the ways in which the number one animal asserts its leadership. This usually takes the sound of "goh, goh, goh." Other adult sea lions emit sounds more like "ga, ga, ga." Pups start out with a high-pitched sound, which deepens with age and growth.

Whales

When a rorqual whale calls as it cruises in the surface waters searching a mate, its voice carries for about 50 miles. If the sound of all the ships at sea were stilled, the rorqual's voice could be heard by others of its species for close to 150 miles.

SOUNDS AND PRESSURE WAVES

When a solid object moves through air or water, it causes the molecules around it to move. The molecules in front of it are pushed together, or compressed. This compression of the molecules is called a pressure wave. The molecules behind the object spread out to fill in the void left by the moving object.

Water molecules are much harder to compress than molecules of air and are also to some extent bound together. This means that less energy is lost in the wave motion than in air, and sound travels about four times farther and faster.

Under some conditions we can feel a pressure wave on our skin, as we do when we are standing near a road and a car or truck whizzes by. Then we are buffeted by the wave of air being pushed ahead of the vehicle. When pressure waves have certain other characteristics, we can no longer feel them on our skin, but sense them with our ears as sounds, which are a form of pressure waves.

If the object does not move continuously in one direction but moves

Sharp-eyed, a California spiny lobster watches for danger, ready to retreat into a crevice in the reef.

The blind cavefish lives in a lightless, subterranean world, and over millennia it has lost its ability to see. Its other senses have therefore become even more important to its survival. Strangely, the cavefish does not have the lateral-line system common to most fish for the detection of pressure waves. Instead, it has numerous sensory papillae on its head to feel vibrations caused by nearby movement. It apparently has difficulty in sensing stationary objects, since in aquariums it continually bumps into objects as well as the walls of its tank.

back and forth about a fixed point, it sets up pressure waves first on one side (as it moves one way) and then on the other (as it moves in the opposite direction). If the object moves back and forth again and again, it creates a series of pressure waves on each side, which travel away from the object. Each complete movement is called a cycle.

If we count the number of cycles that occur in a period of time, we get the frequency of the movement. When discussing sound, the frequency is timed in seconds. When the frequency falls between 20 cycles per second (hertz [Hz]) and 20,000 Hz, humans can hear them. Frequencies in this range are called sonic. If the frequency is less than 20 Hz, it is called infrasonic, and if it exceeds 20,000 Hz, it is termed ultrasonic.

Whales and dolphins have highly developed hearing, particularly sensitive to ultrasonic sounds. Hearing is probably their most important sense, but their ability to hear well was turned against them by early Japanese whalers. These men drove dolphins and whales into shallow bays by beating on the sides of their boats with hammers.

Most fish can best hear sounds that are in the 200 to 600 Hz range. Some species of fish are sensitive to sounds with frequencies as low as 10 Hz, and others may hear frequencies over 10,000 Hz. It is difficult to determine whether a fish hears or "feels" the pressure waves with its lateral line, which appears as a line or series of dots running down the sides of the fish, because sound pulses or turbulences are all basically the same phenomenon.

The lateral-line system is especially sensitive to low-frequency vibrations caused by movement underwater. It gives fish a sort of long distance sense of touch and keeps it informed about nearby animals and objects. Some predators, like the barracuda, use it to detect prey far beyond sight range, and it is of vital importance to fish schooling in dark or murky waters.

The lateral-line sense organs of fish are similar in design to those of its ear. They are contained in canals, which are stretched out along each side of the fish's body, and branches of which encircle the fish's head. The canals are located just beneath the skin, and pores through the skin or scales open them to the water.

The basic components of the lateral-line system are the neuromasts that function in the same way as the cristae of the ear. These are collections of sensory cells, each with tiny hairs projecting from it. When a disturbance near the fish sets up a pressure wave, the moving water strikes the fish and disturbs the mucus in the pores and canal. The mucus, in turn, jiggles the hairs of the neuromasts. The hairs stimulate nerves, triggering the discharge of nervous impulses to the brain.

Pressure waves may also play an important role in the social behavior of a fish. In the ritualized fighting before mating, some fish fan their opponents with their tails to establish dominance. Some other fish fan their mates as part of the courtship procedure between male and female. The vibrations set up by the fanning probably convey appropriate messages to the animals involved.

A dolphin's built-in sonar keeps it constantly aware of its surroundings.

USEFUL ECHOES

To understand how sound can be utilized by man and animals, we must know that sound waves can be bent and reflected. If we remember the alternating compression and expansion of sound traveling through any material, air, or water, we are ready for a new concept—wavelength. If we measure the distance from one compression to the next this distance is called wavelength. The higher the frequency, the shorter the wavelength. Since the

Using echo-location, a blindfolded dolphin retrieves a ring.

speed with which sound travels in various materials is known, by knowing the frequency of the sound, we can compute its wavelength.

Much of the sound will be reflected by an object that is large in comparison with the wavelength of that sound. These reflections are called echoes. Sounds flow around objects that are small compared with a wavelength. The fact that objects will reflect sound has been used by man in navigation to locate icebergs or the bottom, in commercial fishing to find huge schools of fish, and in warfare to pinpoint the position of submarines.

We know that materials of different densities reflect sound, therefore sound in the ocean bounces off the surface, the bottom, and those masses of water of different temperature as well as animals and plants. Layers of water with different temperatures also cause those sound waves that are not reflected to be bent. Thus a sound wave sent through the ocean is spread, absorbed, bent, reflected, and scattered. The higher the frequency of the sound, the greater the effect. For this reason most long-range ship echo sounders operate at frequencies below 5000 Hz. To detect small objects at much shorter ranges, sonars operate at frequencies beyond the range of man's hearing, that is, above 20,000 Hz.

Living Sonar

For almost every invention of man an equivalent system exists in nature that far exceeds man's in efficiency and capabilities. One of these is animal sonar, or echolocation.

In 1938 it was discovered that bats emit high-pitched, inaudible sounds, often called ultrasounds (about 40,000–80,000 Hz), and receive echos that tell them a great deal about their surroundings. About ten years later, observations of a scientist in Florida led to the discovery of echolocation in dolphins. During an attempt to capture dolphins for a seaquarium, the scientist noted that dolphins could be herded into a canal and in the direction of a net. However, within 100 feet of the unseen net, they abruptly changed direction and swam away. But the dolphins swam into captivity if nets with a larger mesh—or water-soaked ones that had no clinging, sound-reflecting air bubbles—were used.

To obtain information about their environment, dolphins emit sounds of frequencies ranging from less than 2000 to more than 100,000 Hz. We hear the audible range as a series of clicks. These clicks may be given as individual sounds or as trains of sounds strung together. The dolphin, and other members of the order of toothed whales, can determine not only the range and bearing, but also the size, shape, texture, and density of objects. It may also be able to perceive more information than we can simply by varying the pitch of individual clicks of each train, and each echoed click, being different, may bring back a different message. Thus one single train of echoes gives a composite mental image of an object. It is the character of those clicks as modified into an echo by the target that informs the dolphin of the object's makeup.

There are at least four types of information in the echo: the direction from which the echo returns, the change of frequency, the amplitude of the sound, and the time elapsed from emission to return. As the dolphin scans, it can determine the direction from which the echoes are returning and thus the bearing of the object under scrutiny. The changes of frequency tell about its size and shape. The sound's amplitude and the time elapsed give clues to the distance.

How the clicks are produced and emitted and how the dolphin perceives the echoes is only now beginning to be understood: the clicking emissions originate inside the dolphin's head. The sounds are produced even while the animal is underwater without loss of air, suggesting that the air is recycled within the dolphin's respiratory tract.

The sides of a dolphin's head and its lower jaw, containing an oily fat, are the areas that receive echoes. The melon in its forehead is very likely the source of echolocation clicks.

When a dolphin is traveling, it usually moves its head slowly from side to side and up and down. This motion is a sort of general scanning; it enables the dolphin to "see" a broader path ahead of it. But then if it gets interested in a small target, such as a fish in murky water, its scanning head motions become fast and jerky. The explanation finds roots in the fact that low frequencies are far-reaching but not directional, and high-pitched clicks are for short-range, high-definition investigations.

Unlike high-frequency sound, low-frequency vibrations are probably received initially in the inner ear. To receive and interpret all these echoes, the dolphin's brain has a much larger auditory lobe than our brain.

There is, of course, no way of knowing what a dolphin hears; we cannot imagine hearing the shape and distance of an object. The dolphin's system is amazingly accurate and gives the animal many more times the information than is obtained by man from his sonar. For example, navy-trained "Dolly" is capable of retrieving three pennies thrown simultaneously in three different directions; the first is picked up while it is still sinking in midwater, the second and third are found in the sediment, in a few seconds, with a few feet of visibility.

Language is a communication of thoughts and feelings. Man is unique in the animal kingdom in being able to communicate through specific, well-defined vocal patterns as well as through written transcriptions of his expressions. The question is: Are there any other animals besides man with a language as we define it?

On land there is no animal equipped with a brain comparable to man's. But in the sea there are several mammals, including orcas, sperm whales,

IS THERE ANOTHER LANGUAGE?

A herd of white-bellied dolphins frolics across a calm sea.

dolphins, and porpoises, that have brains that are at least anatomically equal to man's in size. They are the only creatures on earth to be gifted with the nervous system potentially capable of higher thought processes. The same animals happen to have the ability to produce a wide variety of sounds. This is not the case for the dog (small brain, limited voice), the apes (small brain, limited voice), the parrot (voice but small brain), and so on.

Some captive dolphins have reshaped their sounds to mimic the whistles of men—perhaps attempting to establish a basis for interspecies communication. This ability to manipulate sounds is encouraging, but we should not forget that a parrot can also mimic human sounds and produce them on cue.

Experiments have been conducted in hope of proving that dolphins do communicate and exchange ideas. Two dolphins have been placed in adjacent tanks with a "telephone" consisting of a transmitter and a receiver submerged in each tank. Without the telephone the dolphins could not hear each other, but with it their vocalizations were transmitted electronically and they could talk back and forth. They could not see each other. The dolphins exchanged clicks and whistles for most of the time the telephone was turned on. When one vocalized, the other remained silent. This pattern seems to indicate that the dolphins were conversing, perhaps communicating, but again the meaning of the whistles is unknown. When the telephone was turned off, the dolphins stopped producing a variety of whistles and emitted only "signature whistles," repeating them over and over. These "signature whistles" are personal whistles believed to enable other dolphins to recognize individual animals.

Recent work has led many scientists to think that dolphins are about as intelligent as dogs, and for this reason believe man-dolphin communication will probably never exceed man-dog communication. Some researchers, however, still believe that man and dolphins may eventually be able to communicate on a higher level than this.

Further scientific experiments may give us the answers, but it may be that man is, in fact, alone.

This northern humpback whale off Kaanapali, Maui, belongs to a group of mammals whose songs have awed sailors through the ages.

Songs of the Humpback Whales

Every spring humpback whales pass through the clear blue waters around Bermuda. On calm days, when there are no crashing waves to obscure the sounds, fishermen in silent sailboats may be treated to one of nature's most beautiful melodies—the songs of the humpback whales.

Whales are vocal animals. Humpbacks are the noisiest of all baleen whales, singing their songs day and night during their long migration to and from warm southern waters. The vocalizations are rightly called songs; they occur in complete sequences, which are repeated again and again. A song may be as short as six or seven minutes or as long as 30 minutes. A humpback may put several songs together without a break, thereby giving an extended performance which may last for hours.

At first, observers were confused about the humpback's songs, since they didn't seem to follow any pattern. To simplify their task, the observers recorded the songs and deciphered them with a spectograph. Through analysis they discovered that the songs can be broken down into smaller units. These are called themes, and it was discovered that the songs of any whale, although different in length, always have the same number of phrases in a theme, and a phrase may be repeated any number of times. With each rendition of a single phrase it is changed slightly until, after many repetitions, the final phrase is completely different from the original one.

The whales migrate as a loose community, spread over miles of open ocean. Perhaps the songs help the whales stay together as they move. Since the sounds travel far, and the whale's hearing is keen, the migrants can keep track of their neighbors and keep from falling behind.

INSTINCT
AND
INTELLIGENCE

CHAPTER
7

BIRTH OF CONSCIOUSNESS

Orca is the proud marine mammal that inspired fear in sailors and whalers and was nicknamed the "killer whale." Herbert Ponting, photographer of Scott's expedition to Antarctica (1910–12), reported that he was standing on a large ice floe when eight big orcas, about 25 feet long, broke through the ice in an attempt to throw him into the water. Captain Scott himself reported in his journal: "One after the other their huge heads shot vertically into the air through the cracks they made.... The fact that they could display such deliberate cunning, that they were able to break ice at least two-and-one-half feet thick, and they could act in unison was a revelation to us.... They are endowed with singular intelligence and in the future we shall treat that intelligence with every respect." Ponting, who probably had been mistaken for a seal by the orcas, spoke of them (understandably with resentment) in such inappropriate words as, "Wolves of the sea...their spouts had a strong fishy smell ...little piglike eyes...devils of the sea...."

Orcas are obviously carnivorous, and they display ingenuity in procuring their food. But why should we emotionally judge their manners as "vicious"? Marine animals have to be observed in their element, if possible in freedom, if we are ever to understand their behavior.

In the North Pacific our divers witnessed orcas being captured to be sold to marine parks all over the world in a new kind of slave trade. When surrounded by nets the powerful rulers of the sea quickly understood that there was no escape. They were kept for a while in semicaptivity in a bay closed by nets, where our divers repeatedly went into the water with them. From the very start they were friendly and eager for attention and caresses from the divers. Threats were only expressed if we favored one of them and the others felt neglected.

Day after day at sea we watch orcas and wonder at the complexity of their individual and social manners. Their roving groups constantly communicate extremely complex and even abstract information that could not be transmitted unless they had some form of structured language. On land no animal other than man himself displays such—

let us say—intelligence.

The roots of behavioral studies lie in the automatic responses of organisms to simple stimulation (heat or light) from their surroundings. The way an animal fulfills basic motivations constitutes its behavior. And though the essential drives are only four, they can form an infinite number of combinations, just as carbon, hydrogen, oxygen, and nitrogen combine to form the highly intricate cells that are the building blocks of life.

When behavior is perfectly accomplished without preliminary learning, it is innate and instinctive—all the responses to stimulation from the outside world have in this case been incorporated in the organism at conception.

This definition is theoretical. Every animal, even if it is unicellular, has some ability to learn. This ability activates, to varying degrees, psychological functions including memory and intelligence. What we call intelligence today is much more than the ability to establish relations between causes and effects. Intelligence includes the selective storage of facts in a memory, and the faculty to scan the memory laterally to associate apparently unrelated facts.

Evolution thus appears as progress in psychological aptitudes. From the coral polyp to the starfish, the octopus to the orca, as the share of innate behavior decreases so does the predictability of behavior. The central nervous system increases in complexity as intelligence and learning increase. The parallel progress of brain and intelligence has been followed and checked in animals up to the primates. The gigantic leap between chimpanzee and man is difficult to understand, and it coincides with the all-important birth of consciousness. The brain of a primate is much smaller than that of a man and there had been no intermediate animal to study beyond the chimpanzee until now. Perhaps the answer will come from the sea. The central nervous systems and brains of nearly all toothed whales (porpoises, dolphins, pilot whales, sperm whales, and orcas) are more developed than those of the apes and some are comparable to ours. Our closest relatives are not on land but in the oceans.

Blue damselfish over coral

An Abudefduf *displays normal coloration that will change during the fish's spawning activity.*

COMPLEX MANNERS

Mammal, fish, crustacean, mollusc . . . each animal is faced with the same basic needs for the continuation of life. It must succeed in the face of elements and enemies; it must obtain food, defend itself or its territory, and propagate the species. The manner in which each species performs these tasks is unique, and the combination of these behavioral patterns helps to differentiate each species from all others, just as each species possesses a unique set of chemical entities and systems that makes it different from every other. In fact, some species that are very closely related genetically and could possibly produce offspring, remain isolated because of behavioral differences.

The Family of Demoiselles

The damselfish family offers a unique opportunity to look at behavior patterns in depth. Damselfish are found throughout the world, so generalizations and comparisons may readily be made based on observations of very close relatives.

Damselfish are generally most active during daylight hours and have a tendency to remain localized. Characteristically they cluster around some object, usually a rock or coral head, which they use as shelter and may remain there for months at a time, leaving their comparative safety only to feed. During night hours the damselfish retreat to their shelter rock or coral, with a few solitary individuals remaining quiet but alert. This behavior is not restricted to breeding individuals; juveniles as well as mature males and females have been observed in this pattern. Some damselfish species, however, show a preference for living among anemones rather than coral.

Damselfish generally live in warmer ocean waters, and the food habits of the more than 200 species are far-ranging, with some being herbivores, other zooplankton feeders, and still others feeding on anything from plant and detrital material to small crustaceans and anemones.

Perhaps one of the more remarkable feeding behaviors among damselfish occurs in *Chromis atrilobata,* which has been observed at times feeding on planktonic fish eggs, such as those of the wrasse.

When spawning, the wrasses gather in large groups and swim in the same area for several hours. Periodically, a small group of perhaps five to 15 individuals of both sexes will draw themselves closely together and begin swimming almost vertically, in a rapid fashion toward the surface. Then they make a sharp turn, release their eggs and sperm into the water, reverse direction, and head back down to the larger aggregation.

The hovering damselfish observe this activity and converge on the ascending group to feed on the eggs when they are released. The damselfish has either learned or instinctively knows how to recognize this behavior of the wrasse. Proper timing is important, for if the damselfish arrive prema-

Brilliant color changes in many damselfish occur for a variety of reasons—to show increased aggressiveness or to display readiness to spawn.

turely, while the wrasses are ascending, the wrasses would return to the bottom without spawning.

The complete reproductive sequence of most damselfish starts at the beginning of the season when schools made up of both males and females begin exploratory swimming along the coral reef, hunting for the location of nesting sites. The males head the school and exhibit a color change that varies with the different species. The blue of the *Chromis dispilus* (bicolor damselfish), for example, intensifies, while the *Abudefduf* males change from the usual stripes to blue, a color that becomes more vivid as the spawning activity reaches its climax.

The selection of spawning sites is seen as a ''dive'' from the group by certain males, and this dive halts the activity of the entire school. The nest sites are bare or eroded places on the coral substrate, sometimes the same places that had been used in previous spawnings. It has not been determined whether these locations are recognized by intelligence or chosen instinctively. Continuous threatening, chasing, and biting aids in defining the boundaries, as the coloration of the *Abudefduf* males becomes an intense blue. Aggressive behavior gradually diminishes once the territory has been well defined, whereupon the males begin cleaning the nesting site by nipping off debris.

Abudefduf courtship involves short vertical ''invitation'' swims and returns to the nest, with this activity directed by the males at the females passing the nesting site. The bicolor damselfish exposes his tail fin rapidly so the white margin flashes. During spawning the male and female swim in a circular pattern, and as the mass of eggs being deposited grows larger, the female becomes less regular in her circular swimming, which in turn stimulates aggression in the male. After the egg-laying activity ceases, the male drives off the female. The male's spawning activity may continue, however, with five or six different females.

One of the characteristics of damselfish is the nursery role played by the male, who now begins egg-cleaning and nest-guarding activity. Males care for the eggs by fanning them to give them oxygen, by removing debris, and by driving away most intruders.

The damselfish is so aggressive in territorial defense that it will even

In the highly regulated social structure found among bicolor damselfish, the largest male usually ranks as dominant individual. The smallest female, at the bottom of the pecking order, usually serves as the scapegoat for the whole community.

drive off a larger and apparently better-equipped foe like a crab. This behavior may be instinctive, or perhaps the defender has learned that the crab might prey on damselfish eggs.

The word "territory," however, has a different meaning, according to the intruder. The bicolor damselfish, for example, actually has a number of territories positioned as concentric rings. Certain nonthreatening species are permitted to enter the outer portions of the territory and are attacked only when they intrude on the inner regions. Egg-eating predators such as the wrasse, surgeonfish, parrotfish, and others, however, are driven off upon entering the territory before they have the opportunity to approach the nest too closely.

Characteristically, fish guarding eggs will nip and drive off intruders, even those several times larger, and even ripe females, if the male is busy with nursery activity. An average nest contains about 350,000 pale, elliptical eggs with sticky tendrils at one end to help attach the eggs to the surface of the substrate. The male may remain near the nest after the hatching, but then usually abandons it within hours.

During this time coloration changes back to normal and soon after the hatching is completed, the male resumes his nonreproductive behavior.

Social Status   The reef habitat of the bicolor—and most other damselfish—is unusual in that so many animals of different species live together in such close proximity that they have developed an elaborate and strictly structured hierarchy to cope. It is based primarily on size, and since the males of the species are larger than the females, it is based on sex also. But there is not a straight-line correlation between size and status, since smaller fish on occasion rank higher than larger fish. When the fish are busy establishing territories, the aggression, in terms of challenges and chases, is usually directed at individuals of nearly similar rank. Thus, the No. 1, or alpha, male is more likely to engage, or be engaged, in a chase or challenge with the No. 2, or beta, fish than he is with the lower-ranking individuals. The lowest-ranking fish, the smallest female, is referred to as the omega fish. Though it is usually closely ranked individuals who fight, the alpha fish chases every individual, and every individual chases the omega fish. In other words, the tendency is to challenge those higher in rank; pursue those lower.

The females wander throughout the neutral territory, but the lower in rank are often chased out of the defended areas by high-ranking females. There is no territoriality exhibited by females toward males, and the territories of the females are not very well defined because of their almost constant wandering.

This dominance pattern changes, however, once the territories have been established and reproductive activity begins. The females, for example, while still subordinate to the males, show an increased willingness to challenge each other, and even the males.

This system of dominance has an unusual aspect to it: the omega individual seems to serve as a scapegoat for the entire community. This small fish is chased so much that it spends much of its time in hiding. The chasing in this case is not at all random, for higher-ranking individuals seem to seek out the omega fish for abuse, especially if they themselves have just engaged in some aggressive behavior and have been frustrated. The omega fish, usually a female, is not courted like the other females.

Clownfish's Living Home   A very special interrelationship exists between clownfish and anemones. The clownfish belongs to a branch of the damselfish family that is found mainly in the Indian Ocean and western Pacific. Many of the clownfish species that inhabit anemones are what scientists call "obligate symbionts," which means they probably could not live an entire life without the anemone. Anemones usually prey on small or dead fish, but various species of *Amphiprion,* or clownfish, are able to live among the anemone tentacles even though the tentacles are covered with venomous stinging cells.

When the fish makes contact with the anemone, usually with a fin, the anemone's venomous nematocysts are discharged and the fish jerks back the fin. The anemone's tentacle then sticks strongly to the fish, and probably hurts it, but gradually the contacts by the fish are increased in both frequency and area exposed. It is now known that the clownfish's immunity is not developed because the anemone finally adopts or recognizes its tenant, but rather by a change in the mucous covering of the fish, which inhibits the discharge of nematocysts.

Considerable variation exists in the relationship, with some fish living with only one species of anemone while others show no specific preference. Some live in anemones as juveniles and move to coral as adults. Eggs are laid at the base of the anemone, and the parent often rubs the anemone, causing the tentacles to extend, which in turn forms a protective canopy for the eggs of the fish. It is also common for many clownfish to feed their host, but they sometimes do just the opposite, sticking their heads into the mouth of the anemone and stealing food. The *Amphiprion* may even carry dead fish to the anemone. Some clownfish have been observed feeding on the

An old wreck provides cover for a group of sweepers and silversides.

anemone's fecal material and even on its tentacles.

When clownfish have been artificially deprived of their anemones, they proved they have certain innate behavior patterns. They often dig holes and feed them, defend them, even sleep in them, much as they would behave with an anemone. Instead of digging holes, some clownfish without anemones in an aquarium will settle into algae and direct behavior toward it similar to behavior directed toward anemones. Other clownfish have been observed bathing in bubbles from an air stone and defending the bubbles, probably seeking stimulation comparable to that of the tentacles of the anemone.

Modern man can alter some aspects of his environment to suit his needs. He dams rivers, chops down trees, paves vast areas, burns grasslands and forests. Other forms of life (and even man himself until recently), however, are unable to adjust the environment. Instead they must adjust their behavior, or even their forms and functions. This is especially true in the sea where conditions are harsh and changeable.

Most snails trudge along a solid substrate on the sea floor, foraging as they move at a snail's pace. *Ianthina*, the violet snail, however, has developed the ability to trap air in the cupped end of its foot and secrete mucus to form bubbles. On this raft of mucus-coated bubbles, it drifts before the wind and on ocean currents much the same as its favorite food, *Velella* (the ''by the wind sailor'') and *Physalia* (the Portuguese man-of-war). In fact, if this snail is separated from its raft, it will sink and die, as it would be unable to obtain air to build another bubble raft.

As it drifts, *Ianthina* hangs suspended from the raft with its two-forked feelers and proboscis directed toward the surface where food will be found. When the tentacles contact food the sea snail bites into the flesh with its proboscis and eats as it is carried along on the sea's surface with its catch. *Ianthina* not only seems immune to *Physalia*'s powerful poisons, but seems to mimic it by emitting streams of purple dye.

A fish in the open sea has no place to hide. It must depend on its coloration to become less visible, on its senses to be aware of the approach of predators, and on its swiftness to escape when it detects them. This lack of hiding places may be why fish commonly aggregate beneath floating objects at sea. In one study, darker-colored fish were found close beneath floating debris where shadows are more intense. Their distribution suggests that fish can pick the best hiding places for themselves.

Thor Heyerdahl, sailing across the Pacific Ocean on *Kon Tiki*, reported a population of fish beneath the balsa raft. They ranged from the smallest juvenile forms to large predators like tunas, dolphins, and marlins. Once *Kon Tiki* even attracted an enormous whale shark. Some fish took up residence between the raft's logs where they could be safe from the predators.

The list of floating objects that are found at sea is a long one, ranging from logs, coconuts, and trees washed out to sea from rivers, to sides of tumbledown buildings, shipping crates, and parts of boats. In some tropical seas, huge patches of floating seaweed provide cover for many animals, some living in the seaweed and others swimming beneath it.

When the weather turns cold and the days short, many animals move to warmer climes where procreation will be accomplished safely. When warm weather returns to the temperate zones of the earth, these same creatures move back. These movements are adaptations the animals have made,

A self-made life raft constructed of mucus-covered bubbles enables this sea snail to drift across the ocean surface in search of food.

CHANGE THE WORLD OR CHANGE YOURSELF

Building a Life Raft

Housing Shortage

Migrations

enabling them to be in the most advantageous places for their survival and perpetuation of their species.

Northern fur seals spend much of the year scattered over the Pacific Ocean, but as spring approaches they gather in large herds to move north to their island breeding grounds between Alaska and Siberia. How and why these seals travel 3000 miles or more can probably be accounted for by race memory—knowledge that has led the animals to salutary feeding and breeding grounds for many generations.

Two species of birds sometimes migrate opposite the expected direction in pursuit of favorable climates. Some Heermann's gulls winter along the west coast of the U.S. and Canada and some elegant terns winter in California. In spring they migrate south to their traditional breeding grounds on Isla Raza in Mexico's Gulf of California.

Inhospitable Shores Teeming with Life

Sandy beaches along seacoasts and estuaries seem most inhospitable. Depending on the state and activity of the tide, the stresses of the beach environment include the washing away of substrates and depositing of sediments. Aquatic animals living on intertidal beaches face the possibilities of drying or drowning and of osmotic imbalances as freshwater and seawater come and go with the tides. They may be preyed upon by terrestrial as well as aquatic animals, and they may be subject to poisoning from lack of oxygen or high levels of ammonia or hydrogen sulfide from natural or man-made pollution. Despite this, the habitat supports a considerable population of animals.

Other shoreside environments that house many organisms are mangrove beaches and swamps. These areas are not as inhospitable as barren sandy beaches; in fact, they teem with life. The large population of small creatures attracts hosts of larger predators.

Fiddler crabs often gather in large hordes on sandy beaches that are punctuated with growths from mangrove roots called aerophores, small shoots that pop up through the sand into the air. To survive in such

While most birds migrate north in spring, these Heermann's gulls move south for breeding in the Gulf of California.

environments the shore crabs have had to make many behavioral adaptations including burrowing and burying to protect themselves from dehydration, overheating, and abrasion by sand and rocks in the surf.

Another behavioral adaptation is directional orientation. Fiddler crabs have an uncanny ability to find their burrows after foraging on featureless beaches. Other adaptations help them anticipate tides and time of day so that they know when it is safe to be out of their burrows searching for food.

INBORN SKILLS

Many creatures come into the world programmed to meet the challenge of life. Inborn responses to the events of everyday life are called instincts, the "mechanical" unthinking reactions to the direction of light, the time of day, the threat of predators, temperatures of air and water, and other external stimuli. Instincts are passed on to offspring through the genetic material of the parents, who received them in the same manner from their parents when they were conceived.

One way we can differentiate between instinctive and learned behavior is to raise an animal in isolation, protecting it from others that could teach it and from the means of learning by trial and error. If under these circumstances an animal responds appropriately to a model of a predator, mate, food, or other stimulus, we can conclude that its response is instinctive. Such a stimulus is called a releaser. The animal's reaction is one programmed into it at conception and will manifest itself only when given the proper signal. In addition, there must be a built-in readiness to act or react. The proper food at the wrong time or the competitive male at the wrong season will not elicit a response because there is no inner readiness to react.

Each species has its own set of signals that only others of its species are

Low tide periodically strands many forms of shore life, sometimes exposing them to predation or death by dehydration.

Two views of a hermit crab with a commensal sea anemone. The anemone and crab remain together until the crab outgrows the shell. Then they will move to a new one.

programmed to respond to. It is this mechanism that helps prevent breeding among different species. Instinct helps a species survive under normal conditions.

Mutually Helpful

The Mediterranean hermit crab and a particular type of sea anemone have an instinctive symbiotic relationship even though each is able to live normally without the other. This mutualistic relationship is related to both food and survival. Hermit crabs are a favorite food of octopods, which are very sensitive to the stinging cells of the sea anemone. The poisonous sea anemone becomes more mobile, increasing its chance of food capture, and the crab is protected from octopods.

The hermit crab signals that it will carry the sea anemone by tapping it with claws and front walking legs. The signal is a specific stroking of a particular intensity which the sea anemone will respond to. When it receives the signal, the sea anemone responds by releasing its grip on the substrate and fastening tentacles on the crab's shell. *Calliactis,* the anemone, bends its body into a U-shape, releasing the hold its tentacles have while gripping the crab's shell with its pedal disc. Then it straightens up and rides away atop the hermit crab's shell. When the crab outgrows that shell and moves into a new one, it signals the sea anemone, which moves with it. In other species that have a similar relationship, the crab does not have to stimulate the sea anemone to move. Certain species of sea anemone can take the initiative and transfer their footholds from the old shell to the new one without any action or signal by the crab.

This behavior has been shown to be instinctive. Without the inborn knowledge of the proper signal to the sea anemone, the crabs could not so consistently invite them to attach, nor would the sea anemone shift to new shells with the crabs.

Orienting Toward Light

Animals use their senses both to avoid detrimental environments and to find and remain in favorable conditions. Sea turtles, for instance, just hatched from buried eggs, use their vision to orient themselves and find their way to the sea.

The egg-laying female sea turtles come ashore during their lifetimes solely to dig holes in sandy beaches and deposit their eggs. If the eggs aren't dug up by man or other predators, such as coconut crabs, they hatch in a few weeks. The hatchlings must dig their way out of the sand and then find their way to the sea or perish of dehydration or predation. Hatchling sea turtles are believed today to find their way to the sea through visual orientation toward the brightest horizon. And the horizon of the sea is brighter than that over

land because the sea reflects light more readily than land. If heavy clouds were to darken the sky at sea while sunny weather prevailed over the land, the young turtles might turn inland.

Hatchling leatherback sea turtles circle briefly after digging their way out of the sand to orient themselves, then they home in on the sea. To check the theory of visual orientation by the baby sea turtles, researchers tested them by decreasing or eliminating the light to one eye with light filters and blindfolds. When one eye was completely blindfolded, the turtle circled throughout the ten-minute test period. When the blindfold was removed, the turtle immediately took a seaward course.

One of the strangest fish ever observed is the pearlfish, a slender, elongate, and scaleless fish of tropical seas. These small fish have a symbiotic relationship with sea cucumbers, sea stars, tunicates, clams, and sea urchins. Some consort only with one species of their host organism. Others are less particular. The relationship of the pearlfish to the sea cucumber, however, is the strangest of all. The fish chooses its host while it is still very young, forces entry, and from then on considers the cucumber as its home.

Some species of pearlfish enter headfirst, but the usual way is for the fish to wait with its head by the sea cucumber's anus until it senses that the anal muscle is relaxing. Then the fish whips its pointed posterior end around forming a U with its body and inserts the posterior end into the opening. As it slowly eases its body in, the pearlfish straightens out, and disappears inside, head last. Even very young pearlfish, barely past the second and final larval stage, have been seen entering the bodies of living animals. A few have been found encased in pearl nacre inside clams. Knowledge that they will find shelter in the species their ancestors have always used for protection is almost certainly instinctive. Their ability to get inside their host is probably an inherited skill.

These turtles, which have just hatched beneath the sand and clawed their way to the surface, must now find their way to the sea before predators or the hot sun kill them. Scientific experiments have revealed that hatchling sea turtles of several species instinctively seek the brightest horizon.

Strange Partners

LEARNING BY INSTINCT

Instinct and learned behavior are often erroneously thought of as being opposites. Instinct, we have seen, is inherited, inborn knowledge. Learning is acquired after birth. The ability to learn is inherited, however, and is dependent on genetic characteristics.

Animals are limited in their capacity to learn. Some can learn to respond to one type of situation very easily, while another set of conditions may never be mastered. This capacity to learn is definitely an inborn ability that has been programmed into an animal. As a result, instinct and intelligence are not opposite or even separate but complementary, relating and overlapping in varying degrees depending on the species.

In most animals there is an innate predisposition to learn certain things, whether it be the recognition of prey or the recognition of a type of habitat. Such limitations exist in the learning potential of all species.

Off Point Loma, California, leather sea stars pursue, capture, and consume purple sea urchins. Yet farther north, along the coast of Washington State, the two species live side by side, without such predation. Why do the Point Loma leather stars prey on the sea urchins while those off the Washington coast do not? Availability of prey is a partial answer. Leather stars prefer a sessile prey, that is, one that does not move. Their first food preference is mussels. Off Point Loma, however, few mussels are available and many sea urchins are. The sea stars there, through associative learning, found sea urchins an acceptable and available second choice, while those in Washington waters have no need of a second choice. Not only did the sea star learn to prey on the sea urchin, but the sea urchin learned to avoid the sea star in the Point Loma waters.

Each learned about the other because it had the innate ability to learn. The learning took place only because there was the ability to respond somewhere in each animal's inherited memory. The ability was inborn. The act was learned.

There is a very specialized type of learning that exists in animals that imposes even greater limitations on the individual. It is called imprinting and makes the animal receptive to learning a specific response to a certain object during a very limited period of time. If learning does not take place during that time, the individual will never be able to recognize the object or perform the task. The sensitive period is usually early in life and is very limited in regard to the stimulus and response.

The critical learning period is marked in fish that migrate to spawn. Young salmon learn the scent of the home stream in their first year of life. Then they go to sea to stay until the remembered scent draws them back to spawn in the place of their birth.

In one species of freshwater fish, the cichlid, the parents do not eat their young but will consume the offspring of another species of cichlid. They have learned to differentiate their progeny from others by imprinting. Just

A turtle hatching.

This slender pearlfish maintains a symbiotic relationship with a starfish, living within its host.

after hatching, the parents identify those baby fish nearby as their own and from that time on will not eat them. Even when eggs of a different species are put in the cichlid's nest to hatch, the parents will not eat them because they are imprinted on the little fish from their nest, whatever they look like.

Some forms of learning are based directly on instinct even though they are acquired after birth and are truly learned behaviors. Other types of learned behavior, such as marine mammals' ability to use intellectual capacity, are a different matter.

INTELLIGENCE AND THE BIRTH OF CONSCIOUSNESS

Intelligence begins with the ability to learn from experience. It is, however, more than just the ability to learn. It includes finding the relationship between cause and effect, the capacity to correlate various pieces of information previously acquired and stored in memory, and the ability to respond quickly and successfully to new situations. The ability an animal develops to meet all the complicated needs imposed by diverse and constantly changing environments is perhaps at the origin of consciousness.

As consciousness develops in an animal, much of its programmed behavior is lost. In man such adaptation has reached its climax. Instead of passing genetically from one generation to the next information that quickly becomes obsolete, man transmits consciously selected information through tradition and education. One advantage is that our behavior can be modified quickly to meet rapidly changing situations. A disadvantage is that the young of many ''intelligent'' species are helpless at birth and must undergo extensive education to learn what is needed to survive.

Some young mammals and birds (and probably many more that we have not yet studied) learn the skills needed for survival through play. In such animals learning the necessary skills is acquired and refined by trial and error or, even better, through parental instruction.

Man has evolved as a generalist and is able to adapt to almost any environment. Thus he has successfully lived in some of the harshest habitats—deserts, flood plains, excessively hot or cold areas. It is man's conscious effort to adapt to these varying situations, his intelligence, and his capacity to develop tools that permit him to adjust quickly to changing environments and extreme climatic conditions.

Konrad Lorenz, the father of modern animal behavior studies, once said: ''If a man were asked to perform three tasks—to march 35 kilometers in one day, to climb a five-meter rope, and to swim a distance of 15 meters in four-meter depth and pick up a number of objects in a certain order from the bottom—all are activities which a highly nonathletic person like myself can do without difficulty, and there is no single other animal which can duplicate this feat.''

Octopods are considered to be among the most intelligent invertebrates. Their capacity to learn appears to be considerable. Just how much they can learn and how they learn are subjects of great interest to scientists investigating the mechanics of learning.

Octopods have long-term and short-term memories as vertebrates have. They differ from higher animals, though, in having two separate learning systems. One of these is based on visual stimuli, the other on tactile stimuli. Neither is dependent on the other in any way, and if the ability to use one system is blocked or taken away, it has no effect on learning through the other system.

One of the principal ways in which octopods' nervous systems differ from that of vertebrates is in the topography of the brain and its consequent effect on learning and response. In vertebrates, the centers of the brain that specialize in learning are so closely linked with those specializing in movement that if a portion of a learning center is removed, physical responses may be impaired or eliminated. In octopods, however, an entire learning center of the brain can be removed with no effect on the octopus' mobility. Octopods solve problems through experience and actual trial-and-error experiments. Once it has solved a problem, an octopus remembers and can solve the same type of problem in the same way but faster each time. Unfortunately the octopus lives only two or three years, which is too short a time to take full advantage of its built-in capabilities. The ability of an octopus to learn and to remember has often been demonstrated. *Calypso* divers have observed many octopods building their homes; one of them wedged up a large flat stone with a brick that it had dragged along from quite a distance.

Orcas are engaging mammals that have large, complex brains remarkably similar to man's. Many believe that the orca may be the most intelligent animal on earth.

The only sea mammal known to use tools consistently in obtaining food is the sea otter. Its paws can almost be used as hands, while the dolphin's flippers cannot.

    As evidence of its intelligence, the sea otter has been observed using tools in several ways. Diver-photographer Ron Church planted a rock with an abalone firmly gripping it on the bottom as bait. A wily sea otter simply carried the whole rock away. Church planted another abalone on a much larger rock which had a number of smaller rocks around it. When the sea otter came for this one, it tried rocks of increasingly larger sizes until it found one big enough to jar the abalone loose. The sequence of events demonstrated that the sea otter could differentiate between sizes of rocks and that it knew that larger rocks when used with equal force would produce a harder blow against the abalone. Sea otters are also known to use stones, either as hammers or as anvils, to crack shellfish or sea urchins.

We have seen that the ability to learn is a necessary but insufficient condition for an animal to be considered intelligent. Seals, especially fur seals and sea

Tool-Using Sea Otter

Circus Playboys

To a sea otter mussels are tasty food. Here this clever mammal, floating on its back, uses a rock on its stomach as a tool to open the shells.

lions, are able to learn a variety of tricks and are readily trainable. In nature as well as in captivity they like to play. But their intelligence level remains average, probably slightly inferior to that of chimpanzees. Thus, even though some creatures are brilliant circus animals they should not be considered smarter than more restive individuals.

Craig Kitchen has trained a young female sea lion to respond differently to 32 distinct hand signals. Kitchen was curious to learn whether the sea lion would perform some of these same behaviors in response to a different stimulus. He therefore taught it to respond to seven new symbols printed on a large card. As a third stimulus, he used tones of different pitches. Kitchen found that the sea lion did not become confused but was capable of learning and retaining the same responses to the three different stimuli. The same sea lion solved a problem by itself. In its playtime, it often carried a large ball underwater and released it. Usually the ball would pop to the surface, but occasionally it would become trapped in the overhang of an artificial island in the enclosure. The ball was too big for the animal to grasp in its mouth, so it would hold the ball between its front flippers and exhale, causing its body to sink. Still holding the ball, the sea lion would then swim out from under the island and release it. The sea lion did this as often as the ball became trapped under the structure.

The Gladiator Orca   Among the denizens of the sea the gladiator orca is about the ultimate in beauty and efficiency. The orca is the marine mammal that inspired fear in sailors and whalers and was nicknamed the "killer whale." Herbert Ponting, photographer of Scott's expedition to Antarctica (1910-12), reported that eight orcas, each about 25 feet long, surfaced through the ice and attempted to throw him into the water from an ice floe.

Because of their popularity in marine parks, orcas are prime targets for capture.

A true bond of affection may develop between a trainer and a captive dolphin. Though they are often employed as entertainers, many dolphins also serve in marine biological research.

Captain Scott himself reported in his journal: "One after the other their huge heads shot vertically into the air through the cracks they made. . . . The fact that they could display such deliberate cunning, that they were able to break ice at least two and one-half feet thick, and they could act in unison was a revelation to us. . . . They are endowed with singular intelligence and in the future we shall treat that intelligence with every respect." Ponting, who probably had been mistaken for a seal by the orcas, spoke of them (understandably with resentment) in such inappropriate words as, "Wolves of the sea . . . their spouts had a strong fishy smell . . . little piglike eyes . . . devils of the sea. . . . "

Years ago, whale fishermen on the southeastern coast of Australia reported that orcas made conscious efforts to help them locate pods of humpback whales. While most of the orcas circled the humpbacks, keeping them in a tight group, a few swam toward shore grunting loudly to alert fishermen. The tongues of the slaughtered whales were the reward the orcas received for their services.

Orcas in captivity have given scientists a poor opportunity to study either their physiology or their intelligence: we suspect the validity of studies of animals under shock kept in shallow tanks. How can we learn about the orca's diving ability when in the open ocean it can dive below a thousand feet? And how can we begin to understand its complex behavior in a tank where echolocation is confused by echoes on the walls?

The study of orcas is only fruitful in the open sea, and it is difficult and time-consuming. Each time we have followed a family of orcas with the *Calypso* in combination with motor launches and a helicopter, we have observed the determination of the dominant male to lure us away from his family of at least two generations of offspring even at the risk of his own life. His maneuvers demonstrated imagination, and with his complicated whistling language he was able to direct the route and behavior of his relatives by remote control when they were as distant as a mile from him. On land no animal other than man himself displays such—let us say—intelligence. The long series of chances taken along the line of evolution that was necessary to turn out end products like orcas (and man) staggers the imagination.

Captive Cetaceans

When an exuberant, highly social dolphin has been put through the agony of being captured in a throw net, separated from its family, hoisted brutally out of the sea, confined in a training pen, injected with vitamins and antibiotics, and submitted to weeks of brainwashing sessions, it is turned from a proud raider of the sea into a submissive beggar and clown. It is this now-perverted creature that some behaviorists attempt to analyze.

In captivity a dolphin is a caricature of itself although it has a tremendous ability to learn and to brilliantly execute complex circus tricks. The dolphin's need for affection has been turned into a comedian's appreciation of applause. The dolphin was taught by its mother to catch live fish for food; now a trainer teaches it to consider dead fish a reward. Calling this "overriding learned behavior" or saying that "Pavlovian response" is used for training is an outrage. Behavior forced upon intelligent prisoners is perverted behavior.

Intelligent and responsive to reward, seals are easily trained to perform tricks.

PHARAOHS
OF THE
SEA

CHAPTER
8

THE PACE
OF CHANGE

The very first cruise of my research ship *Calypso* was a diving expedition to the Red Sea. With some of the pioneer diver-biologists and geologists we explored the uncharted mazes of coral keys, islets, and snags of the Far San and Suakin reefs. It was there that I was astounded and entranced by the splendor and the folly of the coral world.

Eleven years later I investigated many Red Sea and Indian Ocean coral reefs down to 1000 feet with a versatile exploration submarine, the diving saucer SP 350. Below 1000 feet I completed the survey with the help of automatic cameras. The picture that emerged from these investigations was very logical, but distressingly plain. The coral reefs, magnificent at the top, were huge barren mounds the size of mountain ridges covered with constantly flowing (often even cascading) sand. I had to admit the hard facts: the romantic coral reefs were gigantic tombstones crowned by a very thin mop of exuberant, obstinate, and complex living communities.

Through the diving saucer's portholes, the size of the monuments and the exiguity of the active layer of the reef suggested to me how important time had been as a construction material. During a lifetime, one witnesses a volcanic eruption, one hears of a few earthquakes, floods, or tidal waves, but the map of the world changes very little. And yet, the face of the earth is constantly remodeled, but at a pace that is not ours.

One single atoll represents a volume of construction several thousand times that of the largest pyramid ever built by the pharaohs. The industrious little polyps have used staggering quantities of both ingredients: calcium carbonate extracted from the sea and time by the millions of years. Buried thousands of feet below the slim bustling coral cities are the fossils of the early ancestors of all reefs dating from about two billion years ago—almost half the life of the planet! Four times the reefs of the world died in all oceans and remained funeral monuments for millions of years before conditions were favorable enough to permit their return. Each time they were reborn they knew a greater diversity. No fossil reef has ever been as rich and as beautiful as those we can study today.

To better realize the mass of time that was needed to build a deep reef we have to use the classic analogy for distances. If you stood close to the Washington Monument, for example, and if you walked back in time 50 years for each yardlong step, two steps would be a century. After 40 steps you would have reached the birth of Jesus Christ. After 200 steps you would find yourself in prehistoric times, maybe looking for a cave. After 15 miles you would witness the very first anthropoids, our ancestors. But you would have to walk all around the earth to reach the earlier reef-building creatures! Such vertiginous explorations of time allow us to size up the difference of rhythm between man's recent hectic developments and the quiet pace of natural changes. The growing concern about the careless destruction that our unchecked technological development is spreading in the ocean is not yet fully understood. Many well-intentioned persons ask such questions as: "Insects adapt to DDT, germs to antibiotics, why would not man adapt to pesticides or to heavy metals like mercury?" Or: "Why should we feel concerned with the possible extinction of some animals as a consequence of environmental deterioration? Dinosaurs have become extinct and they have been replaced!" The answer is given by the patient coral community: "Because men do not allow enough time for such changes to take place. Insects adapt because of the rapid turnover of their generations. Evolution produces a very few new species every million years." If we were to assume that nature could cope with our feverish developments, it is probably mankind that would be submitted to the fate of the dinosaurs.

Flying high over the Red Sea or over the Great Barrier Reef on a clear day, one is puzzled by the abundance and the variety of shapes that coral reefs can take. The impression is one of eternal complexity. Unfortunately the Empire of the Polyps is vulnerable. If we were to bring about its fifth collapse, it may regenerate another time, probably in 10 or 20 million years. Maybe long after man has himself disappeared from the planet.

Destruction is quick and easy. Construction is slow and difficult.

Wuvulu reef flat, New Guinea

Volcanic birth of an island often marks the beginning of a coral atoll. Submerged flanks of the volcanic island furnish a base on which coral can grow in the sunlit upper layer of water.

AN ATOLL IS BORN

Around the world in a band of ocean between the latitudes of northern Florida and southern Brazil lie peculiar islands set in rings around peaceful bodies of water—the coral atolls. Underwater, atolls are huge annular towers that shelter some of the most vibrant and complex collectives of the ocean world. Myriads of tiny marine animals have helped to create each atoll, and now they are crowded and bustling communities of creatures that belong to nearly every major group in the animal kingdom.

An atoll begins to take shape when an underwater volcano erupts and rises above the surface of warm tropical seas. If the peak of the volcano remains above or close to the surface, the larvae of simple marine animals called coral, swimming freely among the plankton of the sea, soon attach themselves in the shallow, well-lighted water along its flanks. Anchored to its new home, each microscopic, pear-shaped larva grows and secretes a limestone cup around itself. The soft part of the mature animal, the polyp, will spend the rest of its life in this external skeleton. The vase-shaped body of the polyp probably will not grow to more than a third of an inch in size. To reproduce, the polyp sends out little branchlike buds, and these grow into new polyps that remain attached to the parent. Each new polyp immediately begins to secrete its own stony cup. In time, each polyp buds in turn, and in this way a closely knit community of coral polyps gradually takes shape. They will also produce larva that settles elsewhere and initiates the formation of new colonies.

The corals, relatives of the sea anemone and the jellyfish, are the chief architects of the reef but not its only builders. Simple marine plants called coralline algae contribute by cementing the various corals together with compounds of calcium. One-celled foraminiferans donate their hard skeletons, as do tube worms, molluscs, and other animals.

Corals need warm water to become reef builders. Even in the tropical zone, reefs will not develop in cold ocean currents. Temperature affects the chemical and physical exchanges that lead to the combination of dissolved calcium with the carbon from carbon dioxide to form calcium carbonate. Also, in warm waters calcium carbonate reaches saturation in small volumes of water trapped near the coral structures and crystallizes to weld together all the reef's components.

For a coral reef to grow, it needs warm, clear water and also abundant sunlight. Microscopic, one-celled algae live within the polyp's tissues. These algae, called zooxanthellae, need the light of the sun to grow, just as much as do the plants that grow out of the water at the coral reef. The precise relation between the polyp and the plant is not yet completely understood, but it is known that the plant helps the coral to calcify its skeleton and helps to remove the waste products of the animal.

The colony grows upward and outward from its point of origin upon the skeleton of corals that have died. Expanding an inch or two each year, these colonies build up the reef over the centuries. Recent measurements reveal that it takes from 250 to 500 years for a reef to grow seven feet—and some are nearly a mile high.

If the sea floor on which the volcano rests begins to subside, the peak that is the island also sinks. As it does, the distance between the island and the reef begins to widen. At some point the reef becomes separated by a vast body of water from the volcanic peak it once hugged. Such a reef is then called a barrier reef.

Finally the last vestiges of the sinking peak disappear beneath the surface. If the descent of the volcano has been gradual, and if the reef has been able to grow upward fast enough, there will remain a circular coral belt all around a tranquil lagoon. Sand beaches and trees turn the reef into an island. An atoll has been born.

Atolls rarely close in a complete circle. They are usually a collection of islets separated by channels of water connecting the lagoon with the open sea. Lagoons would choke and die if there were no tides. The flow fills the lagoon with clean, rich ocean water, and the turbid broth of rotting waste is flushed back to the sea through the flow channels by the ebb currents. The shallows of the lagoon are littered with debris of coral, of green thalassia seagrasses, and of thin, branching segments of a cactuslike growth, halimeda algae. Here and there on the lagoon floor rise knolls of coral that sometimes reach the surface or rise above it. When they do, they are called patch reefs. They dot the green or blue or olive brown of the lagoon waters with their own resplendent hues, providing a most spectacular sight. Sometimes they are merely pinnacles of coral rock, and sometimes they are over a mile in diameter. When they grow to such a large size, they are called table reefs.

Eventually the island becomes surrounded by a fringing reef. Often, the island sinks and becomes smaller, and the channel widens between it and the coral colony.

The moment a coral reef is born, forces set to work to destroy it. The sea begins to dissolve some of the limestone that the polyps secrete. When the reef finally rises above the surface, rain dissolves some of the stone that the seawater can no longer reach. Herbivorous fish come to grind up the coral and get at the algae. Rock-boring worms, sponges, barnacles, and bivalve borers make holes everywhere. This makes it easier for waves to break off big chunks of the rock and to grind it up into fine particles. Shore animals come in hordes to scrape away the rock and feed on the animal and vegetable matter that is exposed.

Waves sweeping across the rim of the atoll, transporting sediments to the lagoon, leave ripples on the surface of the reef. Even more destruction is caused when a tropical storm—a typhoon or a hurricane—unleashes its force. In British Honduras after Hurricane Donna struck in 1961, 80 percent of the coral rock about five miles north of the storm's center had disappeared.

All these destructive forces, however, help make the finished atoll. They create the sand and sediment in which plant life can take root and where even tiny bacteria can find a home.

Wind, flotsam, and seabirds bring seeds to the atoll. Only hardy plants are able to sustain life on the seaward side, splashed by salt spray and buffeted by winds. But in the relatively protected interior, coconut and breadfruit trees are common, and morning glories grow on the lagoon beaches. Atolls in the South Pacific may have only two or three species of plants, and even if they have more, the plants are generally stunted and pale and meager in their production of fruit.

Often plant life comes first to a coral island in the form of the mangrove tree. Looking like a giant spider, the red mangrove stands in shallow water protecting the leeward side of a reef. Farther inland grow the black-and-white varieties. This tough, adaptable tree is a fast grower, climbing about two feet in its first year, and at an early age it sends out green, podlike roots that drop from the trees and hang until they become anchored.

These trees help to build the island. Their roots reduce the water's flow across the reef and help accumulate sand. Bird waste, known as guano, rich in phosphate, changes the chemical composition of the soil, making it more fertile.

Most of the creatures that live on atolls share a peculiar characteristic—a tendency to be smaller than other members of their species that live where food is more plentiful. An exception is one of the most spectacular beings of the atolls—the coconut crab, the largest of its kind. Found throughout the South Pacific and the Indian Ocean, it has large front claws that can work through the hard shells of coconuts. The baby coconut crab develops from eggs in the lagoons or in the open sea off an atoll. Its larval stages are like those of any other crab. Later it lives on land, spending about three weeks inside a borrowed mollusc shell, and then changing itself into a small land crab and burying itself in moist sand. When it emerges, it is colored light violet, deep purple, or brown.

But the coconut crab is by no means the only inhabitant of the atolls. In addition, female turtles come to lay eggs and bats come to drink the juice of fruits. Blue-tailed skinks settle, and some kinds of birds—like the tern and the albatross—come to nest. Others make it a port of call along their migratory routes. The Pacific golden plover is commonly seen on atolls, while other plovers, curlews, sandpipers, and phalaropes also visit to rest.

THE GREAT BARRIER REEF

A barrier reef is like a fortress of rock. Often it rises steeply on its windward side, a great distance separating it from the island it seems to guard. It may be a continuous strip of land, or it may be a series of smaller patches running closely together.

The most spectacular of the barrier reefs is the Great Barrier Reef. This immense coral land strip was first navigated by the British explorer Captain Cook in 1768–69. It was explored scientifically by marine geologists in 1928–29. It is 1260 miles long and from 10 to 90 miles wide. It lies from 10 to 100 miles off the Queensland coast of Australia and is made up of 200 coral-formed sandkeys and thousands of islands with fringing coral reefs. It is the habitat of countless species of marine creatures.

The Great Barrier Reef rests on major oceanic rises and ridges on the continental slope and is a shelf reef. In his ship's log Captain Cook mentioned ''a wall of Coral Rock rising almost perpendicular out of the unfathomable ocean.''

There are no atolls in the Great Barrier system, but there are many platform reefs, patch reefs, boulders, and pronglike or triangular areas of coral beaches between the Australian continent and the reef itself. The reef was first fully mapped from 120 miles in space by the crew of Apollo VII. Because of its staggering extension, the Great Barrier Reef offers an almost complete catalog of coral reef populations.

Thanks to its favorable climatic location, to its deep shelf, and to the high prevailing tides (nine to 11 feet) that generate strong nutritive currents, the reef has flourished for many thousands of years. It might be considered the eighth natural wonder of the world.

EARLIER REEF BUILDERS

The coral reefs we know today are relatively recent structures. There have been other, earlier reefs. The first, and simplest, were created two billion years ago in the middle and late Precambrian times by algae working without assistance of animals; individual colonies of these stromatolites grew upward for tens of feet. Fossils of the structures they built are in the form of trunklike columns or hemispherical mounds.

The stromatolite reefs were enriched about 600 million years ago by stony, spongelike animals called archaeocyathids, Greek for "ancient" and "cup." The archaeocyathids disappeared 540 million years ago in the first collapse of the reefs. Algae continued simple building activity without a partner for the next 60 million years.

A complex new building cooperative formed some 480 million years ago. Fossil evidence identifies its members. In addition to the stromatolite algae, there were colonial bryozoans, tiny mosslike animals secreting calcareous skeletons in shapes ranging from branching trees to convoluted lace; stony sponges shaped like encrusting plates or hemispheres or shrubs; coralline red algae; and the first of the true corals. This third type of reef lasted approximately 130 million years, during which the special relationship between algae and coral was established. Sponges that were to become fundamental partners in ocean reef construction through the ages emerged in both encrusting and free-standing varieties.

Around 350 million years ago all of the reef builders, save the persistent algal stromatolites, were killed off in the second collapse. It was 13 million years before reef development began again. This time two new groups of

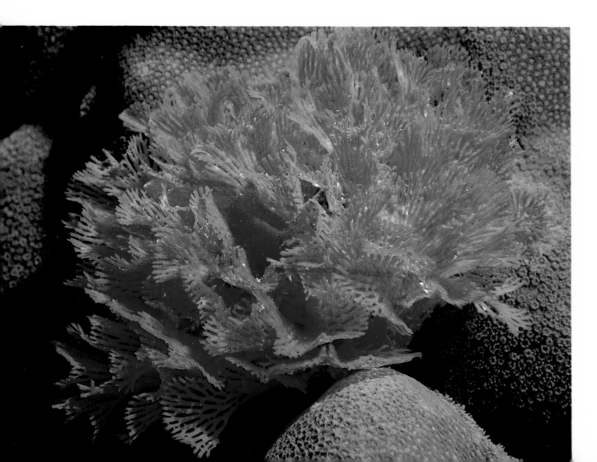

Lacy bryozoans secrete calcium carbonate skeletons that contribute to a reef's structure.

calcareous green algae joined the stromatolites. Bryozoans again became prominent and chambered sponges appeared.

For about 115 million years the reefs evolved until, about 225 million years ago, a third and devastating collapse occurred. Half the known marine and land animals and plants became extinct. Reef builders did not reappear for 10 million years. Then corals made a comeback, this time as six families in the scleractinian group, from which the families of corals we know today have descended.

In the ensuing 130 million years reefs spread from a few scattered patches to flourishing settlements in many parts of the world. Over the period the reef's inhabitants changed. New groups of sponges, sea urchins, foraminifera, and molluscs made their appearance. Roles changed too. Stromatolites faded while a coralline red algae took over the partnership with an increasingly diverse number of coral families. A challenge to the dominant corals came from molluscs called rudists. Along the sheltered landward margins of many fringing and barrier structures, rudists largely supplanted the corals. Their cylindrical and conical shells were cemented into tightly packed aggregates, and many of these grew upward in columns that imitated some coral growth pattern.

Suddenly, the reefs suffered their fourth collapse. The rudists were wiped out, and two-thirds of the coral families disappeared. This happened 65 million years ago, at the time dinosaurs became extinct. After 10 million years the reefs began to take hold once more. They grew vigorously and have maintained themselves to the present time.

The dramatic reef crashes are believed to be connected with cataclysmic periods in the history of the earth itself. For over two billion years vast changes have taken place; ocean basins formed, continents came together

and broke apart, climates that were moist and mild turned severe and dry. Vast though the changes were, they were gradual.

Corals are members of a group known as coelenterates or cnidarians—the flowers in a garden of sea animals. The several thousand types are rich in color and variety, from the tall sea fans and feathers to the small hydroids, from the fleshy sea anemone to the glassy jellyfish.

Among the coelenterates is a class of anthozoans containing thousands of types, including both the soft corals (alcyonaceans) and the hard corals (scleractinians). Both secrete limestone skeletons, as does the group known as hydrocorals. But only the hard or stony coral creates a limestone base that encircles the polyp and forms a colony that can constitute the framework of a reef.

The reef-building coral polyp is much like the simple sea anemone, but its functions are more complex. Both have soft hollow-cavity bodies with an opening ringed by tentacles to snatch food. Most anemones, however, are solitary, whereas the reef-building coral generates a colony with a stony skeleton that is shared by its polyps.

The Growth of a Coral

Just before and during a new moon, a coral polyp sets free hundreds of pinhead-sized larvae from a fertilized egg. For five to nine days these larvae half swim—with the help of hundreds of tiny hairs that cover their bodies— and half drift on the currents of the sea. Almost immediately after the larva is released, the beneficent zooxanthella algae distribute themselves throughout the larva, giving a brown hue to its translucent body. If it survives, the larva thickens and settles onto some smooth surface. Almost at once it begins to secrete its limestone skeleton, creating a shelter and cementing itself to its homesite. The larva is transformed into a polyp. Its center sprouts into a hollow pedestal crowned by tentacles. Some corals develop no further. They are isolated and are joined to neighboring polyps only by thin appendages at the base. With the proper environment, however, a single coral polyp can be the beginning of a whole new colony of individuals.

One way in which coral polyps reproduce is by asexual budding. New polyps bud forth from the cells of this connective tissue and immediately begin to secrete their own stony cups. Polyps also reproduce asexually by fission, putting forth more tentacles and fleshy partitions until they split from the mouth down, and there are two where there was one. All this time the skeleton of each polyp becomes more rigid and complicated.

When a polyp is mature, spermatozoa and eggs begin to develop within its inner tissues. When their maturation has reached a certain point, spermatozoa flow through the mouth of the polyp and float in the sea until they reach other polyps. The other polyps draw them inside with fanning movements of their tentacles. The spermatozoa then fertilize the eggs that have remained attached to the inner linings of the host polyps. Soon after fertilization, the polyp releases the eggs.

According to the environment it lives in, the same species can develop considerable variations in some of its characteristics and can grow in a variety of shapes: some species develop into bushy colonies when they live in shallow water and become flat assemblages in greater depths.

Because stony reef-building corals must have a lot of light as well as pure and warm water for their "working conditions," they are only found in a belt around the equator about 30° north and south, where the temperature of the sea averages approximately 70° F.; the greatest variety is found in temperatures of 77° to 84° F.

At their latitudinal limits reefs include a much smaller number of species; they also grow more slowly and are very vulnerable. Corals have salinity tolerances, too. Water turbulence and wave energy are also important controlling factors on growth, as is the hardness of the substrate and so is depth. Most species stop developing at depths of more than 75 feet and maximum growth appears at depths of less than 30 feet. The critical factor in controlling this is illumination. The polyps are unable to remove great quantities of particulate matter and can easily be smothered.

Influence of Light

Because of the vital presence of algae in the tissues of the reef-building coral, closeness to light for photosynthesis is important. As a result, coral flourishes best near the surface where the water is clear. When water is clouded by silt stirred up by sea action, not only can mineral particles damage or clog the fragile polyp, but turbidity also hinders the penetration of light with an inhibitory effect on the coral's development.

As coral grows farther away from the source of light, it expands horizontally and thins out to best utilize illumination. Away from the surface, the rate of limestone production decreases. Coral types that can successfully compete for light with others in the colony reproduce faster and thus dominate an area.

Corals that Move

Mushroom corals, or *Fungia,* exemplify the amazing ability of coral to adapt to an incredible variety of reef conditions, especially to sandy bottoms.

Disc-shaped mushroom corals may travel on a tidal current. When they feel the need to move, they expand their tentacles like the sail of a ship. If one is turned upside down, it is able to right itself after a long series of

At night the polyps of the fire coral expand to take in food. The tentacles are armed with poisonous nematocysts that produce a fiery sting.

Flexible sea fans orient themselves to face the prevailing currents and bend gracefully in the strong surges. Water flowing through the horny latticework brings food that is caught by the extended polyps of the colony.

Solitary northern corals live in cold waters where they grow slowly and do not build reefs.

contortions. The disc is expanded again and again, but each effort to turn over fails. After each failure, the coral takes a rest. Finally the body is expanded and the disc becomes nearly vertical. With great strain the coral expands a little more, and at last the disc is turned over. The coral then opens its mouth and vigorously spouts seawater. Its tentacles expand again. Then the animal shuts its mouth and hides it among the tentacles. The righting action is completed.

Some mushroom corals creep slowly over the bottom. They are also able to free themselves if they are covered by sediment. Some of these corals seem to be more or less voluntarily associated with a tiny worm that lives in a cavity on its underside. The worm helps the coral by pulling it across the sand. Trails in sand have been seen that were formed when the coral righted itself with the help of a jerk of the worm.

Fire Corals

All corals inflict wounds with their stinging cells, or nematocysts, on the diver who carelessly brushes them, but none so painfully as the stinging, or fire, coral. One of a number of hydrocorals, it secretes massive smooth skeletons of limestone, either in an erect branching or leaflike structure or in a low and encrusting form. The fire corals contribute heavily to the formation of many reefs, mainly in the western Atlantic and the Red Sea. They are white or of a pale fleshy or yellowish hue, and they grow to heights of two to three feet.

Sea Fans

Graceful gorgonians, sea fans made of a horny substance, are amazingly flexible and strong; they grow firmly attached to the sea floor and have the capability of orienting themselves according to the main surge of the water. In studying these colonies of animals, scientists have discovered some interesting points about their growth.

The distinctive feature of sea fans is their placement—most of the fully grown fans stand perpendicular to the motion of the waves. By confronting it broadside, they receive the impact of the ocean's surge more passively and simply bend back and forth. Small fans seem to face in no particular direction, but as they grow taller, their consistent tendency is to avoid the twisting that would result if they were oriented parallel to the surge. As the colony shapes itself into a fan, it acquires a larger surface to catch more planktonic food for its tiny polyps.

Jewelers sell precious objects made of a rare coral species that grows in deep water and has little to do with reef-building varieties. Jeweler's coral is most often red (oxblood), sometimes pink, white, or very rarely "pale flesh" (angel's skin). This prized bounty from the Mediterranean Sea was dredged with leaded wooden crosses in the time of the Phoenicians and early Greeks, and now aqualung divers reaching to 300 feet have practically scraped the Mediterranean bare of red coral.

Brown or ocher trees, shaped like tamarisks, grow to a height of 10 feet in depths of 100 to 200 feet in the Red Sea and in the Indian Ocean. They grow to more moderate sizes in other oceans and at greater depths. They are called black coral, because once their branches are cleaned of their brown polyps and polished, they produce a rich, deep-black precious material. It has long been a treasured raw material for jewelry and objets d'art and is considered a sacred material in Islamic countries, where it is turned into expensive prayer beads. In China, Japan, and Indian Ocean countries it provides the material for charms to ward off bad luck and disease.

As numerous as dandelions in a field are anemones in the sea. About 1000 species have been found in both the Atlantic and the Pacific. Their color and form vary so much that they have been given many appropriate names: brown sea anemone, powderpuff, white plume, dahlia, and so on.

The treelike alcyonaceans are also anthozoans, or flower animals. They often have large, soft polyps, and some pulse rhythmically, making it obvious that they are alive. They have only scattered spicules to stiffen their bodies. Although some soft corals extend into polar waters, they are largely a warm-water group, most abundant in the Indian and Pacific oceans. Some are shaped like leathery brown, great flabby masses; others are like pink or purple branches of celery. If kept in the dark, soft corals show signs of starvation within two weeks. Evidently light is more essential to them than food, and the polyps' association with the algae that live in their tissues is vital to them. Because of the zooxanthellae, they have become dependent on light to function and survive.

Precious Corals

Flower Animals

The Fight for Space and Light

Red coral has been avidly collected over the centuries for jewelry and objets d'art. It is now extremely rare.

CORALS AND THEIR RELATIVES

Corals sometimes kill each other to find space and light in which to live. This rarely happens with corals of the same species but often involves closely related types.

It is generally in the buffer zone of the coral reef, where overpopulation is most likely to occur, that cannibalism is used to solve the problem. No such fratricides have been observed among corals where there is room and light enough for all.

Aggressive acts are set off by physical contact. The stronger "aggressor" polyp attacks by extending mesenterial filaments over the weaker coral, literally digesting its prey outside of its own cavity with the filaments. Thus such battles, although they are initiated as struggles for space and light, revert to the basic, simple feeding response set off by contact between creatures of different species. The victim in the competition becomes a source of food the same as any other unrelated prey would be.

Zooxanthellae are always found in reef-building corals, where their concentration varies among the different species. Their symbiotic relation with their hosts has been an area of particular interest to scientists. It is not always understood whether it is the plant or the animal which is "farming" or "exploiting" the other.

An alcyonacean is a big, rubbery, soft coral common in the Pacific. The treelike body is stiffened by scattered spicules.

One question that has been much debated is the extent to which the algae actually contribute to the nutrition of coral. It was once believed that the purpose of the algae was to provide plant food. But corals have been shown to be strict carnivores.

The majority of them feed at night. Their diet consists primarily of planktonic animals, although at least one kind of coral eats fish. They trap and transport this food to their mouths with tentacles that are laden with stinging structures (nematocysts). Food that is too small to be grasped by the tentacles is trapped on a sticky, mucus-laden area between the polyps of a coral colony. This food is transported to the mouths in moving sheets of mucus. In this way, a colony with its polyps expanded represents an enormous feeding surface in relation to the bulk of its living tissue.

It is known that coral is important to the algae, which make use of the polyp's excreted carbon dioxide, ammonia, and phosphates. Because zooxanthellae are plants, they seek the light of the sun for photosynthesis—the production of organic substances, chiefly sugars, from carbon dioxide and water with the aid of chlorophyll. It is within the upper 200 feet of the oceans that photosynthesis can take place. And yet zooxanthellae have been found living with anemones at a depth of 1250 feet off Antarctica and with anemones and corals at a depth of 1920 feet off Key Largo, Florida. This has led to a suggestion that, at the expense of the host animal, these deep-dwelling zooxanthellae live by obtaining food only from organic material.

But what does the polyp obtain from the algae that might aid its nutrition? One difficulty in answering the question is that the total nutritional requirement of a coral is still unknown. There is evidence, however, that corals acquire soluble organic carbon compounds for their nutrition from the zooxanthellae. Organic material from the algae moves along the metabolic pathways of the coral, but the significance of that movement in terms of quantity is not yet known.

We do know that with the help of the light's energy, the presence of zooxanthellae significantly accelerates the rate of calcification, the process by which the coral's skeleton is formed, although it is not certain exactly how this takes place.

Like other plants, under the influence of light the zooxanthellae consume carbon dioxide during photosynthesis, and the carbon dioxide functions in the acceleration of calcium deposition. This hypothesis is inconsistent, however, with the observation that polyps at the tip of branching corals calcify much faster than lateral ones even though they have relatively few zooxanthellae. It might also be asked how carbonate is supplied for calcification if the algae remove carbon dioxide from the environment.

Two alternatives to the "carbon dioxide removal" hypothesis have been proposed. One is that the transport of metabolic products released by the algae toward the tip of a branch might explain the rapid calcification of these polyps. The other is that the algae remove phosphates, which are also inhibitors of calcification, from the calcifying milieu.

A velvety green coral colony (Monastrea annularis) with its polyps tightly closed.

A CORAL CITY

Coral reefs have been called "nature's cities" and all of its members adapt to and benefit from each other in the crowded area of the reef city.

Reefs have enormous rates of organic production, rivaling the efficiency of man's best agricultural projects. This is particularly remarkable when compared with the rest of the ocean. Most of the bottom is almost bare, dotted with relatively simple communities. It seems that when a bottom ecosystem is locally successful, it manages to eliminate intruding species or to repel them to the next ecosystem—often nearby but different in composition.

Within the jungle-cities of the reefs, however, tens of thousands of species belonging to practically all known phyla coexist in a competitive but successful balance for miles, or sometimes for hundreds of miles. It is primarily through food web interactions that the coral reef achieves its biological stability. All individuals are in some indirect or direct way dependent on all the others in a complex feeding chain.

The coral itself is the center of all activity and is often the main food source. Until fairly recently corals were thought to be virtually immune to parasites and predators. Their stinging nematocysts were assumed to be an adequate defense. But now we know that a surprising number of animals feed on coral. They include fish, asteroids, crustaceans, gastropods, and many different worms. The starfish known as crown-of-thorns *(Acanthaster planci)* is one of the most spectacular predators on coral reefs.

At night, this multirayed starfish, a moving mound of spines about 15 inches across, settles on the stony coral, everts its entire stomach and digests complete polyps, leaving only the skeleton. Algae then settle into the empty skeleton, and borers soon follow.

The removal by collectors of the triton snail, a predator on the crown-of-thorns, might have been the cause of the recent destruction. Water pollution is another possible cause, and so is the testing of nuclear weapons. Blasting and dredging and even spearfishing have also been suggested as reasons for the plague.

A British scientific group based at Port Sudan in the Red Sea, after experiencing a reef invasion of the crown-of-thorns in 1970, increased their efforts to arrive at a solution. Undersea observations showed that the painted shrimp, triggerfish, and pufferfish, as well as the triton, were natural predators; the local plague ended two years later, and it was concluded that it might have been natural. Several remedies were suggested, nevertheless. These included collection of the stars by divers, instructions to fishermen not to remove their natural enemies, and unfortunately the ridiculous suggestion of injecting ammonium hydroxide into them.

The author's opinion is that the crown-of-thorns is instrumental in reef destruction, but that its role has been vastly over-emphasized.

Many herbivorous fish graze on the algae of the reef. Some only scrape the surface with their mouths, but others like the parrotfish bite little chunks of the reef and crush them in their mouths or in their throats, retain the bits of algae, and spit out or reject the crushed limestone in little white clouds before they take another bite. Molluscs and crabs work over the inner reef during the day and disappear into the outer reef at night. In a heavily grazed algal zone, bare whitish patches showing the intense foodseeking activity of these animals can readily be seen. Curiously, wherever herbivorous fish graze, greater varieties of algae grow.

Some carnivorous fish feed directly on the corals, by browsing on the polyps, by scooping mouthfuls from the hard, stony materials of living coral heads, or by feeding on branching coral tips. The triggerfish bites protruding surfaces to wrench loose large sections of coral. It then searches through the fragments for bivalves, gastropods, crustaceans and other things to eat. Some filefish, close relatives of the triggerfish, are very generalized predators on the reef, ingesting just about any form of life they encounter. This includes such seemingly unwholesome things as sponges, gorgonians, hydroids, and stinging coral. Divers can distinctly hear the noises made by all this tearing, chewing, and crushing. It has been estimated that one-third of every year's coral growth is destroyed by predators.

Within this continuously growing and dying city of coral every conceivable strategy of survival has been developed, and every possible feeding method is in operation. Obviously, the herbivores are commonly the most abundant marine animals of any size in a reef community, and all of them may fall victim to more powerful carnivores.

Small fish such as herrings, sardines, and anchovies literally take their choice from the overwhelming quantity of tiny planktonic animals, picking at them one by one. The huge manta ray, however, is equipped with strainers to extract most of the plankton from the water that passes through its gills, and it swallows the aggregate food without discrimination. Fast-swimming tropical predators—tunas, jacks, sharks, squids, or barracudas—prey on the smaller vegetarians or plankton eaters. Open-sea fish accustomed to traveling hundreds of miles in the blue expanses of the ocean find the reef waters an inviting place to gather and breed, to forage for food, or to be cleaned of parasites by specialized cleaning fish. Pregnant hammerhead sharks seek shelter in a reef bay to give birth. The reef provides abundant food for the young sharks. Tuna and jacks from time to time linger around before returning to the open sea. Even whales sometimes venture in.

Among the bottom-inhabiting animals, the sea stars are voracious feeders, preying on clams, which they exhaust and force open by applying continual traction on the two halves. Predacious snails, noteworthy for their long siphons, are able to drill holes into bivalves and other molluscs, insert their siphons and eat the soft contents. Octopods are also capable of drilling holes in shellfish.

In the inner reef crevices there are predators as well. With ease, the mantis shrimp can slash apart crabs in a few seconds. Hidden groupers can use their gill plates as pumps to suck in unwary passersby.

The reef contains its own water-conditioning system. The internal spaces of the reef function as a trickling filter. In addition, many filter feeders, such as sponges and clams and the coral polyps themselves, capture tiny particles suspended in the water.

Sponges have millions of microscopic pores that pierce their surface. Through these pores the sponge sucks the water that contains microscopic plants and animals that it feeds on. A steady stream passes through its body and out the large vents at the top; one estimate set the rate for a Bahamian wool sponge at about two quarts a minute.

All of the filter feeders induce feeding currents that help to ventilate the internal reef spaces, thus perfecting a very elaborate water-conditioning system.

The life processes on a reef produce two sorts of waste: reusable organic material, including bodies of dead animals, mucus, and fecal matter, which must be recycled because it is rare and precious, and inorganic sand, which must be eliminated. The coral community is organized to perform these two functions. Dead animals are eaten by scavengers; crabs, shrimp, and lobsters are the caretakers of the coral city. This category of animal feeds on any corpse, including those belonging to the same species—an interesting case of scavenging cannibalism! As a result, practically all organic tissues are transformed into fecal matter, the main exception being a variable

Sea squirts are the most primitive animals of the phylum Chordata. Their sacklike bodies have one opening to draw in water, another to expel it.

Tunicates, growing in a colony, have openings surrounded with a fringe of cilia.

fraction of the population of bottom crawlers and burrowers.

The resulting waste settles in the cracks, holes, and crevices of the reef, on the bottom of the lagoon, and to a lesser extent outward along the steep slopes where the material is practically lost for the citizens of the reef. The second stage of waste treatment is performed by the deposit feeders (worms, snails, sea cucumbers, and so on), crawling over or under the surfaces of the mud, treating enormous quantities of sediment from which they extract organic debris and fecal matter. The sea cucumber processes as much as 200 pounds through its body in a year. This service keeps sediments from blocking internal passageways, provides ventilation, and resuspends particles in the water to be reprocessed by filter feeders.

The third and final stage is performed by bacteria that teem in sediments. In addition, all calcareous surfaces harbor microbes and fungi which often penetrate skeletons to considerable depths. Bacteria return any organic matter to its fundamental components, basically nitrates and phosphates. Bacteria are oxygen consumers and would asphyxiate the reef if their population grew unchecked. But they are also consumed in great quantities by larger creatures. These organisms allow the reef to remain an almost closed system by eliminating the accumulation of waste. The only waste that remains on a reef is calcium carbonate, which is used to build the coral city itself, the excess being eliminated as coarse sand along the slopes of the reef, which it thickens and reinforces.

MAN AND THE CORAL REEF

Cowries are gastropods with shiny, smooth shells. Some species are known to feed upon coral.

Coral reefs are the most biologically productive of all natural communities, marine or terrestrial. Man once lived in relative harmony with that community. Today a thoughtless use of technology threatens to make that relation a destructive one.

One of the most serious dangers to the coral reefs is sedimentation. This threat is dramatically illustrated in Hawaii. With the arrival of Westerners, upland soil was plowed for sugar cane and pineapple. The impact of the resulting erosion has been tragic.

Since 1897 the shoreline of Molokai has advanced as much as a mile and a quarter across the reef flat. Elsewhere off Molokai, the reef is overlaid with four to 27 inches of red-brown silt.

The use of dynamite to kill great masses of the fish and of bleach to flush them from their hiding places have become very widespread and very destructive practices. On a densely populated coast of Oahu in the Hawaiian Islands, after bleach was used by fishermen, almost no living animal remained.

It is the affluent, jet-propelled tourist who may bring about the fifth collapse of the coral reefs. One by one the islands are being transformed into vacation centers. Channels for cruise ships are blasted through the reefs. Land is bulldozed for airports and building sites. Landfill is dredged from the sea. Untreated wastes poison the waters. Man's ill-conceived constructions could turn the coral castles into an archaeological tomb.

In Florida there is a great reef that runs west from Key Largo for about 220 miles. The reef was once in danger. Curio vendors were tearing it apart, using dynamite and crowbars. Bargeloads of corals, sponges, and shells were piled along the roadsides for sale to motorists. Fish collectors raided the waters, and spearfishermen stabbed everything that swam or crawled.

After much debate, legislation was finally passed that would protect at least 75 square miles of the reef. A preserve was created, to be administered by the governments of the United States and the State of Florida. Now in this one small corner of the pharaohs' realm, the work of these greatest of all builders can be enjoyed without fear of its being despoiled. It is a beginning, at least, and it is a hopeful sign.

MAMMALS AND BIRDS OF THE SEA

CHAPTER
9

SOVEREIGNS
OF THE SEA

Anyone who has seen the early morning feeding at sea on a calm day has also witnessed the superiority of the warm-blooded physiology over the cold-blooded way of life. Sardines, sprats, anchovies, and other schooling fry, chased relentlessly from below by faster, larger fish such as jacks, turn the surface into a boiling kettle, while seabirds, like dive bombers, take their share of the feast. But the aggresive predator-fish—among them bonitos and bass—are themselves easy prey for the second circle of divers: the dolphins. Farther away a few sharks shyly roam, waiting for the end of the feeding frenzy to scoop up the remains. They know that if they venture too close, they will be severely punished by the dolphins. From the air or from below, the fish are no match for the warm-blooded birds and mammals.

The warmer the blood, the higher the efficiency of the living thermodynamic machine. Birds and sea mammals have acquired their power while evolving out of the oceans; their physiology has coped with such problems as keeping their central temperature constant or holding their breath during deep, prolonged dives. The heat-exchanging system displayed in their flukes, fins, wings, or webbed feet is a prodigy of ingenuity. Surprisingly, a few powerful, fast-cruising fish, such as the tuna, have developed a very similar heat exchange, in order to keep their hard-working muscles a few degrees warmer than the surrounding waters. But if the engineering principle used is the same, the purpose is quite different: the tuna is just a little warmer than the sea, though the temperature of the sea varies; birds and sea mammals, on the contrary, must keep their central body temperature constant, whatever the outside polar or tropical conditions may be.

Being gifted with a superior machinery, the warm-blooded lords of the sea have practically no feeding problem, although they have to eat a lot more than comparably sized fish. Colonies of many million seabirds pile up on islands, feeding twice a day in less than half an hour and gorging themselves to the point of being hardly able to fly. In the sea whales fill up with crustaceans in a few daily dives. Pilot whales sound to the deep layers where they swallow hundreds of squid or cuttlefish; porpoises, dolphins, and sea lions spend less than an hour a day to quench their appetite.

Having no difficulty in finding food, these warm-blooded animals have lots of leisure time, which explains why dolphins play almost all day, traveling for no essential reason, performing somersaults many times in the air, escorting boats or ships, outracing a cruiser for a few moments, as if to demonstrate their ability. Even creatures as clumsy on land as sea elephants are amazingly graceful and rapid in their liquid element; and when they come ashore to rear their young, they are so fat that they can afford to fast for several months.

Leisure time has been used by those sea mammals that have a large brain to develop wit, intelligence, communication, and even unnecessary feelings such as faithfulness, tenderness, and friendship. This is particularly true for those that have teeth and are carnivores. But it is not at all certain that the higher degree of intelligence originated from the flesh-eating habit—it may have its cause in the origin of the species. For example, some scientists think that the whalebone whales are descended from some extinct insect eaters such as anteaters, while sea lions came from felines, and toothed whales from an unknown branch of carnivorous land ancestors.

The physiological superiority of warm-blooded animals in the sea has severe limitations. Like complicated clockworks, they are in many ways fragile. Their high intellect produces psychological problems often ending in disaster. If the high combustion furnace of birds and mammals is easily provided with fuel (food), it must also be supplied with large quantities of combustive (oxygen from the air). This means that our warm friends must take in a lot of oxygen, and this in turn limits the duration of their dives. Some birds, such as penguins, cormorants, and puffins, and some cetaceans such as sperm whales, have modified (without radical change) their organs to perform dives of extended duration. The warm-blooded toothed whales are the Sovereigns of the Sea.

ON LAND, SEA, AND AIR

Gray whales make an annual migration to and from breeding grounds in Baja California.

"There's lots of good fish in the sea," wrote Sir William Gilbert in *The Mikado*. Indeed there are. But there are also the warm-blooded animals and birds. In fact, in numbers and varieties of living things, the sea more than equals the land.

The cetaceans are warm-blooded water dwellers that are shaped like fish but are not fish. These mammals—the whales, dolphins, and porpoises—are born live, suckle their young, and breathe air. From the little four-foot harbor porpoise to the enormous 110-foot blue whale, these aquatic mammals have no need to come to land.

The seas are also host, though not total home, to the pinnipeds—seals, sea lions, and walruses—which look less like fish and somewhat more like their land mammal ancestors. They leave the sea to breed, and they go to sea only for food.

Measuring only 56 inches and weighing 80 pounds, the sea otter is the smallest of the marine mammals and is the only member of its family, which includes the mink, weasel, and river otter, to inhabit salt water.

The blue-gray dugong and the purplish-gray manatee, the two living members of the Sirenia order, complete the catalog of marine mammals. These large, sluggish creatures with stout bodies, forelimbs modified as flippers, and no hindlimbs may have been the source of the mermaid legend. One look at them, however, reveals that the sailors who saw human beauty in them had probably been at sea for too long.

No bird, except maybe the penguin, is as at home in water as most pelagic mammals, but many birds spend their lives at sea except for one short period of the year when they come ashore to mate, nest, and raise their young. The prions, or whalebirds, the jaegers, or sea hawks, the albatrosses, and the shearwaters are among these pelagic birds.

Other bird families live in open waters just offshore, coming ashore little more than the pelagic birds but remaining closer to land. Sea ducks like scoters and old squaws and wintering loons and grebes follow this life.

Bay ducks like the goldeneye and scaup ducks remain closer to shore and seek haven in stormy weather on land or in shoreside ponds. The saw-billed mergansers, gulls, terns, skimmers, and a host of other species also occupy these inshore waters.

Two centuries and more ago, when whaling was in its heyday, a successful whaling captain had to be a tough adventurer, but also a superb seaman to navigate poorly charted seas. With the exception of the antarctic, which is a fairly recent hunting ground, what we know about the distribution of the remaining whales today is much the same as what the whaling captains of old knew.

Whales are the most extensive mammalian travelers on earth, barring a few humans. Because of their size, endurance, their reserve of energy, and mainly because they weigh nothing in seawater, they can travel economically, enabling them to migrate even thousands of miles. In the winter, for example, finbacks travel 5000 miles from the Bering Sea and the Arctic Ocean into their breeding grounds in the Indian Ocean, a journey that may take a month or more. However, although whales are migrating animals, they do not cross the equatorial zone, so that almost all species are divided into two different populations: the northern (boreal) and southern (austral). The austral whalebone whales or rorquals—blues and finbacks, sei and piked whales—move from the antarctic to the seas off Africa, the Bay of Bengal, the Gulf of Aden, and the South Atlantic islands of Tristan da Cunha. The boreal rorquals are divided into Atlantic and Pacific groups. The gigantic blues summer in the North Pacific and go as far south as the Indian Ocean in winter. The North Atlantic blues move from the east and west coasts of Greenland southward along the American and European coasts in winter.

The sperm whales, which live a different sort of life because they have teeth and feed on squid and cuttlefish, remain all year in the tropics. According to Melville's classic, Captain Ahab of the *Pequod* hunted Moby Dick, the albino sperm whale, where cooler waters met equatorial currents. The old bulls are the only solitary sperm whales to venture to polar waters in the summer. Some may have been defeated harem bulls. Moby Dick, if he ever existed, could only have been a weak old male.

The little piked whale goes farther into icy seas than any other species and has been seen poking its head up through holes in the ice to breathe. Some of them get trapped too far from the edge of the pack to return to the open sea, and they die slowly of hunger.

Dolphins and porpoises are usually found in an open-sea or pelagic environment, but there are exceptions. Some, like the common harbor porpoise, live principally in bays and estuaries, ascending rivers occasionally.

Other dolphins and porpoises live in specific areas of the open ocean. The lead-colored dolphin lives in the sea off India. True's porpoise stays in the warmer waters of the North Pacific, and other species occur in parts of the tropical Atlantic, Pacific, or Indian oceans.

Next to cetaceans, northern fur seals are the most active migrants among mammals. Although they seldom swim more than 100 miles from the nearest coastline, they may travel 3000 miles along the Pacific coast from breeding territories in and around the Bering Sea to their wintering grounds off southern California, covering the distance in less than two months.

The migration record, however, for any animal, mammal, or bird, is

Where They Go

The enormous proboscis, which prompts the name of the elephant seal, forms an efficient resonating chamber for the loud vocal calls of the male.

held by the arctic tern, a delicate slender winged relative of the larger and stronger gulls. Each year arctic terns travel from the arctic to the antarctic and back—some 25,000 miles, a distance equivalent to a round-the-world trip at the equator. A one-way trip takes up to three months.

OUTWARD APPEARANCES

About 50 or 60 million years ago, for some unknown reason several species of mammals, probably quite different from those we know today, moved to the edge of the sea. With each passing generation changes took place in their physical makeup that were useful to them in their new habitat. As the changes became more functional, they spent more time in the sea until eventually they were living in the sea most or all of the time.

Piglike creatures were probably the ancestors of odontocetes, or toothed whales, like sperm whales, porpoises, and dolphins, while relatives of anteaters may have been the ancestors of whalebone whales or mysticetes.

About the same time, but not necessarily in the same places, a group of hooved animals probably related to the elephant also moved to the shallow waters of coastal marshes and the banks of river mouths. These were to become the present-day sirenians—manatees and dugongs. Twenty-five to 30 million years ago, long after cetaceans and sirenians had developed their ability to live in the sea, another group of terrestrial animals took to the ocean. They probably looked something like wolves or lions and were to become today's pinnipeds, evolving toward a similar form from two separate groups of animals. One of these groups was bearlike and developed into the eared seals—the fur seals and sea lions—and the walruses. They developed paddlelike forelimbs and broadly webbed hindlimbs, which they use to move about on land, however clumsily. The other group developed from an otterlike ancestor into today's true seals—the gray, harbor, monk, elephant, and harp seals. They too have paddlelike forelimbs and hindlimbs, but they are unable to draw the latter forward for use on land. Instead, on land they must crawl like caterpillars and use their front flippers to pull their bodies around. These true seals appear to be more fully adapted to aquatic life.

A thick insulation layer of fat protects the elephant seal in the cold waters it inhabits. The fat is blubber like that of whales and it keeps the mammals warm when they periodically shed their fur.

Streamlining

In the process of evolution toward a streamlined shape, cetaceans have undergone a distortion of their skulls so the nostrils are pushed back atop the head. This enables the animal to breathe at the surface without lunging out of the water. It need only break the sea's surface with the top of its head, open its blowhole quickly and exhale, then inhale quickly, close its blowhole, and submerge. All this occurs in two or three seconds and may be repeated several times before a deep dive. Besides distortion of the skull, cetaceans have all but lost their necks as their cervical vertebrae have become compressed and blubber has filled in the natural constriction behind the head.

As a result of being buoyed up by water and blubber, cetaceans' bones have become lighter in weight. But because of this, beached cetaceans often suffer serious injury to their internal organs. Their own weight, when they are out of water, is too much for their light bone structures.

Among birds, streamlining enables them to pass through air or water with the least effort and greatest economy of fuel. Gannets, loons, and terns are only a few of the bird families that benefit from streamlined body design. Gannets are quick strong fliers that dive into the sea from 30 or more feet above the surface to capture their food. They plummet downward cleaving first the air, then the water, as they pursue their quarry. Underwater, loons and terns are dramatically streamlined with pointed bills, slender necks, and gradually widening bodies. Even the ungainly pelicans are designed so they can switch from what looks like lumbering flight to dive-bomber sleekness as they pursue fish.

Fur, except in the sea otters which manage to keep theirs loaded with tiny air bubbles, is generally useful for insulation only in the air. Therefore most marine animals depend on blubber to keep warm. Blubber is the thick layer of fatty tissue between skin and muscle. The amount of blubber varies with species and season. Those species like the walrus that swim in ice-filled

Something Under the Skin

The powerful flukes of a humpback whale give it enormous propulsive power for deep diving.

waters tend to have thicker blubber than those in warmer latitudes. Blubber is buoyant and helps keep whales and dolphins afloat. The right whales have blubber up to 28 inches thick, and they float when killed.

Besides its insulating and buoyant properties, blubber is also a food reserve. In cold waters whales feed upon the abundant fatty shrimplike krill and store blubber reserves. As they head from food-rich polar seas into warmer waters, food becomes scarcer so they draw on their blubber for energy. Consequently whales in the tropics have thinner blubber layers.

In pinnipeds the layer of fat under the skin is relatively thinner than in cetaceans and is usually about three inches thick. Male fur seals and elephant seals, for example, may go without eating for up to three months during their breeding season. When they first come ashore on their breeding grounds, these seals have voluminous folds of fat hanging from their necks and covering their chests. By the time they put to sea after breeding season, they are relatively slender. During the nursing period, the females draw drastically on their fat reserves to produce huge quantities of milk for their calves.

Birds too have fatty tissue within them that serves as a food reserve when they are migrating. With little time to eat while heading to or from their breeding grounds, the birds use up their fatty reserves for nourishment. Duck hunters know their early season victims are full of fatty deposits, while those they shoot later in the season are not.

Coming to a Head

Once there was a horselike animal with a single, long horn that extended from the middle of its forehead. This mythical creature was the unicorn, and it was said that anyone who possessed the spiral horn of one of these creatures enjoyed great powers. Just how this legend began is not clear, but it may have stemmed from the Basque whalers who brought the tusks of narwhals back with them from their arctic whaling expeditions. The narwhal is a small whale that lives in arctic waters north of Russia and Canada. Its tusk is the left upper incisor that is one of its top teeth and grows in a counterclockwise spiral to lengths up to nearly eight feet in a 12-foot narwhal. It is found only on sexually mature males. No one knows just what the function of this tooth is. Some believe it is used to stir up shrimps and crabs from the sea floor that make up part of the narwhal's diet. Others feel it is a weapon used in ritualized combat between males. A hunter theorized it might be used to punch breathing holes in the ice. The narwhal's tusk is just one of the several unusual features on the heads of the mammals of the sea.

Rorquals, for example, have long, bony plates with hairy edges inside their mouths instead of teeth. These plates are called baleen and are made of the same horny substance as our hair and fingernails. In addition, they have long parallel grooves on the ventral side of their bodies, which run from chin to abdomen and allow their mouths to expand. When rorquals feed, they engulf in their inflated mouths vast quantities of shrimp along with great amounts of water. Water is forced out of their mouth, by the speed of their movement, by a powerful constriction of the grooved crop, or by pushing their massive tongues upward to the roofs of their mouth; thus the shrimp and other food is strained out on their baleen.

Dolphins, porpoises, and sperm whales, on the other hand, are well equipped with teeth. Some species have as many as 300, some as few as 42. They use them to hold their quarry, usually slippery fish, not to chew.

The vegetarian manatees use their lips to gather water hyacinths or hydrilla weeds that make up their diet. Seals and sea lions, with their doglike heads, are well equipped with teeth they use more for holding than for chewing, as dolphins and porpoises do. Among birds, the shapes of bills range from the huge bill and pouch of the pelican to the tiny bills of dowitchers and sandpipers. The broadbilled prion, or whale bird, has strainers on the edges of its bill, comparable to baleen.

Many marine mammals have no external ears, but this is no hindrance. Fur seals and sea lions have small external ears, however, which they lay back while swimming. Small openings mark the ears in birds and true seals. In cetaceans and sirenians, a barely visible crease shows where the ear is.

The most unusual feature on the heads of cetaceans is their nostrils, or blowholes, which have moved from the front of their heads to the top. In the toothed cetaceans both nostrils have merged into one blowhole which remains sealed during dives. The sperm whale has its blowhole located on the left side of its giant head. Other sea creatures have nostrils in more conventional locations, but they have adapted to sea life. Pinnipeds, for example, have nostrils that are closed when relaxed so they, like the cetaceans, must make a conscious effort to breathe.

Limbs, Flippers, and Flukes

Cetaceans have within their flippers, or pectoral fins, a series of bones that compare with the bones of our hands, wrists, and fingers. Flippers of whales are only used as rudders to make short turns or to avoid obstacles. The flippers of humpback whales have an almost disproportionate size, but account for the playful, unpredictable maneuvers of the animal.

For hindlimbs, cetaceans have little or nothing; over the millions of years they have spent adapting to aquatic life, they have all but lost every vestige of hindlegs. In adapting to life in the sea, however, whales and dolphins have added something that might pass for a new limb. Their tails have become broadly flattened with a large area that can push great quantities of water behind them. These are the flukes which are almost exclusively a powerful propelling system. These powerful flukes have been rightly feared by whalers, some of whom have been thrown overboard by the

Seals use their powerful front flippers to propel themselves through the water and steer with the hind fins.

tails of harpooned sperm whales, which have shattered whale boats and the men in them. Sperm whales can, when irritated, reach speeds of 12 knots and more! One reportedly towed an Azores fishing boat at a speed of 20 knots. The enormous blue whale can get up to 15 knots for long runs and 20 knots in sprints: the record seems to be held by the sei whale which can make 35 knots in short sprints.

Dolphins and orcas are swimming machines, as advanced in the warm-blooded category as tunas are in the cold-blooded family. Their flukes are perfectly coordinated with the powerful muscles of their streamlined bodies and the number of fluke beats is almost linearly proportional to speed—it is comparable to a ship's speed being a direct function of propeller revolutions per minute.

The most original use of flippers is demonstrated by sea lions and fur seals; it is with powerful strokes of their long, thin, front flippers that they "fly" underwater, while the soft, relaxed hind fins are mostly used for steering. Sea lions are not quite the fastest, but probably the most maneuverable of the marine mammals. Harbor, monk, and elephant seals swim only with body undulations: their hind flippers are stretched like two parallel caudal fins, beating *laterally* like those of fish. The walrus makes little propulsive use of its limbs, but its thick body is extremely flexible and is rhythmically bent to become an efficient traveling wave.

Like the cetaceans, the manatees and dugongs have no external trace of hindlimbs, only a vestige in the form of a couple of bones imbedded deeply in muscles near the base of the tail. Their forelimbs, like those of other marine mammals, are paddlelike and ideally suited for life in the water. Their tails are large, powerful, and useful to the sirenians in propelling themselves through the water.

Sea otters are recognizable relatives of their land cousins, but their limbs too are beginning to show signs of flattening and growing webs between the fingers. They use their flattened hindlegs for swimming and their tail has flattened to become a third hind flipper. They use their forelegs, with their well-developed fingers, for grasping food. Unlike any other marine mammals, they still have well-developed, retractable claws, which they use in foodgathering.

Floating with its head submerged, a booby bird searches for food.

Wings and Feet

A bird's wings are its front limbs, analogous to our arms, and they probably developed from the front limbs of tree-dwelling lizards some 140 million years ago. Among all the mutations that changed the lizard, some improved their ability to leap from tree to tree. Eventually their limbs became what we now call wings: they had become the first birds. But the aerodynamic design of wings is mostly dependent on feathers. The naked body of a bird seems less made to fly than that of a bat, for example. Feathers, which are an exceptionally light, strong, and flexible material, fill in the wings and give them the various curvatures and shapes needed to take off, fly, or land.

Most diving birds use their wings to fly underwater in much the same way they use them to fly through the air. A few others, on the contrary, like some diving ducks, fold their wings and paddle very strongly with their stretched legs. The body is also stretched out in order to be thinner and to offer the least resistance to propulsion.

Alcids, mergansers, and puffins pursue fish in the way many birds chase insects through the air. But because water is 800 times denser than air, a bird's wings must be stronger to fly underwater. Therefore, diving birds have developed heavier bones than those that do not go underwater. In most cases, the underwater fliers are not strong fliers above the surface, possibly because of this heavier construction. The hindlimbs or legs of birds that swim, such as ducks and geese, have similarly become heavier and stronger than those of birds that spend much of their time in the air or perching.

Storm petrels, called Mother Carey's chickens by sailors, have combined an aerial life with an aquatic one. They spend much of their time flying just above the surface, dipping their webbed feet into the water and dabbling them along the surface, perhaps in search of planktonic food. Because they appear to be walking on the water, they have won the name "Jesus bird" in many places.

Rails, coots, and a few species of shorebirds with lobed or semipalmate feet can walk over marshes and soft mudflats or quicksands along the shores where they live. Probably the most spectacular development of legs among aquatic birds, however, is the long, stiltlike legs of the herons, storks, ibises, and flamingos.

INNER WORKINGS

An aquatic existence means living in a sphere of entirely different physical and chemical makeup from that of the air where most birds and mammals make their homes. Nowhere in any of these aquatic creatures are there any additional or different organs to aid their life in the sea. Instead the properties of their existing organs, found in all birds and mammals, have undergone modifications to suit the needs. These adaptations include the ability to withstand the powerful heat-draining effect of the water, although the temperature variations they encounter at various levels in the sea are less important than in air. Other adaptations are specialized kidneys, respiratory and circulatory systems, and bone structures, as well as the external changes already noted.

Birds too must adapt to life in the sea. The high salt content of seawater requires them to cope with the effects of the chemicals in the ocean. Like some marine mammals, some seabirds need a third eyelid, or nictitating membrane, to protect their eyes and to help them see clearly while they are diving in search of food. They must remain immune to the effects of constant motion as they ride the waves for hours and days at a time. And they require an astonishingly accurate navigational sense to migrate over an environment where there are no landmarks to guide them on their long overwater flights.

Need Amidst Plenty

"Water, water, everywhere,/Nor any drop to drink," wrote Samuel Taylor Coleridge in *The Rime of the Ancient Mariner*. This is precisely the problem facing the warm-blooded creatures in the sea because the salinity of the body fluids of birds and mammals is substantially lower than that of seawater.

First in importance for living in a salty environment is a source of fresh water. Fish-eating dolphins get much of their water from their food because fish flesh, though not pure fresh water, is less salty than the sea. Orcas, feeding on other mammals and birds, obtain water from the tissues of their victims. But both groups take in small amounts of seawater as well as the fluids from the flesh of their prey.

Seals and sea lions reportedly take in no seawater with their food, which is mostly fish. Many whales, the walrus, and the sea otter, however, eat invertebrates whose body fluids are close in salinity to seawater. One of the principal reserves of water for marine mammals is their fat or thick layers of blubber. When this fat is "burned up," or metabolized, water is a byproduct.

Marine mammals conserve the water they gather by various means. Cetaceans have no sweat glands, and because they inhale and exhale in humid atmosphere so close to the water's surface, they lose little water through their lungs while breathing.

Larger, and possibly more efficient, kidneys may help marine mammals dispose of the excess salts they take in. Since the kidneys of some birds are not as efficient as mammalian kidneys, a pair of glands near the nasal passages in the head help secrete excess salts. These salt glands are found in

Salt glands atop the albatross' beak enable the bird to secrete excess salt that would otherwise poison its system.

Most marine mammals, including this crabeater seal of the antarctic, can close off blood circulation to certain parts of their bodies to control their temperatures. Seals, in particular, use their body surfaces, mainly flippers, as a means of heat exchange. They can regulate the flow of blood through their thick skin and in this manner control the amount of heat lost to the cold environment. Remarkably, the arrangement of blood vessels can allow a reduction of heat to practically zero.

all birds, but are not always functional. (Marine reptiles, especially the Galápagos iguanas that feed in the sea, also have salt glands.) The duct from each of the salt glands carries salt solution into the nasal cavity and then to the outside where the solution drips off the end of the bird's beak. The system works only when excess salts accumulate, usually after feeding at sea.

The Heart of the Matter

The elephant seal has about twice the blood volume of man. Yet to pump all that blood throughout its system, the elephant seal has a very ordinary heart. This is because the weightlessness of marine animals in the sea places a far lesser effort on the pump—the heart. The most remarkable part of the circulatory system of cetaceans is the network of tiny blood vessels, called the retia, throughout the fatty tissues as well as a number of other areas. Some physiologists have speculated that the retia's most likely function is concerned with regulating pressure changes within the circulatory system during diving. But at this point no one really knows its functional significance.

Keeping Warm and Cool

Researchers have long been puzzled about how marine mammals keep themselves warm enough even in polar seas and at rest or, just as important, cool enough in the tropics when swimming hard. In the rorquals, some theorize, the ventral folds help regulate their temperature. The bottom of each two-inch-deep fold is rich in small blood vessels. With the folds closed, there would be little loss of heat. If the folds are open to the sea, however, cold water would carry that heat away, cooling the whale. Unfortunately for this hypothesis, sperm or right whales do not have such folds but regulate their temperature just as efficiently.

It is now well established that in whales, porpoises, and dolphins, as well as in seals, sea lions, walruses, and such birds as the penguin, the main role in the complex system of body temperature regulation is played by flippers or forelimbs and by flukes or hindlimbs. These flattened feet, hands, or fins are relatively thin sheets of flesh with no blubber, but they are abundantly irrigated and they allow an important cooling of the blood circulating within them.

Each of the arteries that feeds the flippers and tail with warm blood is surrounded by veins that join to form a sheath through which the blood returns to the heart. The blood in these veins, cooled by having circulated in the cold limb, is warmed by a transfer of heat from the blood in the artery, which is cooled in the process. The venous blood is progressively warmed on its way back to the body and the overall heat loss is very small—the flipper or the fluke is irrigated and oxygenated by arterial blood that is almost cold. If the radiator constituted by the fin is needed at full efficiency, the cold venous blood bypasses the "heat exchanger" and returns through a network of other veins close to the skin that have no insulation.

The rest of the body is insulated from the cold surrounding water by the thick coat of blubber. This heat insulation can be further enhanced by control of the peripheral blood circulation which is automatically reduced during each dive.

Seals also use their body surface as a heat exchanger. A seal's skin is thicker than a human's and is liberally supplied with tiny blood vessels that can be opened or closed. When they are open, the blood in them warms the skin and the heat is carried off by cooler water. When the vessels are closed, the seal's body heat is retained within. In cold water, a seal's skin temperature may be as low as 35° F., while its internal temperature is around 99° F.

The smallest and the least adapted to heat loss of the marine mammals, the sea otter, is unable to protect itself against cold with its fat alone, but has developed a way to use the marvelous properties of a fur coat that is constantly reloaded with thousands of tiny air bubbles. This the otter

A pigeon guillemot about to devour its meal of fish: the bird's digestion will be aided by bits of gravel or stones kept in its crop and used to grind food on the way to the stomach.

maintains by rubbing its fur with its paws whenever it rests on the surface; when the otter dives only the surface of its fur gets wet, and it travels enveloped in a coat of air during dives that are not very deep or longer than several minutes.

Birds generally benefit from the same air-coat protection brought to the ultimate degree of perfection as their greased feathers are even better as a shield against cold than hair, and they are able to control individual feathers in positions that allow precise degrees of "ruffling" to imprison air either before or after a dive.

The bird's feet, especially if they are webbed, can cool off the animal if needed. The feet can protect the animal against cold by simply being retracted within the belly's feathers.

Puffins spend a large part of their lives underwater. Penguins, unable to fly, spend at least half of their lives in water that is often below the freezing point, and they travel hundreds of miles at sea, diving deep, often, and for long times. Their wings have become shorter and stronger and are built very much like sea lions' flippers. They use them as heat exchangers exactly as whales do and have independently developed a thermoregulatory system that is remarkably similar to that of cetaceans.

Rocks in the Stomach

Birds and mammals of the sea swallow their food whole. None is equipped for chewing. Instead, whole fish, squid, shrimp, and other prey are dissolved by gastric juices and possibly ground up by gravel and stones in the stomach or crop. Even the orca, with its powerful teeth, only tears its large prey apart, swallowing large chunks of the victim.

Cetaceans have a three-part stomach reminiscent of cattle and other cud-chewing animals. The first part is actually a great widening of the esophagus, which is the food tube between mouth and stomach. In this first part many whales have stones, or gastroliths. The muscular contractions of this first stomach with its rocks help grind the large chunks of food therein. The second section of the stomach is comparable to the human organ and secretes hydrochloric acid, pepsin, and other digestive enzymes. The third stomach is mostly smooth-walled but has a few glands that secrete digestive

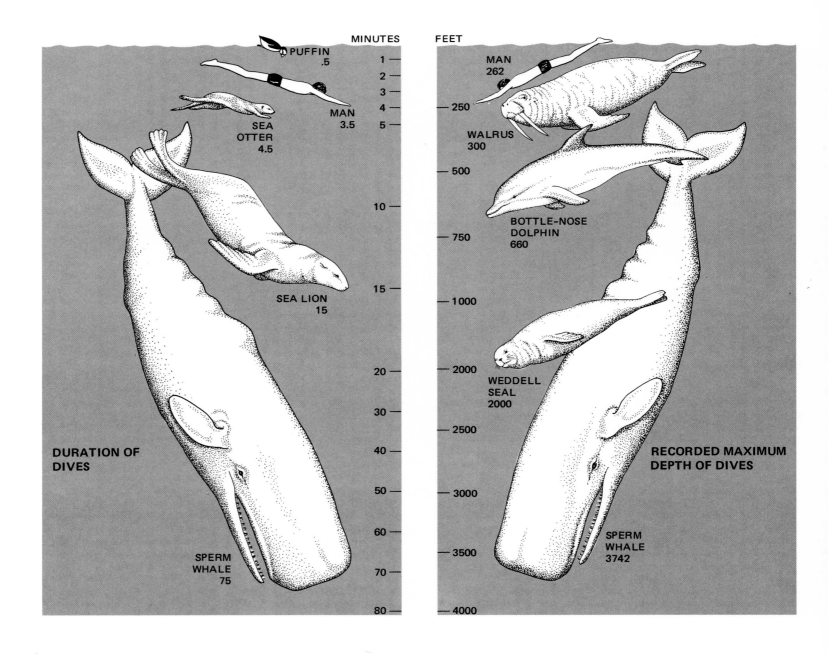

PUFFIN
.5

MAN
3.5

SEA
OTTER
4.5

SEA LION
15

DURATION OF
DIVES

SPERM
WHALE
75

FEET

MAN
262

WALRUS
300

BOTTLE-NOSE
DOLPHIN
660

WEDDELL
SEAL
2000

RECORDED MAXIMUM
DEPTH OF DIVES

SPERM
WHALE
3742

(Above) Star performer among divers, the sperm whale far outlasts all other species. Surprisingly, man does much better than an aquatic bird and nearly as well as a sea-dwelling otter.

(Right) The supreme diving abilities of the sperm whale are nearly matched by the Weddell seal, a large species from Antarctica. Unaided, man can barely penetrate the surface of the sea.

juices. Rorquals can hold up to a ton of krill in the first two stomach sections. From these stomachs, the food passes through the small and large intestines, which in cetaceans are quite long.

Pinnipeds, like cetaceans, often have stones in their stomachs. Again, while the function of these rocks is not completely clear, most scientists believe it is similar to that of the gravel in a bird's crop—for grinding large pieces of food into smaller ones. One of the most puzzling features of the seals' digestive systems is the huge length of their small intestines. Carnivorous animals usually have short intestinal tracts, while vegetarians like cattle have long tracts. Yet the seal's intestines are proportionally three times longer than a cow's. The rocks in the stomach, while they may be for grinding food, could, as in the crocodile, also have to do with buoyancy. Crocodiles are known to swallow stones to alter their buoyancy, so perhaps seals do too.

The Breath of Life

"Thar she blows," was the cry that rang out when the lookout aboard a whaling ship spotted the spouting of whales. The whalers could even tell what kind of whale they had spotted by the character of the blow, because the structure of the blowhole gives the spout a characteristic shape. Right whales, for example, have a double, V-shaped blow. The blue whale has a blow that is shaped like an inverted pear and is quite high, while the blows of finbacks and sei whales have the same shape but are smaller spouts than those of their blue relatives. The sperm whale's blow is pear-shaped but angled 45° forward and to the left because of the blowhole's location on the left side of the head; it is quite long in duration. A blowing whale makes a low moaning sound as it exhales, followed by a softer sucking sound as it inhales before snapping its blowhole shut and submerging. Harbor porpoises, no more than five feet long, sigh as they exhale and sip in a quick breath, then disappear beneath the waves.

As near as anyone can determine, the visible portion of the blow is a combination of two factors. One is the animal's exhalation: warm and humid from its time in the lungs, the air condenses as it strikes the cooler outside atmosphere. The other, according to a pair of British zoologists, is a foamy substance normally found in the cetaceans' trachea that is forced out in particles with the exhalation.

Cetaceans exchange between 80 and 95 percent of the air in their lungs with each breath they draw, while man exchanges only 15 to 25 percent, pinnipeds around 35 percent, and sirenians, which are second only to the cetaceans in their adaptation to aquatic life, about 50 percent.

As compared to man, the exceptional development of diaphragm muscles and the far greater number of floating ribs among cetaceans may give them a more flexible and powerful breathing pump, explaining in part how they can exchange more than 80 percent of the air in the lungs each time they breathe. It may explain why it takes a rorqual only two seconds to exhale and inhale 1500 gallons of air, while it takes man four seconds to exhale and inhale a pint of air.

How Deep? How Long?   Most breath-holding human divers in good physical condition can expect to reach a depth of about 75 feet in perhaps a minute and a half of underwater time. Cruising along the surface of the sea, some whale species may breathe once every two or three minutes, although after a particularly lengthy dive they may "pant" by breathing five or six times in a minute, until they have replenished their tissues and blood with oxygen. At the other end of the scale, the bottle-nosed whale has been observed to remain underwater for two full hours. Sperm whales, which usually remain down 50 minutes per dive, have been clocked at 90 minutes many times. Dolphins and porpoises are far less spectacular, remaining down five minutes during their normally shallow dives. They seldom exceed 15 minutes. They usually sandwich several minutes of rest at two respirations a minute between dives. In emergencies any of these cetaceans may remain submerged below for longer periods of time.

Seals and sea lions, surprisingly, often remain underwater longer than dolphins and porpoises. Their dives are frequently as long as 15 minutes in duration. A Weddell seal made a controlled dive that lasted 45 minutes. The sea otter, less adapted to life in and under water, managed five-minute dives.

Aquatic animals, as does man, experience increased pressure when they dive, but these diving animals never suffer decompression sicknesses when they return to the surface and to lower pressures. How, scientists want to know, can these marine animals dive and dive again? They must accumulate quantities of nitrogen theoretically sufficient to cause the bends eventually. And, in the first place, how can seals, dolphins, and whales remain underwater for such long periods without fresh air?

There is no simple answer to such questions, but we know now that many physiological reactions in the respiratory and circulatory systems combine and make such performances possible. Research has shown that when any air-breathing animal dives, or even puts its face in the water, its heart beats slower, the blood supply to less essential areas of the body is shut off by the sphincter muscles of some arteries. The heart and brain and a few

A blue whale.

other important organs such as the liver, which require a constant supply of oxygenated blood, remain irrigated; other organs, like the kidneys, as well as the extremities, are shut off. Digestion ceases. Thus less oxygen is needed during the dive.

This oxygen is not stored in a gaseous state; the lungs of marine mammals are rather small, and in many cases they are either emptied before a dive, or they are compressed by the surrounding water pressure during dives to such a degree that the remaining volume of air is isolated from the active part of the lungs. Oxygen, in fact, is either accumulated in chemical combination as oxides in blood and muscles or dissolved in organic liquids and tissues.

MAN, THE DESPOILER

There is no such thing as a "balance of nature," because such a term suggests a stable equilibrium. The environment is continually changing; it is influenced not only by natural forces, but also by human activity, often with unforeseen and detrimental effects.

In Scandinavian archaeological digs, the remains of whale bones and stone harpoon heads dating back to about 2600 B.C. have been found. The fishermen of Crete hunted dolphins hundreds of years earlier. The Phoenicians had a whale industry by 1000 B.C., and the Basques had a full-scale whale fishery in the Bay of Biscay by A.D. 1200.

In all the long history of whaling, however, the first scientific study of the life and death of any whale species was not made until 1929. Some excellent but incomplete studies were made in the nineteenth century. The long delay was partly a result of the difficulty of studying these sea creatures and partly the result of man's traditional belief that to survive, he had to fight nature: he was weak and nature was hostile. It is only since man's conquest of fossil energy that things have changed and that to survive, man has to protect nature instead of fighting it. This has affected the lives of hundreds of species of mammals and birds in the sea. Many have been decimated, some have been extirpated, and if they are known at all today, it is only through photographs and old engravings. Man ranks as the most thorough, all-embracing predator of all time.

Whether by shotgun, harpoon, or cannon, whether by poison, over-crowding, or habitat destruction, there are no more Labrador ducks or great auks left in the world. The Eskimo curlew is all but gone. The blue whale and other rorquals may be on their way out. Steller's sea cow, seen in eighteenth-century Alaska, never made it to the nineteenth century. Eight hundred living species have been eradicated by man's criminal carelessness between 1900 and 1960.

There are six ways to kill a baby harp seal for its pelt. Gaffing and clubbing seem to be among the most popular, possibly because the equipment needed is easiest and cheapest to get. In gaffing, the hunter raps the pup on the head with a chunk of wood that has a heavy hook and a spike in it. In clubbing, they use wooden clubs like baseball bats. It may take four or five more blows to render the pup unconscious or dead.

Harp Seal Horror Story

(Left) Harpooned by a commercial fishing boat, a sperm whale is pulled in. (Below) Eskimo fishermen work at the flensing of a narwhal in Arctic Bay, Northwest Territories, Canada.

Young, northern fur seals are herded together, clubbed to death, and stripped of their pelts.

There are other ways like kicking the pup in the face, then quickly rolling it on its back before it can recover and slitting its throat. But this works only with pups; adult seals will try to bite the leg of their assailants. Shooting is used principally for adults trying to get away. Drowning is an effective method of capture, but it takes longer—half an hour or more sometimes. And then there is a long-lining, which is very much like fishing. Baited hooks are lowered in the waters where hungry harp seals may take the fish bait and drown on their own blood after swallowing the sharp hook.

Harp seals are found around the polar seas of the north—off northern Norway, in the White Sea, and in the Canadian arctic and subarctic. The world population of harp seals is about five million. In Canada surveys seem to indicate the seal herds in that country have been reduced by half in recent years. Hunting off Norway and in the White Sea also continues unabated.

Sperm Whales

The sperm whale is one of the most impressive sea mammals. It has a dwarf dorsal fin, always shows its tail when diving, blows its "spout" at a 45° angle to the left and forward, and has a huge, weird-looking cylinder-shaped head. Though its skull is only a very small part of that head, the sperm whale has the largest brain of any living creature. When it is lazing along, its speed is three to four knots; its cruising speed is eight knots; it occasionally swims at 12 knots, and when jumping clear of the water, it reaches 22 knots.

The sperm whale has nothing to fear from the attacks of the few animals that would dare to prey upon it. Its powerful tail would kill or seriously harm the sharks or orcas that would try to approach it.

When disturbed by man, a herd of sperm whales is slow to realize the danger, and it is during that period of surprise that whalers take their catch. But when the whales understand their danger, they display an extraordinary show of intelligence, solidarity, communication, and defensive strategy.

These great whales are largest of the toothed cetaceans. Their large blunt heads are full of a waxlike substance called spermaceti that is probably part of its echolocation system.

212

Spermaceti has been one of the reasons the sperm whale has been a victim of whalers. The waxlike substance, which remains liquid at normal body temperatures and becomes solid when cooled, was very highly regarded as a lubricant and fuel. Happily, it has now been found that oil from the desert jojoba plant has all the qualities of spermaceti and may eventually become a substitute for it.

Blue Whales   Great size is seldom an indication of ferocity. Quite the contrary, it seems as if the larger the animal, the more peaceful it is. Today there is in existence the blue whale, which reaches a maximum length of 118 feet and a maximum weight of 150 tons. At birth, a blue whale is about 23 feet long and weighs close to seven tons. Its mother's milk helps it grow at the phenomenal rate of one and a half inches and 200 pounds every day for the seven or eight months it nurses. During that time, it daily consumes close to 300 pounds of milk which is the richest of all mammals'. By the time it is weaned, the calf may weigh 23 tons and measure up to 53 feet in length.

Once weaned, blue whales feed mostly on plankton and small fish, and during the two or three months they spend in the antarctic annually, on krill, a one- to two-inch shrimplike crustacean which is plentiful in austral polar waters. It is estimated that they are able to consume about five tons of krill per day.

Today there are so few blue whales left it is difficult to learn more about them. Relatively little is known that can help us save these giants from extinction. We know they arrive singly, in pairs, or in small family groups of three on the feeding grounds with layers of blubber thinned out to only 15 percent of their total weight. By autumn, after feeding on the krill with its high fat content, the whales have fattened up until their blubber equals 30 percent of their total weight. During their pelagic migrations, the blues cruise at speeds of up to 15 knots with bursts up to 20 knots for 10 minutes at a time. They can dive to 1000 feet or more but usually remain in shallower water where krill gathers in vast swarms. Krill and other plankton sometimes go deeper, however, and the blues go where the food is.

In autumn, when the whales leave their feeding grounds, they are fattened for their long partial fast while they travel and sojourn in warmer waters. In these warmer waters, a cow gives birth to a single calf no more often than once every three years after a gestation period of ·10 to 11 months. No one seems to know for sure when blue whales reach sexual maturity. Some say at eight years, others say at 25 years. They are known to live to 90 years. Age is determined in many species of whales by counting the layers of ear wax that are deposited each year, a method reminiscent of counting tree rings.

Some persons have claimed the northern race of blue whales is not the same as the antarctic species and thus not subject to the protection they need. The protection given by the International Whaling Commission prohibits taking any blue whales. The job of enforcing the regulations is extremely difficult and there is little doubt that some blues are still being killed each year. Whalers may claim difficulty in identifying the species of whale they have shot because of fog or heavy seas, but blues are usually easily recognized by their pear-shaped blow that billows up some 30 to 50 feet into the air. Their size, their configuration, and their coloration are all additional clues to identification.

Some optimistic observers feel that full protection would enable the blues to increase their numbers to the maximum yield level in the year 2100 which would forestall extinction of their species. But the numbers today are so few in such a vast volume of water that they may have difficulty finding

The elephant seal has made a remarkable comeback after having been considered practically an extinct species.

each other as they travel singly or in pairs rather than in large pods. There may be as few as 1000 in the northern hemisphere and only two or three thousand more in the much broader reaches of the southern seas.

"Thar She Blows"

In the nineteenth century, if a man or boy could put up with maggot-infested food, harsh discipline, and three years away from home, he could go to sea aboard a whaling ship and earn about 20 cents a day. In spite of all these discouraging aspects, in 1846 there were more than 700 American ships hunting sperm whales. (The American fleet then was almost double that of all other countries combined.) New Bedford, Massachusetts, was the whaling capital of the world. More than 100,000 barrels of sperm oil a year were produced by these crews. The ship owners and the captains they hired got rich even if the crews did not. Crews usually signed on for a share of the earnings—often only 1/200 of the take—and they could lose even that pittance as punishment for minor infractions of ship's rules. Bad as these conditions were, the whales fared worse.

Sometimes a skillful crew could make a quick clean kill, but more often the death throes of the whale lasted many minutes or even hours. Once the whale was dead, the ship was brought alongside and cutting in began while the animal lay in the water.

Whaling Today

They don't call "Thar she blows!" anymore. Instead it's a nod or a gesture with an arm. Whaling is a whole different business now, and there are few remnants of the picturesque hunters of 100 years and more ago. Helicopter teams spot the pods of whales and radio information to a mother ship. This is an elaborately equipped floating factory designed specifically for rendering and processing everything that can be obtained from a whale.

A hundred years ago a whaling ship could count on capturing 35 to 40 whales in a three-year trip, netting them 800 barrels of oil. In the middle of the twentieth century, at the peak of industrial whaling, as many whales were processed in two weeks, and each eight-month trip netted a million barrels of oil. Today Russian and Japanese whalers are responsible for nearly 90 percent of the world's whale catch. The principal remaining whaling is carried out by Norway and South Africa. At the current rate, they will shortly be out of business.

MAN, THE RESTORER

For most of his history, man has had to fight nature to survive; in this century he is beginning to realize that, in order to live, he must protect it. Many organizations have been founded to conserve the remaining species of plants and animals, and have had remarkable success. Others, like the International Whaling Commission, usually do too little too late.

In fact, the IWC was founded in 1948 as an association of whaling nations concerned far more with the worldwide glut of whale oil than with the survival of whale species. In 1972, the United Nations Conference on the Environment unanimously recommended a ten-year whaling moratorium. However, the Commission rejected such a proposal under the shameful pretext that such a decision needed a 75 percent majority.

Pressure from the commercial whaling interests on the Commission and its inability to enforce its decisions have meant, in general, that the limits it has set each year have been far in excess of the limits proposed by scientists, limits that would have effectively avoided the probable extinction of most whales in a few years. Nevertheless, the situation is very slowly changing. The Commission is beginning to listen to the scientists and, rather than killing whales indiscriminately, is starting to pick and choose the sex of the animal and its importance to the herd to minimize the ecological impact of the killing.

And the Commission does have a few great successes to its credit. Right after Christmas, boats of all sizes and descriptions anchor off California to watch the stately progression of 9000 or so gray whales traveling to their nurseries in the lagoons on the Mexican coast. Before the foundation of the IWC, thoughtless slaughter of these slow-moving, barnacle-encrusted animals had reduced their numbers to a total of less than 200 on the American side of the Pacific, and today probably the Asian grays are extinct. The Commission's task has been helped by the efficient breeding habits of the gray. Cows can conceive immediately after giving birth; this means they can produce a 15-foot, 1500-pound baby once a year. The United States' declaration of a 200-mile limit in 1977 and the total ban on whaling within those limits will ensure the survival of the shore-hugging grays even if, as the Commission threatens, it takes the animals off its danger list.

The cavorting humpback, breaching and spouting in coastal waters, nearly signed its own death sentence with its playful habits, for it was an easy target for whalers. In 1949 the IWC set a four-day season on humpbacks and in 1963 placed a total ban on their hunting. Today they have begun their long, slow climb back to safety, even though they are still in danger.

Each time a species of marine mammals has been really protected, it has made a dramatic comeback. The elephant seals had perhaps the narrowest escape. By the turn of the century there were fewer than 100 left in northern seas, and only a few hundred in the southern hemisphere. The southern species began to increase once hunting stopped for lack of animals. It was thought that the northern seals were totally wiped out until, in 1900, a small group was discovered living on the island of Guadalupe in the Mexican waters of the Gulf of California. Of these few animals, 14 were killed and shipped to be stuffed and displayed as one of the world's rarest species. In 1911, the Mexican government prohibited any more killing, and established an armed garrison on the beach of Guadalupe Island. Today there are between 10,000 and 15,000 elephant seals living in the area, all descendants of the original 100.

The fur seals are another success story. After nearly destroying the species twice, the Russians in the 1840s ruled that only bull seals on the beach could be taken by sealers. When America bought Alaska from Russia in 1867, American sealers started killing the animals at sea, not knowing whether they were bulls or cows, and losing up to 80 percent which sank. Again the species was threatened. Today, a treaty signed by Russia, Canada, the United States, Japan, and Great Britain permits the taking of male fur seals only on the beaches of their breeding islands. The estimated population of northern fur seals is over two million, and their survival assured.

Birds Come Back

Once the eider duck was hunted for its undercoating of downy feathers, which kept it and its young warm in the cold north of Canada. When the numbers of eider ducks began to decline, geese became the source of down for clothing, comforters, and sleeping bags. The wild geese were reasonably safe from extermination, because the birds were already being raised on farms for food and the down of these farmyard geese could be used as a good substitute.

Egrets, once seriously endangered by the plume hunters who slaughtered the white wading birds for milliners, have long since won protection. The effect has been that egrets are becoming more common each year.

Trumpeter swans are the largest of all American waterfowl. This swan was the biggest and best trophy for waterfowl hunters. Soon there were only a few score of trumpeters left. Rigid regulations were put into effect jointly by Canada and the U.S., but recovery was difficult. Swans still fell victim to

The trumpeter swan, one of the most sought-after trophies of hunters, now enjoys the protection of federal law that gives it a chance for survival.

gunners as well as the usual rats and hawks that prey on waterfowl. But now, through more thorough and expensive enforcement, the trumpeters are better protected from hunters, and their tenuous hold on existence is becoming a little firmer.

The whooping crane was threatened not only by hunters but also by the destruction of marshes and the country the birds need for breeding. Enforcement of laws against a possession of whooping cranes helped a little, but the marshlands along the Texas coast where the cranes breed were continually being reduced in size as tracts were bought up and filled in. A jet airport was built on filled land near the breeding area, and the noise had obvious ill-effects on the birds. During the past 35 years, the whooping crane population has varied between 26 and 60. The population change has not always been upward. There is a good chance we are seeing the last of the whooping cranes, but man has made an honest, if belated, effort to save the species from oblivion.

217

PROVINCES
OF THE
SEA

CHAPTER
10

OUR
WATER PLANET

The continents, as well as the oceans, are in constant evolution, in perpetual motion. The seabeds move, the landmasses collide, peninsulas sink, islands rise from the sea. Oceans expand and contract, continents grow and split. Entire life systems are born or wiped out.

The provinces of the sea, as everchanging as they are, provide the opportunity to study the spectacle of life, as though water were the medium, life its message. The provinces may be familiar, so familiar that they lose their mystery. Or they may be overly romanticized because of their remoteness. But life resists arbitrary boundaries, distributing itself according to the conditions of the environment rather than to man's measure of geography. True, not all animals are found everywhere, but whether it is the tiny life forms of the plankton or the highly selective coral, the impression is one of a world of evolution and change, of profusion and frailty.

The vastness of the oceans, the immense range of waters, is what impressed the early man most. Since he started his difficult career in a hostile world, he wondered—as no other animal ever did—about stars, about thunder and storms, about the immense and impenetrable ocean. Then, as civilization upon civilization developed and replaced each other, the sea remained the obstacle that divided the world. From antiquity to the modern times of the great navigators, the sea was only used as a fishing ground, a highway for coastal trade, and a battlefield. The Americas were first populated during the last glaciation period, about 12,000 years ago, when the sea level was low enough to permit Asians to walk to Alaska and progressively invade the south and the east.

It is only very recently that the oceans no longer are a serious barrier to communication between civilizations. Philosophical concepts were deeply influenced by man's late awareness that all human beings were citizens of the same water planet. Classic humanities have flourished on the assumption that the birth of consciousness in man was the ultimate development of the mind. But the new philosophers think otherwise. Individual consciousness produced a wealth of creative master-pieces in liberal arts but ended up in a junglelike over-emphasis on personal drives and produced an unchecked explosion of technology purely aimed at the artificial creation of individual needs. The process begun in the eighteenth century with the steam engine was bolstered by publicity and drives us to the poisoning of air and water, to the eradication of hundreds of species...and to the assassination of the oceans, with all the consequences for mankind.

Fortunately, thanks to the space programs, a new kind of consciousness is developing in the people's mind: a global consciousness. Space exploration started with the ambitious dream of conquering outer space, landing on the moon, and later on Mars. This great project was a tremendous luxury when a third of mankind was starving, when 15 percent of the population of the richest countries was still in a state of undignified poverty, and when 80 percent of mankind did not have adequate medical care. But space exploration brought back an unexpected, most precious, and most timely gift: a cosmic view of our earth. The photos of the earth materialized and popularized the concept that we were all passengers of the same small, but exceptionally rich and beautiful spaceship. Frontiers and borderlines that show so well on color maps, and for which so many men have died, were not visible on such pictures. There were oceans, continents, lakes, and rivers as a common heritage. There was no replacement for it in the solar system. We obviously had to become earth-conscious, water-conscious, energy-conscious, if we were to survive our population explosion.

From space it will soon be possible to assess the productivity of the oceans; to control forest fires, droughts, and floods; to measure the drift of continents and the height of the swell; to delineate fluctuations of currents; to prevent hurricanes; to pinpoint pollutions; to follow the tracks of whales or other migrating animals; and to interrogate thousands of instrumented oceanic buoys. Monitoring the oceans from manned space laboratories is the only hope we have that the looting of the sea will end and its rational management will begin.

WATER—
THE ESSENCE
OF LIFE

Life originated in the ocean about three billion years ago, and the sea today remains a fountain of life, producing myriads of individuals in hundreds of thousands of different species each year. The chemical and physical properties of seawater, along with the indispensable energy of the sun, made this life possible.

The unique properties of water start with its magic molecule—a simple chemical compound consisting of two hydrogen atoms and a single oxygen atom. This molecule has a slight negative electrical charge at the oxygen end, and a slight positive charge at the hydrogen end. Because positives attract negatives, water then has a tendency to stick to itself and to a host of other chemical compounds, giving it an array of exceptional qualities.

This self-adhesive force is evident in "surface tension," which in water exceeds that of most other liquids. Another manifestation of this property is water's ability to hold large amounts of heat and to lose it very slowly—this helps the oceans moderate the earth's climate. Water temperature varies in response to the summer or winter sun, which affects the distribution of life in the sea. Ocean temperature varies roughly between 85°F. in the tropics to below freezing in higher latitudes. The salt content lowers the freezing point, so seawater is still liquid a few degrees below 32°F., the point at which fresh water freezes. Most surface water stays relatively warm to a depth of 300 to 1700 feet. From there it cools fairly rapidly, in a transitional layer, from 1700 to 3400 feet. Further below, water cools very slowly to an average of 36°F., although it reaches a high of 55°F. in the Mediterranean and a low of 28°F. in polar regions.

Seawater can dissolve gases—oxygen and carbon dioxide among others—that are integral parts of the life cycle. Carbon dioxide is essential for plant life, which is the beginning of the food chain. This gas is often present in greater concentrations in the sea than it is in air. This ability to handle carbon dioxide is so great, in fact, that the ocean can absorb the gas in polluted areas, such as the North Atlantic, transport it, and release it in less polluted areas, such as the South Atlantic.

Oxygen is used for respiration by aquatic animals as well as plants. Even though well-aerated water holds only about one-twentieth the oxygen that an equal amount of air holds, it is sufficient for all cold-blooded animals. Warm-blooded creatures need a lot more oxygen, which is supplied by the air they breathe.

The oxygen in water is also needed for plants during darkness, when they respire like animals. Some oxygen is also used to oxidize organic matter and some is carried deeper by currents.

Salt of the Earth

Imagine the earth's surface being smoothed out and seawater covering the entire face of the planet. Then imagine all the water evaporating. We would be left with a crust of salt about 200 feet thick encasing the entire globe.

This salt is mainly sodium chloride, common table salt, but there are many other chemicals in the sea. In fact most, if not all, of the elements found on earth can be found in the oceans; the most common are sodium, potassium, magnesium, calcium, chlorine, bromine, and sulphur. Hydrogen and oxygen, of course, make up the water itself. The term salt is applied to the resulting compound after an acid and a base react.

When the oceans were formed, they were bodies of fresh water. As weathering of rocks on land took place, salts were dissolved, ran into the rivers, and were eventually carried out to sea. The process continues today, with rain-fed streams leaching the continents of salts and transporting them to the oceans.

The salinity of the ocean, rather than being continuously built up by the continental runoffs, is kept in a state of balance by various mechanisms withdrawing some of the salts. The most observable of these is sea spray, which when it is blown on to land areas evaporates and leaves a salty residue. But there are also slower processes at work in taking minerals from the sea, as evidenced by large salt deposits, gypsum, and chemically deposited limestone. In geologic time, sedimentary beds, such as limestone, have been laid down beneath the sea only to rise later, sometimes to form mountains.

The average salinity of the oceans is 35 parts per 1000 by weight, but in the Red Sea the salinity may exceed 40 per 1000. In contrast, the polar regions have low salinities because of high precipitation, fusion of ice, and

Sea salt, one of the most valuable natural commodities in the world, is mined underground in many nations, but those countries bordering on oceans obtain salt simply by evaporating seawater.

river runoff. One of the least salty seas is the Black Sea, which has a salinity at the surface of 18 parts per 1000.

The importance of salinity is illustrated by its effect on the distribution of life and on the circulation of water in the sea. One classic example of a current controlled by salinity is the flow of deep water into the Atlantic from the Mediterranean. As the surface water in the Mediterranean becomes more saline due to evaporation, it increases in density and sinks. This sinking is balanced by an inflow of surface water from the Atlantic with lower salinity. To complete the cycle, the deeper, more saline water exits to the Atlantic through the Strait of Gibraltar. Because of the rapid inflow into the Mediterranean, almost four miles an hour, ancient sailing ships experienced considerable difficulty entering the Atlantic. The Phoenicians discovered that by lowering weighted sails into the deeper outgoing currents, they could easily be carried against the incoming waters and out to the Atlantic.

Standing on the seashore, the wind rustling your hair, the whitecaps foaming over the rocks, you come to realize that there are two fluids that play a major part in our lives—air and water. These two fluids may seem very distinct, and to be sure, they do possess different properties, but their similarities and interactions are amazing.

For example, these fluids both circulate around the earth, and their flow is caused by their common property of rising when heated and sinking when cooled. The waters of the oceans are constantly moving in a complex pattern of deep-ocean and surface currents so that, however much they might be disturbed or interrupted, eventually every particle traverses the earth's surface. Currents play an important role in influencing the world's climate by moving large masses of warm water to cold latitudes and cold water to

As a wave approaches shore, the drag of the bottom shortens the wavelength (distance between crests) from A to B to C. The friction of the bottom begins to affect a wave when the water depth (D) equals one-half the wavelength (A). As wavelength shortens, the height of the wave consequently increases. Finally when wave height (E) reaches a ratio of three to four with water depth (F), the wave tumbles over on itself or breaks. Upon reaching a sloping bottom, waves spill forward; a steep bottom causes them to leap into a plunging crest.

WATER IN MOTION

A small house on the north coast of Oahu Island, Hawaii, is dwarfed by the tidal waves that battered the shore on Christmas Day, 1964. Swells were estimated at more than 25 feet. Such fury is periodically unleashed, causing extensive damage.

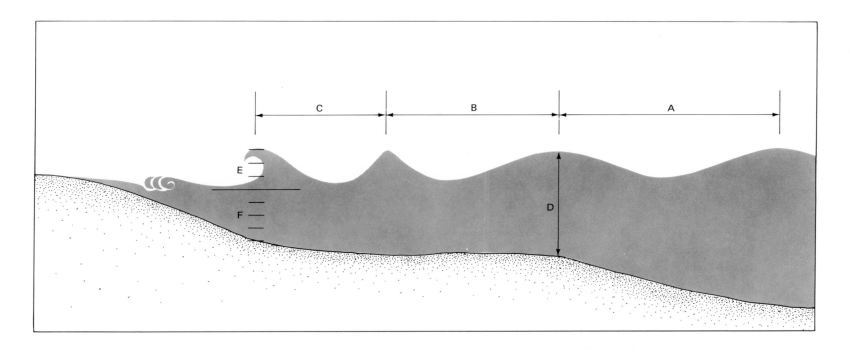

Ocean waves range from gentle swells and curling breakers to powerful shipwreckers fed by the fury of storms.

warm ones. The climate in western Europe is far milder than that in similar latitudes elsewhere because the Gulf Stream moves sun-warmed tropical water past its shores. Fog banks form off the coast of California when warm surface waters are blown to sea, allowing the cooler, deeper waters to rise to the surface.

The movement of deep-ocean water, which is primarily determined by temperature and salinity, is called thermohaline circulation. The principles involved are very simple. As water is heated, it expands and rises. The colder water is pushed away from the warmer water and sinks because it is denser. Increased salinity also increases the density, and this too helps the water sink in the higher latitudes where ice forms; ice is primarily fresh water because the salts are ejected when seawater crystallizes, which increases the salinity of local masses.

In the antarctic region, the dense, cold water sinks and flows slowly northward. The Antarctic Bottom Water, one of the world's densest bodies of water, moves northward very slowly along the bottom, reaching as far north as the northeastern Pacific. The Antarctic Intermediate Water, which is less dense, also moves northward, but at a depth of about half a mile to a mile, and has been followed as far north as the West Indies. The smooth flow of these bodies of water is disrupted by underwater mountains and ridges, especially in the mid-Atlantic region.

In 1812 Alexander von Humboldt detailed the evidence for a flow of cold water below warm tropical seas. He said this showed that, in addition to the general, worldwide circulation of water, there were streams that were distinct from the seas through which they flowed. The currents are distinguished by such properties as velocity and direction, and they have distinct boundaries. Usually currents are linear in form, but they may also be broad or sheetlike.

Most of the oceanic currents are caused either by wind friction on the surface of the sea or by gravity acting on water of different densities. The currents determined by friction or wind generally form loops or gyres matching the wind patterns. In the Atlantic and Pacific oceans these currents are deflected toward the poles after bumping into the continental land-masses. The trade winds, which blow from east to west, form the equatorial currents found in all the oceans, such as the Gulf Stream in the Atlantic or the Kuroshio off the Asian coast. When these equatorial currents are deflected by land, they are accelerated as if in an injector nozzle. As they flow along

the margins of the continents, they become among the strongest and largest currents in the world.

The direction of surface currents, besides being affected by wind and landmasses, is also altered by bottom topography and by the Coriolis effect, which deflects water to the right in the northern hemisphere and to the left in the southern hemisphere.

The surface currents flow slowly: at an average of less than two miles an hour, in some regions the waters seem to stand still. But in a powerful current like the Gulf Stream, the average velocity is three miles an hour and parts of it have been measured at six miles an hour. The width of the Gulf Stream varies from 50 to 150 miles, its depth from 1500 to 5000 feet, and it has a flow of 70 million tons of water per second.

BORDER STATES

The wind is the source of most of the ocean's waves whether they are ripples formed by gentle breezes or the savagely high seas piled up by gales blowing over large stretches. The size of the waves raised by wind depends on its velocity, on the amount of time it blows, and on the extent of the surface it acts on. These are the factors that start the sea rolling, a movement that can reach to the very edge of the ocean basins. Oceanographers have tracked waves generated by winter storms in the antarctic region all the way through the Pacific Ocean until they have broken on the coast of Alaska.

How big do waves get? One of the largest ever recorded was reported by the American tanker U.S.S. *Ramapo* in 1933. Lt. Cmdr. R. P. Whitemarsh estimated the height of the wave, which hit well above the mainmast crow's nest, to be 112 feet.

Waves do not always meet a beach head-on. Often they approach at an angle, but there is a tendency to swing around toward the shore, called wave refraction. This angle adjustment toward the beach is not complete, and the result is a longshore current which flows along the beach. This current increases until it can overcome incoming waves, and then it flows back into the sea in a rip current. These rip currents, as fast as two miles an hour, can easily trap an unwary swimmer.

When the ocean is at its most spectacular, it is often at its most dangerous. The awesome and destructive tsunamis are often mislabeled tidal waves, but they have nothing to do with tides since they are caused by volcanic eruptions or some movement of the ocean floor, such as an earthquake.

Tsunamis can have extremely long wavelengths of over 100 miles and can travel at speeds over 350 knots, so when they begin to drag on the bottom in shallow water and break on the shore, they can achieve incredible height and destructiveness. Like all breaking waves, tsunamis are affected by the topography of the bottom, breaking highest near submarine ridges and breaking the least near points of land surrounded by deep water. Since the force of a breaking wave can be controlled by the sea floor, in some areas the full force of a tsunami might be channeled and concentrated into a small area, while in other places it might be dispersed.

Another type of catastrophic wave is the storm surge, which consists of a steady rise in the water level rather than a rapidly rhythmic rise and fall of water. Strong winds, such as those associated with hurricanes and typhoons, can pile up water on the coast. The high-water levels due to these storm surges can be especially damaging if they coincide with high tide in low coastal areas.

INS AND OUTS OF TIDES

Tides are waves that occur every 12 hours and 25 minutes and that have a wavelength of half the circumference of the earth, approximately 12,600 miles.

Simply speaking, they are caused by the gravitational pull of the moon, and to a lesser extent, the sun, on the surface of the earth. As the earth rotates, the point of the ocean closest to the moon is more strongly attracted than water farther away, and the result is a "bulge" in the ocean. At the same time, on the other side of the globe, the centrifugal force caused by the rotation of the earth-moon system causes a bulge at a point farthest from the moon. High tides are thus generated on opposite sides of the earth.

As the earth continues to rotate, water recedes from our original bulge and moves continuously to swell other parts of the ocean. When the sun is aligned with the moon, at times of a new moon and of a full moon, the gravitational forces of these bodies act in concert to produce the highest

Wave action, the flow of currents, and the erosion of rivers continually alter the shorelines of continents. In this small area on the Mexican coast, sand and sediment are trapping a small lagoon behind the beach.

tides, called spring tides, which can occur in seasons other than spring. When the moon and sun are at right angles to each other, when the moon is in its first and third quarters, the gravitational pulls of the two bodies work against each other and the tides are lower than usual. These are called neap tides.

Another factor affecting the height of tides is the elliptical orbit of the moon, which at times brings it closer to the earth. Twice a year the moon passes very close to the earth causing very high perigee spring tides, and twice a year, when the moon is farthest away, there are apogee tides of minimal height.

The influence of the moon on tides being much greater than that of the sun, the time cycle of tides is 24 hours and 50 minutes, a lunar revolution, rather than the 24 hours of a solar revolution.

The rhythm of life has often been related to extraterrestrial events—like the cycle of the moon. Some thinkers saw a connection between the menstrual cycle of women and the lunar cycle. Charles Darwin took the connection of the moon, tides, and menstrual cycles to show that life had indeed originated in the sea and that women were still showing the effects of the tidal cycle common to many ocean organisms.

Animals that live near the shore must do daily combat with the tides. Mussels and barnacles, for example, simply close up tightly and wait for the water to return. Certain gastropods are able to secrete a mucous coating that effectively prevents desiccation, or drying out. Other animals burrow into the muddy or sandy bottom to await high tide, while some organisms remain in the shallow tide pools that are left by the outgoing sea. Some shellfish display this instinctive behavior even when placed in aquariums.

A more bizarre response to tides is displayed by *Convoluta*, a small flatworm which lies exposed on the sandy bottom during low tides at daylight hours. The flatworm uses this opportunity to bare itself to the sun so that the greenish algae that live symbiotically in its digestive tract may use the light to grow. But as soon as the tide starts to return, *Convoluta* returns to cover in the sand.

The slope of the continents from the highest points of land, the mountain ranges, down to the abyssal plains is characterized by broad, relatively flat areas separated by abrupt declines. Just beyond the coastlines lie submerged shelves of land—the continental shelf—that descend to the deep ocean floor. Upon reaching abyssal bottom, the continental slope flattens into plains, punctuated in some ocean regions by the deep oceanic trenches. Some coasts lack a continental shelf; instead, nearby mountains plunge into the deep ocean bed. The continental slope is the true end of the continent.

Action and Reaction

Free swimmers of the continental shelf area include the squirrelfish.

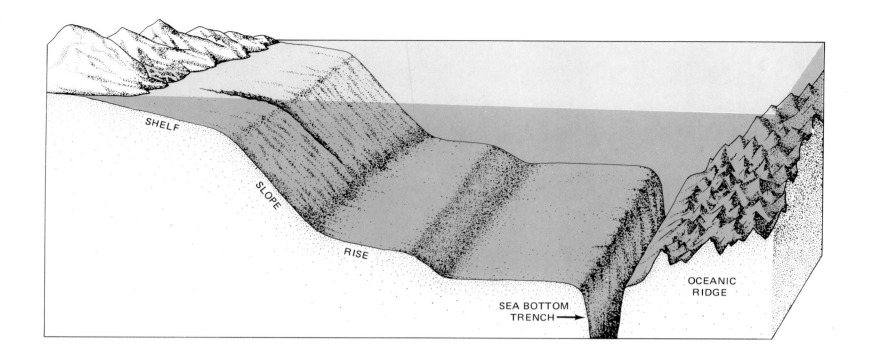

SHELF

SLOPE

RISE

SEA BOTTOM
TRENCH→

OCEANIC
RIDGE

THRESHOLD OF THE SEA

Land and sea appear to be in almost constant conflict, each seeking to overtake the other's domain. Yet they are in a state of dynamic equilibrium, for the total sum of land on the earth changes little, however dramatic the gain or loss might be in any specific place.

The profile of the shoreline is constantly changing. Much of the Atlantic coast of the United States, from Long Island to Florida and through the Gulf of Mexico, for example, is characterized by barrier beaches—constantly shifting sand islands, which are able to survive because of their malleability.

An important feature of East Coast beaches is the berm, that area from the base of the dunes to the high tide lines, that is formed by waves. This is an unstable feature and may be removed by storm surf or just as easily reformed by less turbulent waves. On Fire Island, a 30-mile barrier island off Long Island, the western tip is being extended at a rate of 212 feet each year. As waves approach the shore at an angle, each wave picks up some of the surface grains of sand and washes them a short distance parallel to the beach. The succeeding waves do the same and result in a drift of sand along the beach, called littoral drift. A lighthouse built on the tip in 1858 is now five miles inland.

BELOW THE DEEPEST TIDES

The relatively smooth surface of the ocean, broken by the rhythmic waves and the occasional arc of a leaping fish, actually conceals the true contour of the continents as the sea floor slopes from the landmasses toward the abyssal depths. Below the water are the true edges of the continents. The edges may be at the end of a broad, slightly sloping plain known as the continental shelf, as occurs in the Atlantic Ocean, or the edges may be very close to the present shoreline, the continent sloping steeply from the top of a coastal mountain range to the deep ocean trenches not far offshore, as off the Pacific coast of South America. Often these underwater continental shelves are scarred and marked from past glaciers. Or they may be terraced, revealing the successive lower levels of the ocean at various periods in the geologic history of the world. In the Gulf of California, the transition is very abrupt and there is little or no trace of the continental shelf and waters get very, very deep extremely quickly.

At the edge of the continental shelf, the incline becomes sharply steeper

in the area known as the continental slope, which drops rapidly toward the bottom. In many cases, before the ocean floor is reached, the slope flattens out again in a formation called the continental rise, which is really the buttress, the counterpart of the landmass.

Burrowing marine organisms range from those that live in the intertidal zone, to those that spend all their lives under water in shallow waters, to those that live on the bottom and never see the light of day. These organisms that live in, or attached to, the ocean floor are called infauna. The more mobile creatures that live just above the floor, often stirring the bottom for food, are called epifauna. Together they comprise the benthic, or bottom, life.

The benthic zone ranges from the deepest parts of the ocean up to the areas affected by the tides, but the most productive in terms of organisms is that region over the continental margin which is unaffected by tides. Here is found a great variety of animals from many groups including some so small they can live between grains of sand. There are worms, sea pens, crustaceans, brittle stars, tiny starfish, and protozoa, just to name a few.

The larger organisms include lugworms, surf clams, and the exotic feather-dusters. These infaunal forms usually feed by extending filters into the gentle current or by straining the muddy sediment, taking whatever food they can find.

Bottom dwellers depend for much of their food on dead and decaying organisms dropping from the shallower waters above. Many of the benthic organisms, then, are scavengers, especially among the mobile epifaunal creatures. In shallower areas, where light can penetrate to the bottom and

The queen angelfish, commonly found off the continental shelf, is a member of the highly mobile group called nekton. In the environment of the continental shelf these free swimmers move up, down, or in any lateral direction over great distances.

Life on the Bottom

where plants can grow at all depths, life is much more plentiful. The deeper the water, the less life there is likely to be. But as long as there is some rain of debris from above, there is some life on the bottom to take advantage of it.

Fancy Free

The waters above the continental margins are the most fertile as far as commercial fishing is concerned, for it is here that most of the world's free-swimming organisms, or nekton, can be found.

Nekton includes everything from free-swimming molluscs, like squid and octopus, to all varieties of fish, to mammals like whales, and to man with his aqualung, submarines, and bathyscaphes.

Nektonic life helps provide the organic material needed for life on the bottom, and the decomposition of dead nekton by bacteria is a source of raw material for the producers when elevated to the upper regions by upwelling currents.

THE OPEN OCEAN

For many years it was assumed that the only vegetable life in the ocean was the seaweed attached to the bottom or floating in places like the Sargasso Sea or the algae that covered the surfaces in well-illuminated shore waters. It wasn't until early in the last century, long after the discovery of the microscope, that plankton was discovered.

The plant plankton, or phytoplankton, are not rooted to land and float about in the ocean, providing food for both zooplankton and larger organisms. The productivity of the open ocean is limited because the nutrients are in short supply and without this fertilizer the floating microscopic plants cannot populate in the vast numbers seen in coastal waters. The average productivity of an acre of open ocean is certainly inferior to that of rich soil, but higher than that of a desert. And, though this productivity may vary greatly, it is far more constant all over the seas than land's output all over the world.

Start of Something Big

Phytoplankton, those tiny free-floating plants that directly or indirectly provide energy from the sun to so many ocean creatures, are mostly water. Over 80 percent of their weight is water, with the rest of their tiny mass consisting of protein, fat, carbohydrates, and mineral components—either calcium or silica compounds—making up their shells or skeletons.

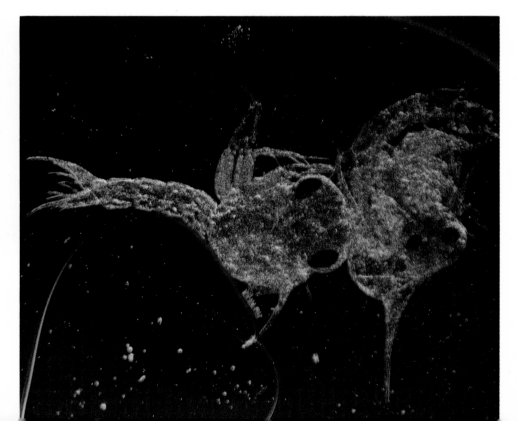

Plankton, which float and drift at the mercy of currents, include diverse life such as the ctenophore (above), decapod larva (right), and copepods or sea fleas (overleaf).

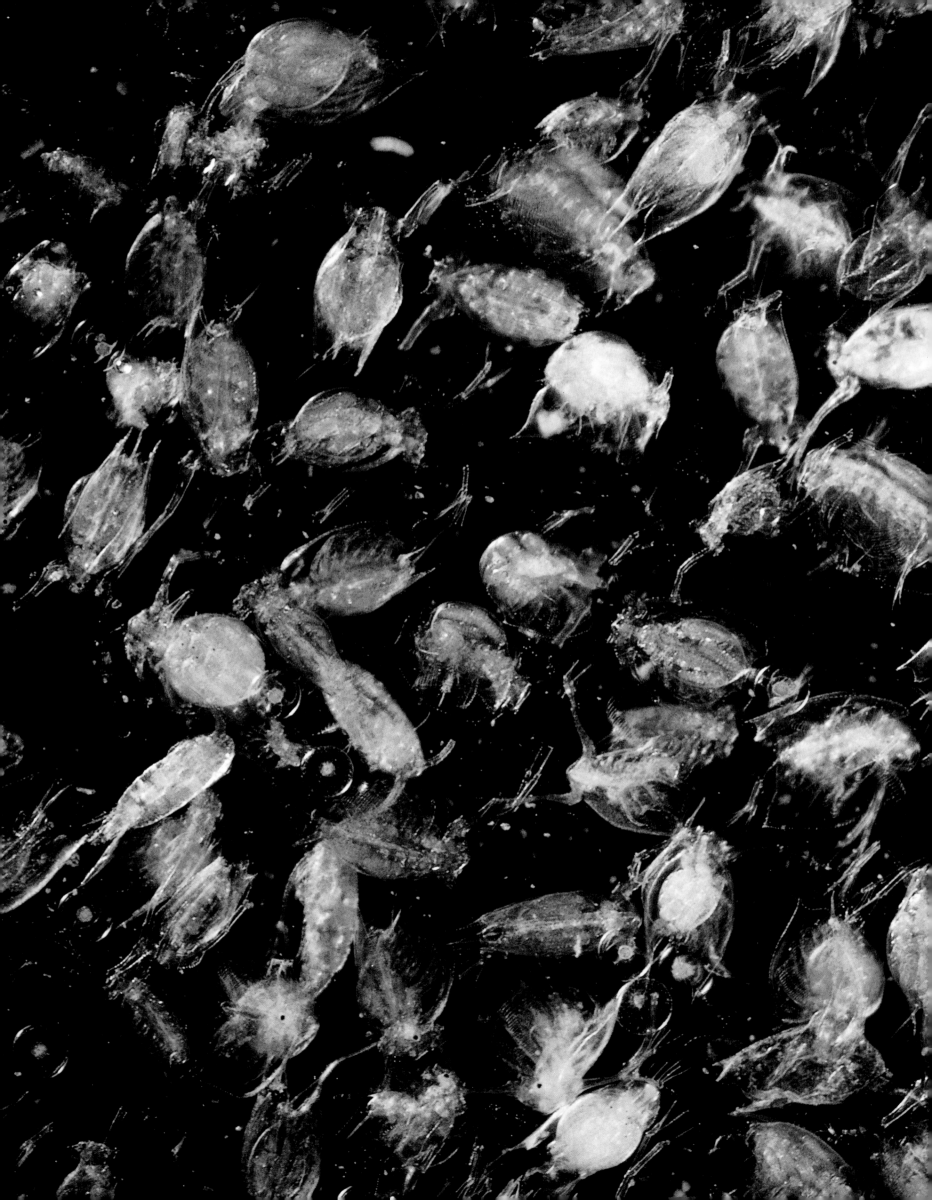

Wherever there are dissolved nutrient salts, carbon dioxide, and radiant light to trigger photosynthesis, that is where phytoplankton can blossom. The sun penetrates in sufficient quantity for photosynthesis to perhaps 300 feet deep in the ocean, and this defines the photic zone.

Probably the most common algal plankton is the diatom, whose name—meaning two atoms—is derived from the fact that each cell wall consists of two nearly equal overlapping halves. All diatoms are single-celled, but some reproduce and form complex chains.

Diatoms have a hard, glassy shell made of silica, which is pitted and grooved. When these plants die, their shells litter the ocean bottom and eventually become part of the sedimentary rock being formed. This rock layer is called diatomaceous earth and is used commercially as a filtering material.

Coccolithophorids and dinoflagellates are the two other major types of phytoplankton. The coccolithophorids are distinguished by the calcareous plates that cover their external surface. Dinoflagellates are the most complex of the phytoplankton. Some may have traces of chlorophyll and others do not, indicating they are part animal as well as vegetable. One factor that enables dinoflagellates to succeed is their motility. With their two flagella, they are able to propel themselves vertically. They have been found to descend below a shallow thermocline to absorb nutrients at night, then ascend to the surface for light during the day. Consequently they are able to bring nutrients to the sunlight and carry on photosynthesis in a nutrient-poor environment.

Drifting Animals

Just as there are plants that live a floating life, so are there planktonic animals, or zooplankton. In most cases, they feed upon the phytoplankton. Many of the zooplankton are protozoans, or single-celled animals, and are microscopic, but others are quite large, such as the salps. Most planktonic species possess some mechanisms for propelling themselves through the water, but these are usually very weak and the organisms are subjected to current and wind.

As with phytoplankton, zooplankton contribute skeletal matter and debris to the sedimentary layers on the ocean floor. Foraminifera are usually snail-shaped with shells made of calcium carbonate, while the closely related Radiolaria have shells made of silica.

Among other types of zooplankton are the larvae of larger organisms, such as fish, crabs, and lobsters; several small forms of creatures that have pelagic or sessile relatives; arrow worms, which have fins to aid them in moving through the water, jawlike bristles, and teeth for eating; comb jellies like the sea gooseberry that capture food with sticky tentacles; and the Portuguese man-of-war, a siphonophore, which is a colony of specially adapted individuals living as one adult. Juvenile and adult crustaceans are an important part of the zooplankton. These include ostracods and copepods, various kinds of shrimp.

Wide Open Spaces

If one word could be used to describe the dominant swimmers of the open ocean it would be ''speed.'' The foremost residents of the area are the tunas, sharks, billfish, and dolphins. They all have streamlined bodies and the ability to produce short bursts of extremely high speed. Some sailfish have reportedly been timed at 70 miles an hour. And these fish use their speed to hunt down slower fish that share their environment.

There are many thousands of species of sizable creatures in the open sea, but some of them have been far more successful than others. The basic food for the giants is essentially made of sprats, sardines, flyingfish, squid, and cuttlefish. And though they are relentlessly hunted and caught, they still thrive in great numbers.

Squids generally stay very deep during the daytime when only the mammals can dive to chase them. At night the squids have to come closer to the surface to feed (also upon the poor flyingfish that can know no sleep). Cuttlefish were believed to be shore animals but they are also pelagic, and recently huge concentrations of them have been spotted at a depth of 1500 feet in the mid-Atlantic. Pilot whales and orcas feed on them.

The oceans are deep, very deep, but far below the last glimmer of sunlight life goes on. There must be special adaptations in a world with no light, no green vegetation, and low water temperature.

The deep sea, or abyssal region, is described as the single largest environment on earth, since it covers 85 percent of the areas we know as ocean basins. For vast reaches the abyss is a smooth plain, covered with thick sediments, much of it moved by turbidity currents that occasionally rush down gorges in the continental slope called submarine canyons. These currents can move as fast as 50 miles an hour, and have snapped submarine cables as though they were thread.

The sediments along canyons are stirred up by the currents and slide, gaining momentum and more sediment as the mass proceeds all the way down to the abyssal plain. There the turbidity current flows like a river on land, carving its bed in the almost flat sediment of the plain. When the velocity begins to slow down, the heavier particles begin to settle out first. In a process known as grading, the particles get finer and finer farther from land.

Occasionally the abyssal plains are broken by volcanic mountain peaks, which periodically protrude above the surface to form an island. More often the peaks are totally submerged and are called seamounts. Some of these seamounts are flat-topped, having been leveled, perhaps by wave action, when they encroached upon shallow waters; they are called guyots.

Life in the abyssal zone has its pelagic types, like the squid, rabbitfish, deepwater cod, anglerfish, rays, and eels. But life is mostly concentrated on the bottom. The sessile and burrowing creatures, the crawling and weak-swimming organisms all compete for the organic material—living or dead—that comes their way from above or is brought by currents. The increased pressure of the lower depths is not an excluding factor for life, since most organisms are made up largely of water, which is almost totally incompressible.

The soft oozes and sediments are often streaked with the trails of organisms such as the brittle stars which sometimes live in very large aggregations.

There are also sea urchins, heart urchins, sand dollars, and deep-sea cucumbers, the holothurians. With their tubelike bodies, these holothurians gather in organic matter, living or dead, by means of tentacles near the mouth end.

There are several kinds of worms living in the abyss, such as the acorn worm, which may be the creature responsible for some of the intricate spiral patterns that have been photographed on the sea floor.

Two of the most frequently encountered open-water fish are lanternfish (Myctophidae) and hatchetfish (Sternoptychidae). Both groups grow only to a few inches in length and possess bioluminescent photophores along their sides and ventral surfaces.

Another weird fish of the bottom is the tripodfish, which supports itself on two long ventral fins and an extended tail lobe. It sits facing the current with elongated ray fins directed ahead, possibly to detect an approaching victim. Although there are some monsters in the deep sea like the giant squid, most life is very small, possibly because food is so scarce. In the

FACING THE ABYSS

Like a miniature waterfall, this sand fall off Cape San Lucas in Baja California shows how erosion continuously builds up the abyssal plains.

Among the unusual creatures that inhabit the deep is this batfish.

236

Many creatures live in the solitude of the deep sea including brittle stars and sea urchins.

absence of light, the more traditional methods of predation, like chasing, are useless. Feeding is more dependent on trickery and chance encounters. As a result the strong bones and firm musculature of the rapidly swimming fish above are generally not seen.

Life in the ocean contributes its share to the deposits on the bottom, whether it is whitish yellow or brown chalky globigerina ooze, the straw- or cream-colored diatom ooze, or the pale greenish-yellow radiolarian ooze. The siliceous and calcareous skeletal material of these plants and animals falls to the bottom, sometimes playing a determinate role in the type of rock layer which will form in that area of the seabed.

A Striking Profile

One of the more outstanding features under the sea are the seamounts. The spectacular Great Meteor Seamount in the northeastern Atlantic is a totally submerged peak that rises 13,000 feet above the sea floor and is 68 miles in diameter at its base.

The world's greatest mountain range, 40,000 miles long, is the mid-ocean ridge. This is a continuous chain of mountains that stays beneath the surface for the most part and extends down the middle of the Atlantic, around the tip of Africa, and splits as it heads east. One spur travels up the axis of the Indian Ocean, while the other continues east through the Pacific Ocean until it approaches the coast of South America, where it splits again. One branch continues almost up to the South American continent itself, while the other heads north until it ends off the coast of California. In the Pacific, the name mid-ocean ridge system is misleading, since it is located to the south and to the east of the central axis of the ocean. Great lateral fracture zones typically interrupt these chains of undersea mountains. In some places the mid-ocean ridge extends above the surface to form islands, such as the Azores and Tristan da Cunha.

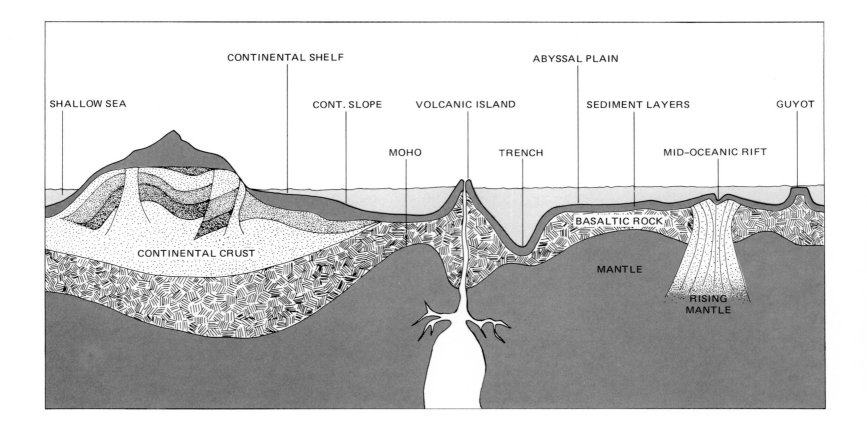

The very deepest parts of the ocean, which have been viewed only a few times through the windows of bathyscaphes, are called the hadal zone. The total darkness, the intense pressure, the constant cold, conjure images of a Hades quite different from the fire and brimstone of the Hell of fundamentalist preachers. But the fire and brimstone are there, provided by the earth itself when molten rock bursts through the thin crust and spews forth with a grisly glory.

Some theorists propose that the trenches are carved along the edges of large plates that slip and slide around the globe. These deep-sea trenches are the places where two plates meet, one descending under the other and thus creating the mammoth depressions that can extend 36,000 feet below the level of the sea. The deepest such trench is the Marianas Trench in the northwest Pacific, where Jacques Piccard and Don Walsh reached the depth of 35,800 feet in the *Trieste* in 1960. Among the first things they saw at that depth were a flounderlike fish and a shrimp, which helped soften the hadal connotations of the trenches. There are several of these large depressions in the Pacific, always adjacent to a volcanic arc. There are trenches in the Atlantic, such as the Puerto Rico Trench in the Caribbean Sea, but these trenches are much shallower than the Pacific trenches.

In the Pacific, especially, mountain ranges or ridges are usually found on the island arcs and continents flanking the trenches. It is on these island arcs and on continental coasts that many of the world's active volcanoes are found. Earthquake activity is also associated with trenches.

The shapes of creatures living in the trenches are probably not very different from those to be found in other parts of the deep ocean. The trenches probably do not form a separate environmental niche as do coral reefs, estuaries, and surface waters.

The hadal zone is not lifeless, although there certainly seem to be far fewer species and individuals in the trenches than there are in other parts of the sea.

Among the adaptations most deep-sea creatures have made to abyssal life is the reduced use of eyesight. Most of the fish of the bathypelagic region do have eyes, often enormous eyes, indicating there are enough flashes of

THE HADAL ZONE

PROVINCE OF THE UNKNOWN

bioluminescent light to keep the eyes from degenerating totally, as has occurred in some terrestrial cave dwellers, but there is one bristlemouth species, *Cyclothone obscura*, that is believed to be without light-sensing organs.

Many of the animals which live on or near the bottom have reddish or pinkish coloration. This may be because red light waves are absorbed in the uppermost layers of the ocean, and as a result red-tinted creatures appear no different from black creatures because no red light is available to be reflected.

When a child asks, "Where did I come from?" the answer deals most often not with "where" but "how." The same is true of the earth. Questions of where the continents or oceans came from are usually answered with the hows of continent building or ocean formation.

Since the very symbolic and somewhat misleading explanation in Genesis, throughout most of history men have accepted the idea that the earth has been as it is today since the time of creation.

Then, particularly since the dawning of the age of science in the seventeenth century, men began thinking about the possibility that the continents were not always as isolated as they are now. In the second half of the twentieth century, this theory of the drifting continents has been combined with revolutionary new findings about the composition of the earth itself in a dramatic hypothesis. It explains not only how the continents arrived in their present positions, but also why certain areas are subject to earthquakes and volcanoes, why continental rock is so much older than the sea floor, and describes what part the mid-oceanic ridges and deep trenches play in the creation of the earth as we know it.

Molten lava flows into the Pacific Ocean at Kealakomo, Hawaii, following an eruption in 1971. The Hawaiian Islands are of volcanic origin but are not part of the "ring of fire" that girds the Pacific. Volcanic activity is not confined to the Pacific; in 1963 the volcanic island of Surtsey rose out of the Atlantic near Iceland.

Journey to the Center of the Earth

In order to understand the geography of the earth, it is necessary to understand the composition of the globe itself. At the very center is the intensely hot core composed, it is thought, mainly of iron with lesser amounts of nickel and silicon. The inner core, a little larger than the moon, is solid because of the immense pressure. It is surrounded by a transitional layer, and then the 1330-mile-wide molten outer core.

Surrounding the core is the 1800-mile-thick mantle. Its upper portion is fairly well determined, since earthquake waves speed up considerably when they reach the mantle. The measurable alteration in wave pattern is called the Mohorovicic Discontinuity, after the Yugoslavian geophysicist who discovered it. The Moho, as it is called, is the frontier of the mantle. It lies about 25 miles below the continents and six miles below the sea floor.

The upper layer of the mantle is fairly rigid and, with the crust, forms what is known as the lithosphere. Below it lies another layer about 95 miles thick, called the asthenosphere, where temperature and pressure are much higher and rocks are believed to flow like molasses. This is of crucial importance in understanding how the solid surface of the earth can move.

Coating the earth like the paper-thin skin of an onion is the crust, composed of two kinds of rock. The ocean floor and the underlying rock of the continents is called sima, a term derived from the chemical symbols for silica and magnesium, its major constituents.

Dark-colored basalt typifies sima. Resting on the sima are blocks of less dense sial, made up of silica and aluminum. These blocks are the continents, typified by light-colored granite.

Because sial is less dense than sima, it is lighter in weight. Thus the landmasses are buoyed up above sea level because they float on the heavier material, a principle called isostasy.

The Spreading Sea Floor

Geologists have developed a time-scale of the earth's history, beginning with the pre-Cambrian deposits, those rocks that are 600 million years and older. Until the 1950s it was assumed that the sea floor was composed of these oldest rocks. However, the development of sophisticated measuring devices during World War II led to the discovery that the oldest parts of the sea floor appeared to be not much more than 140 million years old.

The assumption made by scientists today is that material from deep within the earth wells up through a rift in the central zone of the mid-oceanic ridges and spreads toward the continents on either side. Since we know that the amount of land remains roughly the same, the older parts of the sea floor must go somewhere. Geologists today think that these parts are wedged under the edges of the continental shelf, as though the sea floor was a conveyor belt, pushing up in the middle and heading down at the continental margins, which, being lighter, override the denser sima. It is thought that the downwelling of the ocean floor occurs at, and actually creates, the deep-ocean trenches.

In support of this theory of sea-floor spreading, scientists use the magnetism of the earth. There have been occasions when the magnetic poles have reversed themselves. Why, or how, this occurs no one can say for sure, but as molten rock solidifies, the magnetic particles within it record the magnetic orientation of that time. If new material appears at the mid-oceanic ridges, then strips of rock with identical magnetic polarity should be found parallel to and at equidistant points from the mid-oceanic ridges. This has proved to be so. As further proof, rocks farther away from the mid-oceanic ridge have been found to be older than the rocks near the underwater mountain chain.

In this Apollo 7 photograph the Red Sea appears in the foreground; the gulfs of Suez and Aqaba run northward. The eastern Mediterranean lies below clouds on the horizon. The earth's crust here, where plates are diverging and a new ocean may be born, is both rugged and complex.

The Drifting Continents

The evidence supporting the idea that the sea floor spreads gave an enormous boost to the various theories proposing that the continents had not always been where they are today. In 1628, Francis Bacon had noted the similarity between the outlines of western Africa and eastern South America, and the theory that they were once joined continued to be proposed sporadically until the twentieth century, although never accepted by the majority of scientists.

In 1912, the German geologist Alfred Wegener envisioned the continents as "ships of sial plowing through a sea of sima." He argued that a super-sea, Panthalassa, once surrounded one super-continent, Pangaea, which began to break up about 225 million years ago.

Arguments raged, and are still raging, about exactly how the continents achieved their present shapes. The most likely solution follows Wegener's theory of the super-continent. Pangaea existed 280 million years ago and 80 million years later had split east-west into Laurasia (North America and Eurasia) in the north and Gondwana (Africa, South America, Australia, Antarctica, and the Indian sub-continent) in the south.

As Pangaea was breaking up into two parts, Gondwana itself was being subdivided by the rift known today as the southwest Indian Ocean Ridge. This rift separated West Gondwana (South America and Africa) from East Gondwana (Australia, Antarctica, and India).

One hundred and seventy million years ago, South America and Africa began to split, as India was drifting away from the Antarctica-Australia complex toward Asia.

Then some 65 million years ago, Africa and Eurasia were rotating toward their present positions, while India was heading for a collision with the Asian mainland. Australia began moving east and South America west, where it would eventually rejoin North America. Greenland, not yet an island, linked North America with Europe.

During the last 60 million years, Antarctica rotated slightly to the west; New Zealand split from Australia as it shifted its course from east to north;

and the Atlantic and Arctic oceans were joined as Greenland became an island. Africa moved slightly to the north to its modern location, and India crashed into Asia, with the impact throwing up the Himalayan mountains.

When the evidence became strong enough that the continents had split and had widely changed their respective positions, the question that remained was: how? The theory of plate tectonics seems to provide some answers.

The Crustal Plates

A NASA infrared satellite photo shows the area of the San Andreas Fault in California, where the edge of the Pacific plate slips past the edge of the American plate.

Duxbury reef at Bolinas, California, runs along the San Andreas Fault. It has been predicted that part of southern California will eventually break off from the mainland and form a separate island off the coast.

According to this theory, the lithosphere—the crust and the hard upper part of the mantle—is divided into as many as 20 or as few as six blocks. These slide around on the plastic asthenosphere. Plate boundaries need not coincide with continental margins; some contain both continental and oceanic material while others consist only of oceanic rock. These plates, ranging in thickness between 43 and 143 miles, move at a rate of between 0.4 inch and two inches a year, and warp, crack, fold, and fault when they come into contact with one another. It may be a violent collision such as the meeting of the Indian and the Asian plate. Or it may be a more gentle meeting, when one of the plates folds under the other and dives back toward the center of the earth at a steep angle. When this happens, geophysicists say the plate is being "consumed."

This theory of moving plates explains the existence of belts of seismic activity and vulcanism in different parts of the world. At the mid-oceanic ridges, forces from deep within the earth create a fracture along the axis. Through this fault molten magma wells up to create the new floor and also rushes up through cracks to form volcanoes. In 1963 Surtsey Island was created as it rose out of the sea near Iceland. It is one of the few visible portions of the mid-Atlantic Ridge. The ridges are also the site of shallow earthquakes, originating at depths of less than 45 miles.

Where one plate dips down under another, shallow earthquakes are caused by the downgoing plate breaking, and deeper earthquakes (down to 450 miles) result from one plate grinding past another. As the downgoing plate edge melts, molten magma is forced up into faults in the overriding plate, causing volcanoes like those in the Andes of South America.

The Mechanics of the Motion

What force causes the plates to move? It is generally accepted today that the answer is giant, circulatory movements of the rock within the earth's mantle, caused by convection currents. To imagine such a phenomenon, it helps to consider the earth's mantle as liquid heating in a pan. As the pan is heated from below, the hot liquid moves to the top, pushes the surface water to the outer rim where the cooler liquid descends, moves to the center, and the whole process repeats itself. This circulatory movement is known as a cell of convection. Essentially the same thing may be happening to the earth. Huge convection currents deep within the mantle are generated when the hotter, plastic material nearer the center rises toward the earth's surface. There the material is deflected and flows laterally until it has cooled enough to descend once more into the bowels of the earth.

The ocean ridges, then, are areas of upwelling, corresponding to the rising of convection cells in the mantle. The leading edges of the continents, either slopes or deep-sea trenches, or both, are areas of downwelling, where the convection cells are descending.

The Future

What does the future hold in the light of plate tectonics? Is a new ocean being formed in Africa? Will California slip into the ocean? The East Africa Rift Valley extends from Mozambique through Ethiopia to the Red Sea and has become six miles wider in the last 20 million years, leading some people to believe the continent is breaking apart and a new ocean is being formed. But this movement is slow, even by geologic standards; in the same 20 million years the Red Sea and Gulf of Aden have spread 200 miles.

Southern California is another story. It rests on the edges of the Pacific plate, which is slipping past the American plate along the San Andreas fault. If the present motion continues, part of southern California would be carried in a northwest direction into the Pacific and could become an island in the way Madagascar was separated from Africa.

Many questions remain without an answer, and one of the provinces of the sea has to be the Province of the Unknown.

MAN REENTERS THE SEA

CHAPTER
11

MAN'S REENTRY INTO THE HYDROSPHERE

Are human beings about to return to the sea as some mammals did a few million years ago to become seals, porpoises, and whales? In the absence of drastic anatomical and physiological mutations, it is very improbable indeed. Our silhouette, our limbs, our lungs, heart, veins, and arteries; our fat and our liver; our kidneys, our skin, our blood—all would have to be modified radically before we could stay submerged for weeks or months at a time, without dying from exposure to cold, losing our skin, or being compelled to go back too often to the surface for air. In spite of the new popularity of skin diving, there is no indication that in the grand scheme of evolution men were programmed to become marine creatures.

However, in his own way, which is an artificial way, man is preparing his reentry into the sea. He makes up for his lack of blubber by developing better diving suits. He struggles both to perfect breathing devices and understand diving physiology. He has already lived for an entire month in undersea settlements. He is the proud owner and user of dozens of exploration submersibles, and he has gone down in bathyscaphes to depths beyond those ever reached by the sperm whale. Man cannot fly. Man cannot dive very well. Yet he has conquered the air, the moon, and the deepest ocean trench.

Man's futile efforts to penetrate the alien element that nursed his ancestors have been traced to the earliest times. In seaports of the eastern Mediterranean, in the warm waters of the Persian Gulf and the Indian Ocean, in scattered islands in the Pacific, and even in the icy waters of Tierra del Fuego, man was diving before there were scribes to record the event. These primitive divers were at the same time practical and mystical. From the mysterious shallows, they brought back food and treasure, as well as fantastic tales that kept mythology alive—pearls, coral and monster stories, sponges and legends of mermaids. The Sumerian hero Gilgamesh found and lost the fabled seaweed that provided eternal life.

Through empirical knowledge these pioneers perfected naked diving, reaching depths of 150 to 200 feet, in dives lasting two minutes—sometimes longer. Their techniques were passed on from generation to generation. It was only at the end of the nineteenth century that technology and science were able to improve on these methods and develop the equipment necessary to open up a new world for man. Progress in diving was explosive, and coincided with the population and the industrial explosions.

Why have men always been motivated to dive? Was there a subconscious yearning for the element that had produced life, for the mother-sea they came from? Maybe. But the conscious drive was for freedom and adventure. Freedom from their own weight, adventure in exploring a world that can barely be imagined from the surface.

Today the motivations are more materialistic. If diving is to progress, it has to demonstrate that it is practical and economical. In the realm of air breathing (surface to 200 feet or slightly more), divers have proven to be irreplaceable: salvage, rescue, coral and pearl collecting, oyster farming in Japan, geological, biological, and behavioral research, speleology, and undersea archaeology have all demonstrated the efficiency of the human presence under water. Deeper diving, however, involves gas mixtures, elaborate habitats and decompression chambers, voice unscramblers, power lines, power tools, and communication devices, and each working hour on the bottom becomes extremely expensive and often dangerous. For this reason, the main customers (offshore oil exploration companies) have developed alternate methods of intervention, using either completely automated tools controlled by closed-circuit television, or capsules lowered from a surface platform and housing men working at atmospheric pressure. Deep divers are in competition with robots.

It would be equally wrong to assert that either deep divers or robots were the best solution in all cases. The same question has been debated endlessly: could direct observation be replaced by instruments in outer space exploration? Experience has proven that both were useful—in inner as well as outer space.

A diver examines stalactite formations in a submerged cave. Undersea cave habitats play host to unique animal species and house unusual geological features.

THE CALL OF THE SEA

Today the seashores attract human populations as magnets draw iron filings. This "call to the sea" is, however, very recent. Early man was more of a hunter than a fisherman.

It has been speculated that man's evolutionary roots matured in ocean water and that his recent return to the sea is just as natural as that of the sea mammals a few million years ago. Nothing can substantiate such a romantic statement. After all, the sea was not really very inviting to men. Storms took a heavy toll among sailors. Typhoons and hurricanes came from the open ocean and were included in the overall awe and suspicion. The water itself was cold and infested with stinging and biting creatures.

Much respect was earned by fishermen who did daily battle with the sea and by the navigators who dared traverse the ocean. But the superheroes were the divers like Gilgamesh, who 4000 years ago plunged to the bottom of the sea looking for a weed that would provide eternal life, and Glaucos, the Greek mortal who became a god after eating a weed and diving into the ocean. Gods and animals, apparently, were the only ones at home underwater.

This, however, did not prevent a small number of human beings from diving. The best-chronicled achievements are those of the Mediterranean, where the Greeks have been gathering sponges since immemorial times. Aristotle described the value of sponges to soldiers who used them to cushion heavy armor.

It is only in the last few decades that science has exposed the myths of the ocean and that masks, fins, suits, and breathing apparatus have made it possible for modern man to answer the call of the sea and to enjoy her wonders in thrilling weightlessness. Divers can truly speak of their adventure in the sea.

Down for Pearls

In ancient times potentates (and more recently Western jewelers) created a demand that motivated naked divers in the Red Sea, the Persian Gulf, and the South Pacific to risk their lives in collecting thousands of oysters to find perhaps one good pearl. Today the oyster beds of the Pacific have been depleted to such an extent that they are protected by law and diving is permitted only every other year on a small scale.

Easily the most romantic of the traditional divers are the pearl divers of Tuamotu Archipelago in the South Pacific. The divers begin their ritual with a prayer asking protection from sharks and moray eels. Then they hyperventilate, usually emitting a low whistle when they exhale the quick, deep breaths. Then the diver wraps his toes around a rope with a weight at one end. When the weight is thrown overboard, the diver is pulled along, feet first, with little or no effort on his part. His only tools are a thick glove, which protects his hand from cuts and scrapes, and a collecting basket.

The diver surfaces on his own. If the area is productive, he may make as many as a dozen dives before moving on. He averages 40 or more dives a day, ranging from 100 to 140 feet down to collect 150 to 200 shells, and experiences severe physiological stresses.

Women: The Perfect Divers

Ama divers have been practicing their art in Japan for 1500 years. Long known as pearl divers, today these women dive almost exclusively for food. They work throughout the year with no special equipment or clothing and only recently have they used goggles or face masks.

Japanese women are among the most liberated when it comes to the sea. The ama divers have been practicing their skills in Korea and Japan for at least 1500 years, and while they may have been pearl divers at one time, the 30,000 or so practitioners today dive almost exclusively for food. In times past, both sexes engaged in diving, but now there are few males practicing this art. Because women have additional layers of fat beneath their skin, which insulate them against the effects of cold water, men are mostly relegated to the role of assisting the women divers by manning boats at the surface.

More important perhaps than the physiological difference, however, is the training the women receive. It may start when they are girls of 11 or

12 and they may continue diving until they reach their 60s.

Diving requires very little muscular strength but great litheness of the body and great resistance to cold. Women thus should be the perfect divers, today as ever.

Diving Warriors

The earliest dives made by men, or their anthropoid ancestors, were probably in pursuit of food. They first collected clams or shells at low tide, then stepped into the sea and ventured progressively deeper. Later, shells were also used for decorations. Such artifacts are among the oldest recovered by archaeologists. As early as 4500 B.C. divers were collecting mother-of-pearl as proven by the ornaments on the walls of Bismaya.

Who were the first military divers? We don't know, but, by the fifth century B.C., two famous divers from antiquity, Scyllias from Sione and his daughter Cyana, went diving to cut the ropes from the anchors of the warships of the Persian king Xerxes. During a terrible storm the ships ran aground and sank. Later Scyllias and Cyana dove again to plunder the wrecks. During this same period the Greek historian Thucydides relates how Spartans used divers to ferry supplies past Athenian ships blockading the island of Sfaktiria. The Athenians eventually caught on to the idea and later employed divers themselves.

One of the most spectacular uses of diving soldiers was during Alexander the Great's siege of the island stronghold of Tyre in 332 B.C. The divers were used to destroy the Phoenicians' submerged boom defenses, and Alexander reportedly watched the operation by descending in a glass barrel or diving bell.

FROM AIR TO WATER

The two most important components in the world, at least as far as life is concerned, are air and water. These two components behave according to the same physical laws concerning pressure but because their physical and chemical makeups are so different, the results differ greatly and are of vital importance to the diver.

Pressure Points

The physical laws of pressure were first stated by Robert Boyle in the 1660s, and they applied to all gases.

Boyle's law says that the more pressure exerted on the gas in the container, the less volume there is; and the less pressure that is exerted, the greater the volume. This means that a diver taking a deep breath before plunging into the sea finds his lungs being "squeezed" as he descends and as the water pressure increases on the air trapped inside the body. As the air becomes smaller in volume, his chest subsides and his belly grows hollow as if he were exhaling. When he swims back to the surface, chest and belly come back to normal.

We are so accustomed to the air around us that we do not always notice its density and pressure. The pressure of air is uniformly exerting a force on everything at sea level of one atmosphere, or 14.7 pounds per square inch (usually abbreviated psi). Rising from sea level, the column of air that weighs on us becomes shorter, and atmospheric pressure decreases, so that on top of a 16,500-foot mountain, it is only half as much as it is at sea level. Correspondingly, the atmospheric pressure increases at points on land that are below sea level.

Water is 800 times as dense as air; thus, when descending in water below sea level, the pressure increases very rapidly. At the water surface the water pressure is zero, but at 33 feet down, the pressure of the water itself is 14.7 psi, and since the air above the water is also exerting a force of 14.7 psi at the surface, they combine to put pressure on a fish or rock or human of "two atmospheres," which is 29.4 psi. At 66 feet down, a diver

An archaeologist-diver fills a salvaged, ancient wine jar with air from his tanks. Buoyant, the inverted amphora will float to the surface.

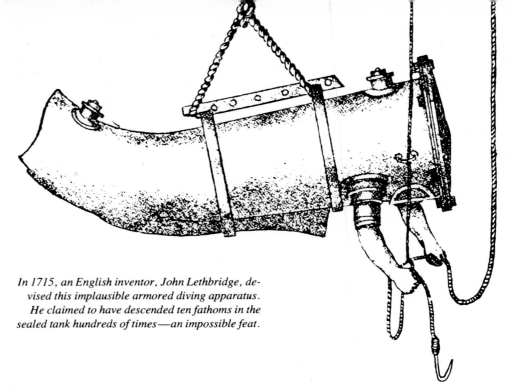

In 1715, an English inventor, John Lethbridge, devised this implausible armored diving apparatus. He claimed to have descended ten fathoms in the sealed tank hundreds of times—an impossible feat.

is subjected to slightly more than 44 psi, or three times as much pressure as at the surface. If man, with all the air cavities inside his body, wants to reenter the sea, pressure is a force that must be dealt with.

Gases in a mixture exhibit a unique property. Each exerts a pressure that is independent of all other gases. The force exerted by each gas is called partial pressure, and the force exerted by the mixture—called total pressure—is the sum of all the partial pressures of the constituent gases. Air is composed largely of about 21 percent oxygen and about 79 percent nitrogen. If air were trapped in a container and all the oxygen removed, the pressure would drop from 14.7 psi to 11.6 psi. And if the nitrogen were removed instead, the pressure would drop to 3.1 psi. Thus, the partial pressure of oxygen is 3.1 psi; that of nitrogen is 11.6 psi; and their total equals 14.7 psi, the absolute pressure of air at sea level.

MAN AND MACHINE

Man is a technological animal, and over the centuries he has devised many contraptions to supplement his limited ability to dive and his total inability to breathe underwater.

Until the invention of the aqualung, devices for breathing air underwater had many drawbacks. If air was to be breathed at the same pressure as at the surface, then the diver had to be enclosed in a carapace, protected from water and the effects of pressure. With the invention of the compressed-air pump, man could move around on the ocean floor at greater depths for longer periods, but he was still tied to the surface vessel by lines and chains; he was still a tethered animal.

The Human Lobster

Most of the earliest designs, even those of Leonardo da Vinci, were as fanciful as they were unworkable. In 1715, for example, John Lethbridge described his leather case with armholes in which he claimed to have descended, "ten fathoms deep many a hundred times"—an impossible feat.

Theoretically the problems of working under water could be solved by an articulated diving dress. Early designs used leather pleats like those on an accordion for the joints, but these stiffened and shrank under pressure. The first successful suit, patented in 1913, combined ball-bearings and ball-and-socket joints to increase flexibility. Nevertheless, during the first successful deep-water (400 feet) salvage operation in 1931, the divers quickly abandoned the diving suits. Instead they sat in diving chambers and directed manipulators lowered from above to recover 95 percent of the gold carried by the S.S. *Egypt.*

On August 18, 1973, 42-year-old Enzo Maiorca from Syracuse, Italy, descended to a record depth of 265 feet with no fins, no mask, no goggles, and no breathing apparatus. Preparing by hyperventilating for eight minutes, he reached the record depth, dropped the 50-pound weight he was holding, and ascended rapidly. The dive lasted only two minutes and 18 seconds. The photograph shows Enzo surrounded by his companions immediately following the dive.

Underwater observation chambers have venerable ancestors, not counting Alexander the Great's—in 1865 Ernest Bazin tested one of his inventions in 250 feet of water, and in 1890 Balsamello was lowered in his sphere down to 430 feet. The Italian engineer Roberto Galeazzi has constructed both articulated armored diving dress to be used at a maximum depth of 800 feet and observation chambers effective at depths below 2000 feet, where the pressure exerted by the water exceeds 1000 pounds per square inch.

Galeazzi's observation chamber is tube-shaped and had limitations compared with a spherical shape. Galeazzi knew this, and his roughly cylindrical design is in fact made of many short spherical sections welded on top of each other. Dr. William Beebe showed the superiority of the sphere shape in the 1930s when he used a bathysphere to exceed depths of 3000 feet. Beebe's observation chamber had three windows for vision and was tethered to a mother ship at the surface by means of a steel cable, containing electric and telephone lines. The bathysphere had its own self-contained breathing system of oxygen; it used soda lime to absorb exhaled carbon dioxide and calcium chloride to remove moisture from the enclosed atmosphere.

The credit for designing the first true mobile submarine usually goes to David Bushnell, who designed an egg-shaped boat that was powered by means of manually operated cranks attached to screw-type propellers. Bushnell's submarine, the *Turtle*, was used in the Revolutionary War to move beneath British warships and plant explosives.

Recently we have been using minisubs launched from the *Calypso*. We call them ''puces,'' French for fleas. These minisubs give the diving pilots artificial arms through external manipulators and visual memory through cameras mounted on their backs. The pilots are protected from the high pressures of the underwater environment, even down to depths of 1600 feet, and are able to breathe air at atmospheric pressure. Perhaps more important is the ability of these vehicles to be used in tandem, one being able to rescue the other if such a need were to arise.

Obviously diving armor will always be the only tool with which human beings will ever be able to reach the ocean's depths. On January 23, 1960, the bathyscaphe *Trieste* reached the bottom of the Marianas Trench, 35,800 feet deep. The Americans had conquered the Everest of the sea with a vehicle designed by a Swiss professor in Belgium, built in Italy, and which included a German sphere. The ocean was demonstrating how effective international cooperation can be.

Compressed Air

The rigid diving bell is a very old device, and one that has been used with success, especially in shallow water. The greatest difficulty is that water pressure compresses the air inside the bell as the depth becomes greater; at 33 feet down, air inside a bell will be reduced to half its original volume. In 1689, the inventor Denis Papin, a pioneer in the development of the steam engine, had the idea of a crude pump to supply a diving bell with a continuous flow of air. The air would thus be forced down the rim of the bell whatever its depth. The constant flow of fresh air would greatly extend the duration of the stay. It was a revolutionary concept in diving techniques.

This concept is used in what, for many years, was the standard equipment of a diver: a miniature diving bell over the diver's head (called a hard hat) connected to an air pump at the surface by a hose. Heavily weighted, the diver remained in an almost vertical position and had to walk clumsily on the bottom.

The modern outfit, which weighs nearly 200 pounds, consists of a helmet, breastplate, weightbelt, flexible watertight dress, and weighted shoes. Those boots are made of either brass, which is good for walking around in mud, or lead, which provides better traction on solid surfaces like ships' hulls. The brass shoes weigh 20 pounds a pair; lead, 35 pounds a pair.

The umbilical connection to the surface includes the air hose, a lifeline, a telephone line for intercom between the diver and the tender. The air hose and the lifeline remain separated for safety reasons; the telephone line is either attached to the air hose or included in specially braided lifeline ropes. The depth of the diver is known by the crew above by reading on a gauge the pressure of the air that is pumped down.

FREEDOM IN THE SEA

Attempting to become free as fish in the sea, we had to forget our fears and to adapt our minds to an environment that had little to do with ours. Obviously we had first to get rid of those hoses and lines that turned us into captive animals kept on leashes. Then we had to abandon walking on the bottom. Therefore we had to develop a practical self-contained breathing apparatus; and we had to move by swimming in a longitudinal position and to be neutrally buoyant.

Four years before the publication of Jules Verne's classic *Twenty Thousand Leagues Under the Sea*, Benoit Rouquayrol and Auguste Denayrouze unveiled a device which allowed a diver to store a small amount of compressed air on his back, disconnect his air hose to the surface, and walk free on the floor of the ocean. The freedom was short-lived and the apparatus primitive, but those first steps had been taken in 1865. The key to the Rouquayrol-Denayrouze device was a regulator that helped control the flow of air from the underwater reservoir to the diver's mouth.

There were other steps, and missteps, along the way. One was the oxygen-rebreathing apparatus, invented by Henry Fleuss in 1878. These devices feed only oxygen to the diver, and his exhalations are filtered through a chemical agent to purge them of carbon dioxide. In World War II the frogmen used oxygen rebreathers and probably pushed them to their limits, learning that pure oxygen under pressure causes dangerous convulsions and that prolonged oxygen dives were only safe near the surface and no deeper than 23 feet.

OVERLEAF: *The aqualung gives divers a great amount of mobility and freedom to explore.*

The first air regulator—needed to deliver steady, breathable quantities of air to a diver—was developed in 1865 by French aquanauts Benoit Rouquayrol and Auguste Denayrouze. The air reservoir on the diver's back was kept full by a pump above.

Ten years later, another Frenchman, Georges Comheines, tested a semiautomatic regulator attached to a compressed-air container. It was a modified version of an air-breathing apparatus used by firemen in toxic atmospheres. Unfortunately, Comheines died during one of his first dives. By that time, Emile Gagnan and I were already working on our fully automatic aqualung, which supplied air upon demand to the diver at exactly the appropriate pressure. We were on the threshold of true freedom beneath the sea.

The Aqualung

The aqualung is based on the principle of open-circuit breathing apparatus, which allowed the exhaled air to escape into the water; this is a waste of oxygen but is only the price of simplicity and safety.

In the system, the compressed-air cylinders are carried on the back. The air passes from the container through a control valve which then brings the pressure down to about 100 psi above the surrounding pressure. Then the air passes through the demand valve operated by a membrane which is subjected, on the outside, to the surrounding water pressure. The air on the inside of the membrane is soon equalized with the hydrostatic pressure. Each inhalation applies a small depression on the large surface of the membrane, which acts as a force multiplier and pushes the valve open: air is fed to the diver. As he exhales, the membrane falls again, a spring closes the valve, and the air flow stops. The exhaled foul air escapes freely into the water through a one-way exhaust valve and never mixes with the air to be inhaled. A simple warning system alerts the diver when the air supply is low and assures him of a reserve sufficient for a return to the surface.

Nitrogen and the Bends

Nitrogen is considered a neutral gas in the air since it does not perform any essential function in the life process and does not react with the human body. At sea level, the body is penetrated and saturated by the surrounding air. Approximately one quart of nitrogen is dissolved in the blood and tissues at normal atmospheric pressure.

When a diver descends, the amount of nitrogen dissolved in his body increases proportionally to the increase in pressure. And when a diver breathing compressed air for some time and at some depth rises, enough nitrogen has been dissolved in his system to form bubbles as the pressure decreases—in much the same way as bubbles form and stream to the surface when a bottle of soda water or champagne is uncapped. These beverages are bottled under pressure and are decompressed when opened.

Such bubbles present special hazards when a diver has been submerged for a relatively long period or has dived to relatively great depths. There are several kinds of accidents that can occur by not following decompression tables and procedures. The most serious involve a release of bubbles into the blood as the diver ascends, bringing on a blockage of pulmonary circulation, which can be fatal within a short time. Other severe damage can happen as a result of oxygen being cut off from the brain by partially interrupted circulation, such as hemiplegia, paraplegia, or other extended paralysis.

Other, less severe effects of decompression accidents include localized pain in the joints, which often forces the diver to double up—the bends. A minor affliction called the ''fleas'' is an itching sensation on the skin.

The treatment for all these accidents, without exception, is immediate recompression. This return to a higher surrounding pressure reduces the size of the bubbles, reestablishes circulation, and restores the function of nervous tissues. If the starvation of these tissues was not too serious and if recompression was accomplished in time, the accident's effects disappear entirely within several hours. If these conditions are not met, however,

In 1920, Benjamin Franklin Leavitt proposed to use his diving dress to salvage the Lusitania, *sunk during World War I and resting in over 300 feet of water. However, water pressure at that depth would have collapsed the suit's flexible limbs and the salvage was never attempted.*

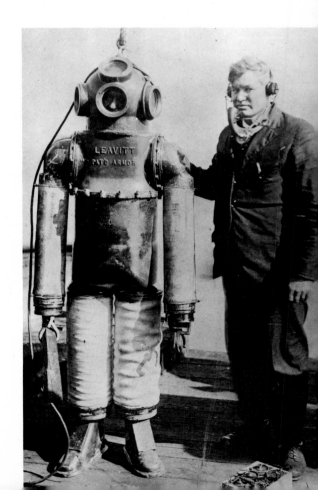

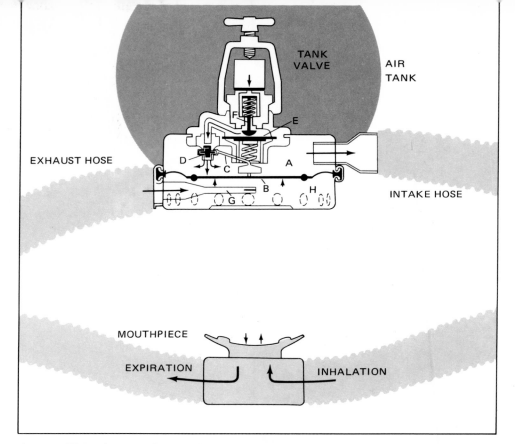

How the demand valve regulator works: Inhalation reduces pressure in chamber (A) above the membrane (B), causing it to rise. The lever (C) activated by this membrane allows air to flow through the low-pressure valve (D). A reduction in pressure is caused within the chamber (E), opening the high-pressure regulation valve (F), and allowing air to flow to the diver. Expired air flows through a one-way valve (G) into the chamber (H), which is open to the water through holes (dotted lines).

TANK VALVE

AIR TANK

EXHAUST HOSE

INTAKE HOSE

MOUTHPIECE

EXPIRATION

INHALATION

there will be irreversible lesions and permanent damage. And even divers who have been involved in only minor accidents may develop, with time, necrosis of the bones, especially joints at the shoulder and hips.

Such accidents can be avoided only by the diver making a very controlled ascent, slow enough to eliminate excess dissolved gas gradually so that oversaturation does not reach the critical point where dangerous bubbles form.

The best-known method of ascent, called stage decompression, was proposed by J. S. Haldane in 1906. Stage decompression requires the diver to stop rising every ten feet for specified time periods, which are determined by the depth and the duration of his dive. The length of time of each of these stops, which increases near the surface, is listed in decompression tables.

Other Dangers

The diver who breathes compressed air underwater introduces an additional quantity of air into his lungs. If he stops breathing during his ascent, his glottis acts as a tight valve, trapping the expanding air in his lungs causing lesions and forcing gas under pressure into the bloodstream, bringing about embolisms. These, as do nitrogen bubbles, cause severe damage and lead to quick death.

This may happen during ascents from as little as eight feet. Many victims of air embolism are neglected or improperly treated because of the assumption that their dives were not deep enough or long enough to qualify as decompression accidents. The whole problem can be alleviated by ascending slowly and exhaling during the rise to the surface.

Heat loss is another, often underestimated, hazard. When a body is immersed in water it loses heat by conduction much more rapidly than on land because water absorbs a greater amount of heat than air. As a rule of thumb, a protective suit is necessary for diving in water below 64° F. It may be a "wet suit," which allows some water to seep in, or a "dry suit," which has tightly sealed openings at the wrists, face, and neck.

A little understood danger to the diver breathing compressed air is nitrogen narcosis. At 100 feet down his efficiency begins to diminish; at 150 feet down, his faculties are obviously impaired, and below 200 feet there is a feeling of being drunk. Below this level, the intoxicating sensation may become violent enough to bring on an epileptic fit followed by loss of consciousness.

We called this effect "rapture of the deep" because it felt like drinking a little too much alcohol or smoking marijuana. I am susceptible to it, and yet I fear it like doom because it destroys the instinct of self-preservation. There are no after-effects, however; to escape its effects one simply ascends 30 or 40 feet.

LONGER AND DEEPER

Although man and his technology have greatly improved his diving equipment over the centuries, there are severe restrictions on the depth and length of time divers can spend in the world beneath the sea because of the property of the gases he breathes.

Finding a Substitute

The ideal breathing mixture appears to be a combination where nitrogen is absent and the partial pressure of oxygen kept as low as possible but still sufficient to sustain life.

Experiments have been conducted on various substitutes for nitrogen as an inert diluent for oxygen. The most encouraging were helium and hydrogen. Helium, on the one hand, is plentiful mainly in the United States but was fairly expensive since the government had a monopoly on it. Hydrogen is very plentiful and relatively inexpensive, but is dangerous to use since it can react explosively with oxygen when mixed in certain proportions.

Decompression is still a problem that all deep divers must face. At one time, ascending in slow stages with long waits at specified depths was the only safe way to return to the surface; it was both exhausting and time-consuming. Now divers can enter submersible pressurized chambers which can be sealed and hauled onboard ship where decompression is continued in comfort and safety.

As early as 1917, Elihu Thomson proposed helium as a substitute for nitrogen. Laboratory experiments on men have shown that there are no psychological problems that develop when using a helium-oxygen mixture (heliox) down to at least 1300 feet. The low density of helium is of considerable advantage in facilitating gas exchanges and pulmonary yentilation. Studies such as those at the Center for Advanced Marine Studies in Marseilles, France, have demonstrated that the limits of human diving with heliox are in the vicinity of 2000 feet.

But even helium has shortcomings. The hope for shortened decompression times, due to faster diffusion out of the system, for example, is negligible in tests conducted at 225 to 300 feet. Helium mixtures require sophisticated, expensive equipment and complicated manipulations. Helium is a good conductor of heat and exacerbates body heat loss.

Advances

The conventional open-circuit, self-contained compressed-air breathing apparatus has opened the world of the sea to millions of people. Of all the breathing devices available it is by far the easiest and safest for the novice diver to use. But in addition to the restrictions imposed by oxygen and nitrogen in a compressed-air system, there are limitations in the device itself. It is bulky, noisy under water and, because of the open circuit, a large volume of gas is used in relatively short periods when the diver is at great depths.

The Krasberg unit is one of several systems that electronically control the mixture of gases that divers may safely breathe.

The latest devices generally use helium as neutral gas. The breathing mixture is automatically controlled and monitored electronically so that the diver knows if the oxygen partial pressure varies from tolerable limits. Oxygen is automatically introduced by an electrovalve operated by sensors. Carbon dioxide from respiration is scrubbed in a chemical absorbent canister. The units operate silently and leave no bubbles. The diver can remain underwater from four to six hours at any depth, provided his decompression is made with the help of a chamber or of another unit.

These advanced closed-circuit systems have allowed longer and deeper stays in the sea, but they still carry some risk. The increased capabilities allow a diver to get into trouble by going too deep or staying down too long. Therefore these devices must remain, for the time being at least, professional or military tools. Aqualungs are still the only systems that leave no doubt about the exact nature of the gas that is taken in.

LIVING UNDERWATER

Since decompression takes such a long time, it is obvious that reasonable productivity in great depths can only be obtained by staying down for several days. On September 6, 1962, Robert Stenuit stayed 24 hours in a small, submerged recompression chamber at a depth of 200 feet in the Bay of Villefranche. He breathed a helium mixture and dived several times out of his cocoon using a breathing tube. From September 14 to 21, Albert Falco and Claude Wesly lived in Diogenes, our Conshelf I station, which is 35 feet deep; they worked several hours a day at depths to 85 feet.

Underwater Colony

For Conshelf II, our second underwater habitat located off the Sudanese coast in the Red Sea, we were planning a true colony, where five men would live 32 feet down for a month, in addition to the two men who would inhabit the Deep Cabin 82 feet down for a week. The Conshelf II complex included a main unit we called Starfish House because of the branching workshops and living quarters off the main chamber; a wet garage for the scooters and tools; and a dry domed hangar for the diving saucer. The various buildings were linked together with telephone and television lines, and cables from the surface ship *Rosaldo* provided power for communications and electrical equipment and the compressed air which kept the pressure inside at two atmospheres.

We wanted to test man's capacity for prolonged living on the sea bottom and mainly his efficiency when working extended hours from a pressurized habitat. The oceanauts' psychological stamina was as important as their health. They were examined each day by our diving doctor, Jacques Bourde. Soon the men living in the Conshelf station realized there were subtle effects of the pressure. Beard growth was retarded. Small cuts and abrasions healed much faster than similar injuries suffered by our other divers living on the surface.

In spite of the initial lack of an air conditioner, five relative strangers adapted well to living in very close quarters. Pierre Vannoni wrote in his diary, "I am beginning to be aware of time passing, which I had rather forgotten about. I feel I may rise to the surface next week without having seen and experienced absolutely everything." Vannoni was an interesting case: he had been buried alive by an explosion during the war and had suffered claustrophobia ever since. But in Conshelf his fears vanished; he also found that once his routine was established, he slept more deeply than usual and found it difficult to wake up.

One of the disappointments came midway through the stay when I took down some bottles of champagne to toast the milestone. In the high pressure of the habitat, the bubbles failed to come out of solution; the wine was as still as the world outside.

The Deep Men

An important part of the Conshelf II plan was the Deep Cabin, where Raymond Kientzy and André Portelatine were to spend a week in a helium-oxygen atmosphere.

Deep Cabin was equipped with two chambers: the lower one, called the wet room, containing the diving gear, tools, and sea hatch; and the relatively dry living quarters above. One piece of apparatus that was missing was an air conditioner. We had originally planned the experiment for March, when we thought the water at that depth would be sufficiently cool to make Deep Cabin livable. But in July, the Red Sea outside Deep Cabin was 86°F. and the humidity was always 100 percent.

The morale and psychological well-being of the Conshelf II aquanauts is maintained by providing some of the comforts of home. Here, two men relax over a game of chess.

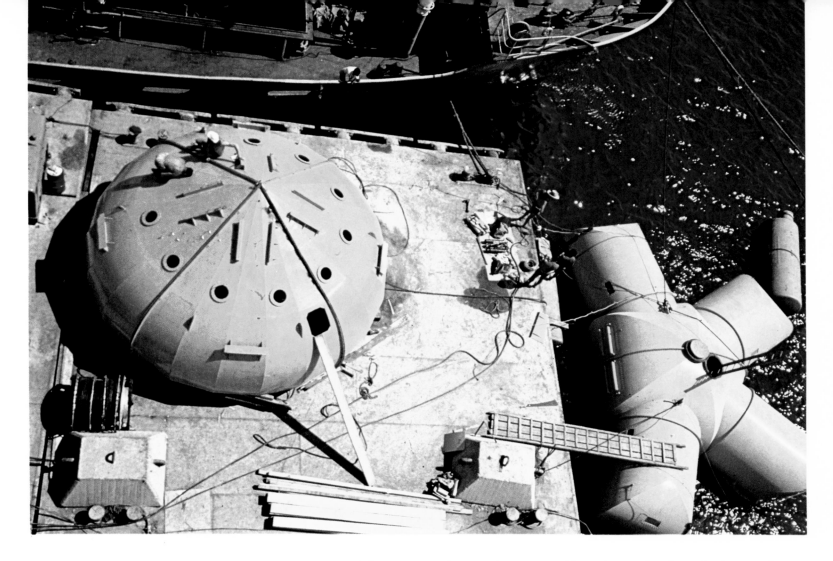

Starfish House and the diving saucer's hangar are made ready for their epic adventure below the waves.

There were also equipment malfunctions—in the telephone, the refrigerator, and the backup oxygen system. Living in a 24-hour sweat bath also caused Kientzy's and Portelatine's appetites to malfunction. But Dr. Bourde visited them every day and said that other than being uncomfortable the two were in good health.

On the sixth day of their stay, Kientzy and Portelatine descended toward the edge of the reef on which Deep Cabin rested. Then they dropped to 300 feet where the pressure is 11 times that of the atmosphere at sea level—161.7 psi. They felt no drunkenness, no rapture of the deep that ordinarily accompanies air-breathing men at that depth. And though they entered their achievement in the log as 330 feet, they later revealed that they had reached 363 feet—a record for the deepest compressed-air dive in history, but one which they did not want to claim because setting records was not their purpose.

The United States Navy

The United States Navy designed two underwater habitats, Sealabs I and II. The Sealab II habitat was larger than its predecessor, 57 feet long and 12 feet in diameter and less tapered at the ends—resembling a railroad tanker car. The plan was for 28 men to stay down in 15-day shifts, with astronaut Scott Carpenter and Navy doctor Robert Sonnenberg remaining down for 30 days. The project was conducted in 1965 in the Pacific, where there was no danger of hurricanes, less than a mile off the coast of southern California.

Collecting all physiological data from the inhabitants was one of the primary goals of the project, and each day the aquanauts sent up samples of breath, blood, urine, and saliva in a pressurized capsule. In addition to chores like placing monitoring instruments on the sea floor, the men performed strength and manual dexterity tests before and after each dive. Every evening there were daily activity and mood checklists to be filled out and occasional "brain teaser" quizzes and arithmetic tests to determine the possible effect of the high-pressure helium-oxygen atmosphere on the brain. The breathing mixture was 77–79 percent helium, 18 percent nitrogen, and 3–5 percent oxygen. Lithium hydroxide was used to absorb carbon dioxide.

The bodily changes that the aquanauts experienced were transitory: some decrease in strength, manual dexterity, and coordination was observed. These conditions abated when the divers returned to the surface after their 56-hour decompression. There was no measurable change in mental test results before and after the stay. Humidity, as always, was a problem. But as Scott Carpenter later said, it was "the most richly satisfying experience of my life."

An experiment was conducted with Tuffy, a porpoise trained to respond to sound signals. At first he was shy and irresponsive, but then rose to the task by delivering mail between the surface and Sealab II and by finding men outside the habitat who signaled that they were in danger and acted accordingly.

Cutting the Cords

While aquanauts of Sealab II were down on the floor of the Pacific, we were 6000 miles away in the Mediterranean working on Conshelf III. This was a radical departure for us, since the habitat would settle 328 feet below—too far for support divers to reach it from above. If man was going to inhabit the continental shelf, he had to sever the umbilical cord to the surface, and this was a decisive step in that direction.

Conshelf III consisted of a spherical chamber 18 feet in diameter resting on a chassis 48 by 28 feet. An engineering characteristic of the habitat that proved essential for safety was that the sphere could resist inside as well as outside pressures higher than the 11 atmospheres to be met

Deep Cabin was part of the Conshelf II project. It was set 82 feet below the surface of the Red Sea.

Conshelf III rested over 300 feet below the surface and proved that man could perform difficult maintenance jobs at that depth.

The United States Navy's Sealab II was larger than its predecessor. On its first mission it was towed into position and submerged off the southern California coast.

at the operating depth. The men were breathing a mixture of 98 percent helium and only two percent oxygen. Helium is so light that the oceanauts' vocal cords vibrated much faster than they do in air. The result is a squeaky, high-pitched voice. Voices from topside or music from loudspeakers were, of course, undistorted.

While working outside Conshelf III, the oceanauts were tethered to the habitat with 200-foot breathing hoses that pumped heliox through one tube and sucked exhalations through another; the foul air was scrubbed in the habitat's recycling system where the helium could be used again. The oceanauts also wore aqualungs filled with heliox, good only for a few minutes, in case something went wrong with the breathing hoses.

The major task of Conshelf III divers was to prove that men could perform practical tasks at great depths. One of the oceanauts, Christian Bonnici, worked seven straight hours, an unheard-of achievement at that depth. Among the things Bonnici accomplished was threading a stiff wire

through a series of pressure-proof seals—a task considered almost impossible underwater.

The three-week stay of Conshelf III reaffirmed our experience that surface vessels and research equipment are more vulnerable to damage than the underwater habitat and that machines fail more often than men.

Industry Adapts

Industrial divers, though, found that in many cases habitats were impractical or unworkable for such reasons as expense or location. As a result, the cachalot system was developed for divers at dams and underwater construction and offshore oil well sites. The divers are transferred to and from the surface in a special submerged decompression chamber. On the dock or the deck of a surface ship the divers are transferred through an airtight lock into a more comfortable pressurized chamber, where they live between work shifts under the same pressure as the bottom. This method allows saturation diving to a depth of 1000 feet and maybe soon to 2000 feet, by teams

of two, three, or four men.

There are drawbacks, however. The tender ship has to be large and specially equipped, its cost will be running all year, even if it is sparsely used. And handling a heavy weight onboard a ship that is pitching and rolling most of the time has always been a hazard. Also, helium divers need a lot of local support, gas, heat, and comfort, and a cramped SDC (submerged decompression chamber) is far from affording such necessities. The list of casualties in the past few years is unfortunately long.

Properly trained men supported by the appropriate equipment and technology can perform underwater almost any task they have been familiarized with at the surface. One thousand feet is already an operational depth, but we know that as long as the human body will need to breathe gases, even the lightest conceivable mixture will be too heavy to ventilate man's lungs at depths greater than 2000 feet.

In 1965, after Conshelf III, we drew the lessons from the success that had needed so much money, so much surface manpower and equipment. It became obvious to me that all the support complex had to migrate from the surface down to the immediate vicinity of the diver's habitat.

Our studies ended in the construction of the ultimate in deep-diving equipment, a submarine bearing the name of our water spider, the *Argyronete*. This absolute weapon for ocean reentry is designed to service in safety four oceanauts working up to eight days, several hours a day, at a 2000-foot depth. The 300-ton *Argyronete* has a "dry" crew of six, a surface speed of seven knots, an underwater speed of four knots, and a cruising range of 400 nautical miles. It eliminates the need for a tender vessel.

The government suspended construction when it was more than three-quarters built. Leaving aside the confusing lobbying and the political motivations, the real reasons for the suspension are that the oil companies (the only industry rich enough to fund such a project) were convinced that

A Water Spider Made of Steel

266

they could successfully achieve their deep jobs without divers, using remote-controlled tools. The concept, however, remains above criticism, and the *Argyronete* family of steel water spiders will multiply in the future.

Whatever ingenuity man uses to facilitate his reentry into the sea, his respiratory system provides him with few alternatives. He can either enclose himself in a pressure-resistant, watertight shell in which he maintains his normal atmosphere or he can breathe more-or-less elaborate gas mixtures delivered to him at ambient pressures. In the first instance, he will have very poor contact with the ocean environment and, in the second, the physiological effects of breathing exotic gas mixtures under high pressure put a limit on his drive toward great depths.

The only breakthrough appears to be breathing without using gas at all. Experiments conducted on test animals have shown that liquid can be pumped through the lungs of mammals and sustain life. But the liquid is a special concentration of salts with much, much more oxygen dissolved in it than normal ocean water. In addition, because the fluids in the human body are generally less saline than surrounding seawater, the resulting osmotic differences would cause the lungs to lose water and become dehydrated. For a number of reasons, then, man could not sustain himself by breathing seawater directly. As we explain later, because of the incompatibilities in seawater and the physical capabilities of his lungs, there is no way that man can breathe seawater directly.

Hopes have been raised that perhaps man could "breathe" water indirectly, since small animals can live without difficulty for several days in a watertight enclosure made of Teflon membranes and completely submerged in an aquarium. Carbon dioxide accumulates inside the enclosure until its partial pressure becomes higher than that of carbon dioxide in the water. The gas then diffuses through the semipermeable membrane into the water and eventually into the air. In the same manner, as oxygen inside the enclosure diminishes, it is replaced by oxygen from the water. Thus the respiratory exchanges are assured. The atmosphere of the enclosure is just a little poorer than the air outside the tank.

Nevertheless, the cage can contain *only* air at atmospheric pressure. If any gases other than oxygen and nitrogen were introduced into it, they would escape to the atmosphere, and would be progressively replaced by air.

Pressure inside the enclosure being that of one atmosphere, the membrane walls must withstand the hydrostatic pressure of the surrounding water, an easy task in an aquarium, but positively impossible at any diving depth. Any conceivable wall construction material cannot at the same time have the required mechanical strength and porosity.

Sports and games, science and engineering have outlined routes to take in man's reentry into the ocean. Now the pragmatists take over. The petroleum industry has been the first to really exploit the world under the sea. Those soon to follow include mineral-mining interests. Sea farms, cultivated from beneath the surface, can't be far behind since the need for food is a major driving force.

Science fiction writers dream of entire populations moving permanently to underwater cities, developing diverging civilizations, and making war against their land brothers. No. Settlements will be built on the sea floor and inhabited as long as a definite task justifies the cost and the sacrifices. But *Homo aquaticus* himself, after a few weeks of work in platinum mines 7000 feet deep, will gladly rejoin his native village, celebrate a friend's birthday, and smile in the warm sun.

THE UNDERWATER MAN

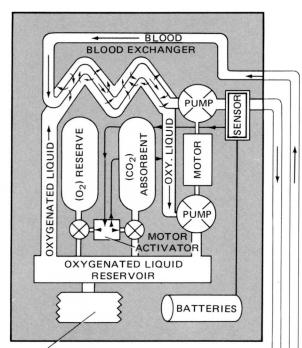

PRESSURE
REGULATION
CHAMBER

Still on the drawing boards, a lightweight lung substitute machine will one day enable divers to descend to any depth and return without decompression. Blood will be pumped through a highly oxygenated medium; the exchange of gases will take place through a semipermeable membrane. A computerized motor activator will respond to sensors detecting the gases dissolved in the blood and adjust the flow of oxygen (O_2) and the absorption of carbon dioxide (CO_2).

Invasion of the Sea?

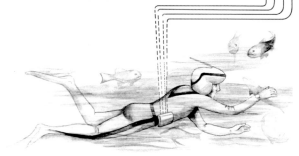

WET MUSES

In the clear tropical waters of the Red Sea, along the exuberant coral reef of Port Sudan, a great iron ship ran aground and sank many years ago. It has since been coated with a medley of graceful polyps, sponges, and gorgonians; it was invaded by scores of gaudy reef fish, feather-duster worms, starfish, and sea snails that surpass in brightness and gorgeous colors the butterflies and flowers of our gardens. On board *Calypso*, André Laban dons his diving equipment and slowly walks down the ladder into the sea, his arms burdened with unusual impediments: a tripod, a metal stool, a large rectangular canvas, and a bag full of brushes and paint tubes. Soon he is comfortably installed, facing the wreck's crumbling and festooned bridge, and he begins to paint the scene. Back on deck, Laban carefully flushes his "masterpiece" with fresh water from a hose. The first known undersea artist was Pritchard, who painted in a heavy helmet suit at the end of the nineteenth century.

Photography developed as soon as the peculiar technical problems raised by the physical and optical properties of seawater had been solved. Today it is not enough to make a well-exposed, color-corrected, well-focused picture: the art of photography also implies tasteful composition and the choice of the most effective angles.

Off the coast of the Italian Riviera, a large bronze Christ has been installed in the depths, and a statue of the Virgin watches over the undersea world in Spain. Miró made a ceramic sculpture of a strange creature meant to be the Spirit of the Deep, which was set up in a cave, 120 feet down in the Mediterranean. Aside from these spectacular pieces, the art of sculpture is often inspired by such streamlined shapes as the fluid, graceful lines of the dolphin.

Architecture is also faced with new problems. Engineers had early solved the problems of designing structures to withstand water pressure. They were either spherical, cylindrical, or ellipsoidal, but they had to be tailored to the needs of men. Conshelf II and Conshelf III were the first to be influenced by designers and decorators. Undersea settlements of the future will certainly be drafted by teams of architects and technicians such as those that are responsible for the most beautiful and efficient buildings on land.

Poetry and music have hitherto been very slightly influenced by the ocean, but poets and composers are today discovering new sources of motivation in the realities of the undersea world. A full symphony inspired by the melodious songs of the humpback whales recently marked the beginning of a new field of music.

Writers and playwrights have long been fascinated by the mysteries of the ocean and by information available in the legends of the sea. But today facts are just as extraordinary as tales, and quality books of science fiction or science anticipation are written in increasing numbers.

The popular films of the aquatic ballets of Esther Williams were an early attempt at including the sea in some aspects of the art of dancing. But there remains a wide fascinating field to be explored. Humans are prisoners of their weight and thus sentenced to move around and struggle in a two-dimensional world. Dancers symbolize the efforts of the body and the dreams of the mind to escape gravity. Weightlessness, experienced by divers in their three-dimensional realm, could be a totally new departure for a different choreography.

A more subtle but overwhelming influence is rapidly spreading and slowly penetrating our daily life. Studies of the behavior of marine animals show that even in such a different environment as the sea, the motivations of all creatures are basically the same. The concept of the unity of life is every day more obvious. The oceans, in the same manner as space, are contributing to link the nations of the world; oceanographers of all countries work hand in hand in an extraordinary atmosphere of understanding and collaboration. The obsolete laws governing the high seas and the exploitation of all provinces of the ocean are about to be revised, and new rules will establish that the seas are the common property of mankind. The damages caused to the world by the mismanagement and pollution of the sea call for action on a global scale and in a way will help a stronger solidarity of our species. In fact, our entire philosophy is about to be changed by the sea.

Gerald Arpino's ballet Sea Shadow

In the Revelation of Saint John the Divine, the last book of the Bible, John imagines a utopia in which there is earth and sky but no sea: "Then I saw a new heaven and a new earth, for the first heaven and the first earth vanished, and there was no longer any sea." John's fear of the sea was what every man felt. The sea was great, and it was a stranger. It was said that it led to the edge of the world, perhaps to the land of the dead. This fear inspired the creation of a mythology to answer the unanswerable, to explain the inexplicable, and to assuage, a little, man's terror. And the first questions man asked about the sea were: "How was it created, and why is it like it is?"

THE WATERS UNDER THE HEAVEN

The Shaker of the Earth

The powerful god of the sea, Poseidon, is believed to have come to the Greek peninsula as the chief deity of northern Aryan tribes that settled near the Mediterranean around 2000 B.C. He had created horses for them, and because the stampeding animals caused the earth to tremble, he was also identified with earthquakes. As a symbol of his dominion over the seas he usually carried a trident, which he used to smite the earth to cause it to quake or to bring forth fountains and horses.

When the Greeks named Zeus as the first god of Olympus and ruler of the sky, his elder brother, Poseidon, was assigned command of the sea and all that belonged to it. He lived in a palace deep in the lagoon with his wife Amphitrite, eldest of the Nereids—daughters of an earlier sea god, Nereus. Homer's description of Poseidon revealed him standing upon his chariot, clothed in gold with a golden whip. His golden-maned horses bounded across the waves while all the sea monsters gamboled at the coming of their king.

Poseidon could be a formidable enemy. His anger at the Trojan king Laomedon, who failed to pay a promised reward to him and Apollo, caused him to send a terrible monster to ravage the city. He favored Troy's enemies, the Greeks, but later he turned against the arrogant victors and made their return home a disaster. Storms destroyed many of their ships and blew others far off their course. But the great god of the sea could be kind as well. Protector of seafarers and fishermen, his sacred dolphin was a symbol of a calm and peaceful sea.

A Boy's Best Friend

Today many a schoolboy would like to come out of school to find a dolphin, his best friend, waiting to take him home on his back. Once, very long ago, this really happened. At least, that is what some authors of ancient times assert.

If we are to believe those writers, it is certainly true that at least one boy's best friend was a dolphin. According to Aristotle, the great Greek philosopher, "among the sea fishes many stories are told about the dolphin, indicative of his gentle and kindly nature, and of manifestations of passionate attachment to boys in and about Tarentum, Caria, and other places." Somewhat later, the Roman historian Pliny related how "a dolphin that had been brought into the Lucrine Lake fell marvelously in love with a certain boy, a poor man's son." The boy and the dolphin became fast friends and "when the boy called to it at whatever time of day, although it was concealed in hiding, it used to fly to him out of the depth, eat out of his hand, and let him mount on its back, sheathing as it were the prickles of its fins, and used to carry him when mounted right across the bay to Pozzuoli to school, bringing him back in similar manner, for several years, until the boy died of disease, and then it used to keep coming sorrowfully and like a mourner to the customary place, and itself also expired, quite undoubtedly from longing."

In ancient legends dolphins were not only boys' best friends but guides for sailors lost at sea. Dolphins understand men because, according to legend, they were once men themselves. The gods, it was believed, had

Many legends throughout the world tell of the friendships between boys and dolphins or other sea creatures. This Japanese woodcut by Kuniyoshi is entitled Kintoki Swims Up the Waterfall.

273

transformed them into dolphins. In the *Homeric Hymns* we find just such a legend. The child god Dionysus was captured by pirates of the Tyrrhenian Sea. Thinking that they would sell the boy, they put him on a ship to take him to the slave market. But the child was possessed with divine powers, and he began to transform the ship. He made red wine run over the decks and leaves sprout from the mast. He transformed himself into a lion. The pirates were terrified and leapt into the sea where they were immediately changed into dolphins. Not only was their form changed, but their character as well. Henceforth they began to aid mankind.

It is of little wonder then that there were more admonitions against the killing of dolphins. Oppian, the Greek poet, declared that "the hunting of dolphins is immoral" and that the men who killed dolphins "would not spare their children or their fathers and would lightly slay their brothers born." Sailors today believe that dolphins will not play about a ship that has caused the death of one of their kind. The dolphins can tell, they say, and they stay away.

The Sea that Roars

One of the most dangerous passages for the ships of old was through the Strait of Magellan at the southern tip of South America. An ancient myth known to the Yamana Indians of Tierra del Fuego helps us to understand the terror that those waters held.

The myth relates that in very ancient times women commanded men. Since the women knew that their power would one day be taken from them, they attributed that power to Tanuwa, the supreme goddess. They constructed a sacred dwelling where they met to carry out the most complicated initiation rites. Then, to further convince the men of their right to rule, they disguised themselves with masks and painted their bodies and presented themselves as incarnations of the goddess.

At this time the sun was not yet in the sky but lived with the men of the island. One day, when the sun was taking a walk in the woods, he discovered the women's secret, and he hurried to unveil their deceit to the male members of the tribe. The men were furious. They armed themselves with arrows, harpoons, and lances, and they marched to the sacred spot. In the struggle that ensued between the men and the women, the women were

A colorful fabric mola *from Panama illustrates the Cuña Indians' skillful use of sea imagery.*

274

In the bowl of a Greek cup of c. 535 B.C., Dionysus, god of wine, is shown crossing the sea.

transformed into sea animals. The sun threw a pail of water on the temple, and it turned into a gigantic wave that carried off all the animals. From that day on, the sea roars when it gets angry.

The Ocean of the Cuña Indians

Among the many stories of the creation of the ocean to be found around the world is this one from the Cuña Indians of Panama.

Ibelelus, who was their demigod, saw a woman pass by, chanting these words: "Tree of salt . . . tree of salt . . ." Ibelelus followed her to see where she came from.

The next day he came back to his men and said: "I have discovered where the woman comes from. She lives in a tree of salt. At the top there is water and fish and flowers and other animals. At the foot of the tree there is a lot of gold." His nephews decided to go and attack the big tree, and they set to work the following day. But they did not even manage to cut the bark off. Ibelelus was surprised at this, and in the night he posted himself in the brush to find out who was keeping the tree from falling. He saw a jaguar, a snake, and a frog come to lick the trunk's wounds, which immediately healed. He resolved to get rid of these magic animals, and the following morning his nephews began again to cut the tree. The chips flew, and when they touched the ground, they were transformed into fish; but the tree remained standing because the clouds themselves supported it. Then Ibelelus commanded the squirrel to cut the clouds that held the foliage of the tree, and at last the tree fell. When it did, the water that was at the top spread everywhere and formed the sea.

Why the Sea is Salty

According to a Danish story, the source of all the salt in the sea is way down at the bottom of the ocean. There once was a beloved king of Denmark who had among his treasures two enormous grinding stones. The

275

stones could grind jewels and anything the king wished for. But they were too heavy for anyone to turn.

Once, when the king was in Sweden, he saw two women who had been captured in the land of the giants. The king of Sweden gave them to him as a gift. He took them home and put them to work turning the grinding stones. He commanded them to grind gold and silver and peace and joy, and the giants obeyed. But one day they got tired and asked the king if they could rest. The king insisted that they grind on, and so the women decided to play a trick on him and they began grinding out soldiers for the king's enemies. When his enemies were strong enough, they attacked the kingdom and carried off the magic grinding stones and the two giants. The king's enemy had a great need of salt in his country, and while they were still aboard the ship carrying them to that land, he commanded the giants to grind salt. And they did. They ground and they ground, and the salt filled up the ship. Finally there was so much salt that the ship sank to the bottom of the sea. All of its company was drowned except the two giants, and they went right on grinding salt. Even today no one has told them to stop, and that is the reason the sea is full of salt.

Every part of the world has its legends of life in the sea. The Eskimos tell of a girl who became Sedna, the goddess of the sea. She was a girl who refused her suitors and married a bird. The girl's father, outraged at this, killed her husband and took her home in his boat. On the way, a storm arose, and the father threw the girl overboard. She clung to the gunwale until he chopped off her fingers, one by one. She then sank to the bottom of the sea, and there she still lives, keeping guard over all that dwell there.

But her severed fingers were transformed into the fish and mammals of the sea, and these, her children, ate up the girl's father. The girl became the chief deity of the lower world, and each autumn the people of the Far North hold a great feast and festival in her honor. To some of the Eskimos she is known as Sedna, but she goes by a great number of other names in various parts of the arctic, and her story has many variants.

Nowadays her companions are a dwarf and an armless woman with whom she shares her husband, a sea scorpion. Sedna feels no kindness toward humanity, but she does not act arbitrarily. She never moves about of her own free will but is rooted to her stone dwelling. Her sinister appearance would kill an ordinary man, and only a shaman, or priest, is able to stand the sight of her. Huge, voracious, and impotent, with a wild temper, she keeps watch with her one eye over the mammals of the sea.

When a hunter kills a mammal of the sea unnecessarily, Sedna feels resentment and great physical pain in the place where that animal originally sprang from her body.

Civilization first appeared in the valleys of ancient rivers. Since all these rivers led to the sea, man was eventually tempted to venture out onto a dangerous ocean.

The boats of the early challengers were not very seaworthy and this made the danger very real; but they also had to overcome their imagination. Early man peopled the sea with monstrous and terrifying beings. Only by meeting the challenge could these myths be dispelled as they proved to be imaginary.

When Jason was 20, he came to Iolcus in Thessaly to claim its throne from his uncle, the usurper Pelias. Pelias told his nephew he could have the throne if he would first travel to Colchis and bring back the Golden Fleece which was guarded by a sleepless dragon. Jason accepted the challenge and

Sea Goddess of the Eskimos

THE CHALLENGE

Jason and the Argonauts

A temple dedicated to Poseidon, god of the sea, stands high above Cape Sounion, Greece, welcoming seafarers home. Doric in style, the temple dates from the time of Pericles, about 440 B.C.

so began the epic sea voyage of Jason and the Argonauts.

The goddess Athena personally directed the construction of their ship, seeing that its stout pine planks were bolted and fastened with a tight girdle of well-twisted rope to withstand the battering of the waves. *Argo,* as it was named, had benches for 50 rowers and a mast for sail. The crew of Argonauts under Jason's command was made up of 50 heroes from all over Greece.

They had many adventures during their voyage. At Salmydessus they aided blind Phineus by ridding him of a pair of loathsome Harpies. He repaid them with wise advice on how to get past two floating rocks that guarded the Bosporous. They freed a dove as they approached. The rocks clashed together as the dove flew between them, nipped off some tail feathers, then recoiled. At that very moment the Argonauts stroked hard and got through the narrow guarded passage before the rocks came together again. They lost only a stern ornament.

Aeëtes, king of Colchis, agreed to give up the Golden Fleece only if Jason would perform tasks thought to be impossible. He was able to do them with the help of the king's daughter Medea, who had fallen deeply in love with him. But in the end the Argonauts had to steal the prize anyway. Medea sailed away on *Argo* while Aeëtes followed angrily in pursuit. Legend has it that Medea cut up the body of her little brother and threw the pieces in the sea so that Aeëtes would be delayed when picking them up.

After many more adventures they finally returned to Thessaly, where Jason gave Pelias the Golden Fleece. But the crafty old man revoked his promise: Jason would never occupy the throne at Iolcus. Medea squared matters in her usual way by tricking his daughters into killing Pelias; after that the newlyweds wisely found refuge in Corinth, where they beached *Argo* and consecrated it to Poseidon. One day when Jason was old, he crawled beneath the ancient hulk to rest. The stern broke off, fell, and killed him.

A portrayal on a famous Greek vase of another well-known myth shows the type of boat that Jason and his crew of Argonauts might have used while seeking the legendary Golden Fleece.

The Infancy of Marine Biology

One of the most industrious of scientists was Pliny the Elder, who lived in Rome in the first century A.D. His great work was a 37-volume encyclopedia of natural science called *Historia naturalis,* which dealt with the whole

This shadowy figure, photographed in 1961, was offered as proof of the existence of the elusive Loch Ness Monster.

Monsters of the deep, both real and imagined, have been portrayed throughout history. This drawing of a Minoan stirrup jug bears a credible rendering of an octopus.

MONSTERS OF THE DEEP

physical universe. Most of Pliny's information came to him secondhand and his work is quite useless as science, but this does not take away from his achievement, for he had made a beginning. Others would follow along the path.

Pliny the Elder devoted Book IX to fish, and what he wrote reveals how much natural science was intermingled with tales and imagination: "The mullet and the wolffish are animated with a mutual hatred; and so too, the conger and the murena gnaw each other's tails. The crayfish has so great a dread of the polypus, that if it sees it near, it expires in an instant; the conger dreads the crayfish; while, again, the conger tears the body of the polypus. Nigidius informs us that the wolffish gnaws the tail of the mullet, and yet that, during certain months, they are on terms of friendship; all those, however, which thus lose their tails, survive their misfortune. On the other hand, in addition to those which we have already mentioned as going in company together, an instance of friendship is found in the balaena and the musculus, for, as the eyebrows of the former are very heavy, they sometimes fall over its eyes, and quite close them by their ponderousness, upon which the musculus swims before and points out the shallow places which are likely to prove inconvenient to its vast bulk, thus serving it in the stead of eyes."

Over the centuries there have been thousands of reports of the sighting of gigantic and fantastic creatures in the sea and on the shore.

There were monsters whose backs were a mile wide. Others had arms long enough to pluck a sailor from the crosstrees of a mast. There were even stories of a gigantic serpent that circled the earth at the equator. Investigations of these reports in recent times have led us to believe that most if not all of the sea monsters described are nothing more than water-logged trees, masses of seaweed, whales, basking sharks, giant squid, walruses, manatees, dugongs, or seals. Nevertheless, some of the descriptions given have not been all that easy to explain away, and it is not possible to state categorically that strange, still unknown creatures of the sea do not exist. We know that there once lived in the sea animals that we would call monsters if we were to see them today.

Monsters of the Middle Ages

It took extraordinary courage to be a sailor when the earth was still thought to be flat, for if one ventured too far out to sea he was taking the risk of being swept over waterfalls at the edge. Even if that did not happen, there were always those places in the sea where the water was boiling and where

In the Middle Ages some people believed that everyone on land had a counterpart in the sea. The sea bishop (left) and his attendant, the sea monk (right), are drawings that appeared in L'Histoire entière des poissons, *published in 1558 by the naturalist Guillaume Rondelet.*

fearsome monsters lay in wait to crush his ship and devour him. The Great Wall Snake was a peril the sailor believed he might have to face at any time. And then, of course, there was always the great serpent that encircled the entire globe at the bottom of the sea with its tail in its mouth.

Naturalists of the Middle Ages did their part to people the sea with imaginary beings. They reasoned that every land animal must have its counterpart in the sea. Thus there must be sea horses, sea dogs, sea cows, sea elephants, and sea lions that looked like their terrestrial cousins, and they were partly right. These early scientists also believed that since the sea was the home of mammals like the whale, the porpoise, the manatee, and the seal, counterparts of human beings might also be found in the sea. Among the species they created and envisioned as living in the sea were the sea bishop and his attendant, the sea monk. The sturgeon might have been the inspiration for these beliefs because this fish has a body covered with what looks like scaly armor. Perhaps it was a member of the ray family, such as a skate which has on its underside markings that resemble a human face.

The Kraken

In 1555 a Swede named Olaus Magnus wrote a history of Scandinavia in which he described a peculiar monster called a kraken. It appears that in reality it was a giant squid or octopus or cuttlefish. None of these animals is imaginary. But the imagination of the sixteenth century made the kraken something far more gigantic. Engravings of the time show squids so large that they can snatch a sailor from high up in the rigging of a ship. A famous drawing of a kraken by a Swiss artist who had never even seen the ocean gave the animal a body like a catfish, legs like an alligator, and tentacles that resembled ribbons.

Probably the first recorded mention of something like a kraken was in the *Odyssey*. The picture that Homer gave of the monster Scylla clearly suggests an octopus or a squid enormously magnified.

The Loch Ness Monster

No monster past or present has had more newsprint devoted to it than Nessie of Scotland's Loch Ness. Monster sighting at Loch Ness is a perennial sport that reaches its highest pitch during the holiday season each August. It is then, during what they call "the silly season," that the editors trot out the Loch Ness monster story.

Those who believe in Nessie explain its presence in landlocked Loch Ness quite convincingly. During the last ice age when the weight of the ice

According to legend, the grouper has gill openings on the side of his head because he once swallowed a beautiful girl that he loved who was then forced to cut her way to freedom. The grouper adjusted to the flow of water through these openings and retains them to this day, but he vowed never to fall in love again.

had pushed down on the earth, Loch Ness was connected with the sea. As the ice melted, the sea level rose, but without its heavy cover Scotland rose too, like a cork in water. Loch Ness was cut off from the sea, and many marine animals were trapped in it, among them Nessie's ancestors. The long, narrow loch, averaging 430 feet and plunging at one point to 754 feet, could support a large animal in its deep water.

Nessie's notoriety goes back a long way. In 565 A.D. it had killed one boatman and was chasing another when the abbot of Iona, Saint Columba, came along. The abbot raised his hand and moved it to make the sign of the cross. Then he commanded the monster to cease and desist, which the terrified beast immediately did. The abbot's spell was a lasting one, for there have been no reports of molestation from the animal again.

A lot of people say they have seen Nessie, who likes to bask in Scotland's thin sunlight and might be mistaken for a floating log. She is estimated to be a fast swimmer, reaching 10 to 15 knots. In recent years Nessie has been sought by divers and even by men riding in a yellow submarine and by people peering out of a submerged capsule. Probes have been made electronically with echo sounders. But Nessie is resistant to discovery; it lives in water where visibility is measured in inches—a perfect habitat for bashful beasts.

Scotland is not the only country that can boast of such a creature. Ireland has a great variety of them. Iceland has its *skirmsl*, a 46-foot lake monster. Lake Victoria in Africa has its gigantic *lau*. Canada's Lake Okanagan allegedly harbors a monster called *ogopogo* by the Indians, and to this day they refuse to cross the lake in certain places.

The national epic of Finland, the *Kalevala*, a legend that is well known to all Finns, explains why the fish do not speak.

There was a time, the legend goes, when no one knew how to speak. The animals had no cries. The birds had no songs. The waters flowed and the winds blew, but they made no sound. Even man made no sound.

One day Vainamoinen, the Master of Song, commanded them all to take for their own the language that suited each best. The wind chose the loud roar and rattle of Vainamoinen's big boots as he mounted to his seat. But the thunder got first choice, so its language is much louder than the wind's, although it never talks for as long a time as the wind does. The river

MAN'S AFFINITY WITH THE SEA

Why Fish Do Not Speak

decided that the rushing swish of Vainamoinen's cloak made a delicious sound. The trees thought the rustle of Vainamoinen's sleeves was best for those who had leaves for lips. The birds found no speech pleasing until Vainamoinen played a little melody on his harp. As for man, he learned all the different sounds that Vainamoinen's harp made and that his garments made as he moved this way or that. And he learned to sing better than the birds themselves.

But while everything on the earth and in the sky had been listening to the great Master of Song and choosing a specific language for itself, the fish had been quite helpless. They knew that something of great importance was going on, but they had no idea what it was. The fish could see all the creatures of earth and sky opening their mouths and shutting them but, being underwater, they couldn't hear a sound. Nevertheless, they made up their minds that they should behave just like the others. That is why you can see the fish opening and closing their mouths and not making a single sound.

Why Fish Have Gills

There are many legends that tell why fish have gills, and a favorite in the islands of the South Pacific concerns a grouper that fell in love with a beautiful girl.

The fish saw her one day as she was weaving on the shore, and he fell in love with her at first sight. Twice he asked her to marry him, and twice the girl refused, and he swam out to the deep waters of the reef and lay there mourning for his unrequited love.

Then he decided that he could not give up so easily, so he swam back into the lagoon right up to the shore where the girl was weaving. He stretched out a fin, snatched the girl, and swallowed her. When the girl realized where she was, she demanded to be let out, but the grouper refused. "I love you," he cried, "and I cannot let you go."

Then the girl had an idea. She still had with her the sharp shells that she used to pattern her tapa cloth. With them, she cut two slits in the fish's body, one on each side, and then she slipped out and swam to shore.

As for the grouper, once he got used to the water rushing in and out of his throat through the cuts, he rather liked the feeling, but he vowed never

The beautiful, but soft and boneless, jellyfish.

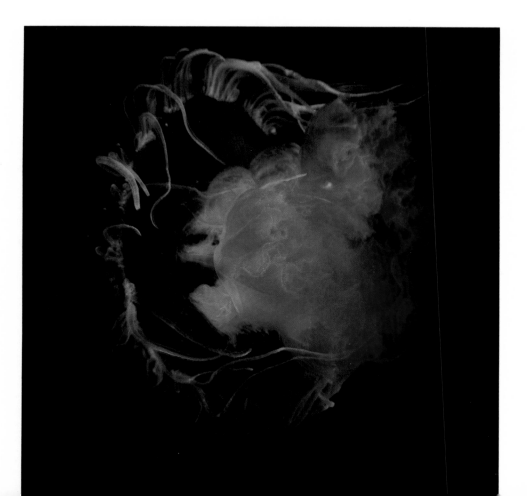

again to fall in love with a human girl. And that is why all fish have gills in their throats today.

Why Jellyfish Have No Shells

From Japan comes a tale that explains why jellyfish have no shells. A long time ago it happened that the Sea King's wife fell ill, and the doctor told him that the only thing that would cure her was the liver of a monkey. One of the few denizens of the sea that was able to walk on land was the jellyfish, and to him was given the task of journeying to Monkey Island and enticing one of its inhabitants to return with him. Obediently, the jellyfish set out. When he landed on the shore of the island, he spied a monkey in a tree and fell into a conversation with him. Soon he was telling the monkey of the splendors of the palace of the King of the Sea, of its trees of white, pink, and red coral, and of the fruits that hung from their branches like great jewels. The monkey was so entranced that he agreed to go with the jellyfish to see these wonders for himself, and he climbed on to his back and they set out over the sea.

Presently, however, the jellyfish asked the monkey if he had brought his liver with him. The monkey was perplexed at such an unusual question and demanded to know why the jellyfish had asked. The jellyfish, suddenly feeling sorry for the monkey, told him everything. The poor monkey was horrified at what he heard and angry at the trick that had been played on him. He trembled with fear at the thought of what was in store for him. But he thought that the wisest plan would be not to let his fear show. He told the jellyfish that he had not brought his liver with him but had left it hanging on the tree where he had been found. So they turned around and swam back to the island.

If it were not for the faithful turtle, the island of San Cristobal in the Solomons would long ago have sunk into the sea. Legend says that when an earthquake occurs the turtle tightly clasps the rock that holds the island up.

Just as soon as they landed on the beach, the monkey scampered up into the branches of the tree and jeered at the jellyfish and admitted that he had deceived him. There was nothing the jellyfish could do now except repent of his stupidity at letting the monkey dupe him and return to the King of the Sea and confess his failure. Sadly and slowly he began to swim back. The last thing he heard as he glided away was the monkey laughing at him.

The king's wrath was great, and he at once gave orders that the jellyfish was to be severely punished. His bones and his shell were taken from him and he was beaten to a pulp with sticks. Then his limp and battered body was carried out beyond the palace gates and he was thrown into the water. Ever since that day, the descendants of the poor jellyfish have been soft and boneless.

The Origin of the Coconut

The South Pacific Turtle of Tamarua legend tells of an 11-year-old princess and the Prince of the Turtles, who saw her bathing one day and fell in love with her. The Turtle Prince wanted to marry her, but the girl insisted that she was still too young. Even when the Prince of the Turtles transformed himself into a handsome young man, the girl refused, for she could not leave her dear old father. The prince wept large turtle tears at hearing this, but he was not angry. In fact, he made the girl a present. He told her that when she got home, it would begin to rain, and it would continue until the waters were up to the doorstep. In the morning, the girl would find a turtle outside the house. She was to take her father's ax and chop off its head. Then she was to bury the head and the body together on the hillside above her father's house.

"Will the turtle be you?" asked the girl. "No," answered the Prince of the Turtles, a bit haltingly. "It will only be one of my messengers." When the girl got home, it began to rain, just as he had promised. And in

the morning there was the turtle, and the girl killed it as she had been instructed and buried it on the hillside.

A few days later a short green shoot sprouted from the grave. Day by day it grew taller and taller, and soon the people of Tamarua could see that it was to be a plant unlike any they had ever seen before.

When the shoot had grown into a tall tree, it flowered and bore fruit, and the girl knew what her gift from the Turtle Prince was to be. It was a coconut—a truly wonderful gift, not only for her, but for everyone in the South Pacific. Ever since that time, the people have eaten its meat and drunk its milk and used its tough leaves for weaving mats, baskets, fans, and thatched roofs.

If, perchance, the reader does not believe in the truth of this tale, he has only to take a look at a coconut to see that its shell is so hard as to be almost unbreakable—just like the shell of a turtle. And the milk inside the coconut is clear and limpid—just like the Turtle Prince's tears.

The Mermaid of the Magdalens

A Canadian legend tells of a young girl taken from her home by a lobster and turned into a mermaid. This happened as a result of a slaughter of sardines by fishermen. The sardines asked the bigger fish to help them. In answer to their appeal, a meeting of all the fish in the sea was called. The big fish took an oath to help their small cousins in their struggle with man and to punish anyone who ate or fished the sardine family.

One day a large ship loaded with packed fish was wrecked on the sunken rocks of the Magdalen Islands, off the northeast coast of Canada. That evening, after the sea had calmed, a young girl walked along the shore to view the wreckage of the broken ship. On the beach she saw one of the boxes of sardines, and she resolved to eat them. She tried to smash the box against a rock, but it would not break. She began to sing a mournful song. Beneath the rock lay a large black lobster, sleeping quietly. The tapping on the rock had awakened him, and he remembered his oath at the great gathering of the fish, and he determined to punish her. He came out of his hiding place and, waving his claw politely, asked if he could open the box. But when the girl held the box out to him, the lobster grabbed her by the wrist with his strong claw and swam with her far out to sea. It is believed that the lobster sold the girl to a merman, and that she is still slowly being changed into a fish.

On the first day of May, however, she always appears on the water looking into a mirror to see if she is closer to becoming a fish than she was the year before. She combs her long hair which is now covered with pearls, and she looks with longing eyes toward her old home. Sometimes on moonlit nights, fishermen hear her strange, plaintive song across the water. When they do, they stay on shore, for they know that she is lonely and that she might seize them and carry them off to be her playmates in her home of bright shells far under the sea.

The Mermaid of the Fairy Tale

In the imagination of Hans Christian Andersen the mermaid became a thoroughly lovable girl—a girl who gave up her life in the sea in the hope of winning the love of a mortal and of gaining an immortal soul.

Andersen's little mermaid was the youngest of six girls who lived in a castle of coral with a roof of mussel shells filled with pearls. They were friends of the fish, who ate from their hands and allowed the girls to pet them.

Many an evening the five older sisters rose arm in arm to the surface of the sea to sing, and they had voices more beautiful than any mortal. Whenever a storm was approaching and they thought a ship might be wrecked, they swam ahead of it and sang to the sailors about how beautiful it was at the bottom of the sea. The little mermaid always felt very sad at being left all alone at home, and she looked as though she were going to cry. But she did not cry, because a mermaid has no tears to shed. So she suffered all the more.

In Herman Melville's Moby Dick *a murderous white whale that no seaman could kill drove Captain Ahab and the crew of the* Pequod *across all the oceans of the world in its pursuit.*

When the little mermaid was grown up enough to go to the surface of the sea, she rescued a prince from a shipwreck and fell desperately in love with him. Back at home she pined for him, and one day she made the dangerous journey beyond the roaring maelstroms to the den of the sea witch. Here she made her fateful bargain. In the hope of winning the prince's love, she agreed to give up her life in the sea. But alas, the prince married another. The poor little mermaid was just about to turn into foam when the kindly spirits of the air rescued her.

THE IMPOSSIBLE MISTRESS

The impossible love of the fisherman for the mermaid is only the legendary illustration of the violent and contradictory feelings man always developed for the sea. Immense, but limited, powerful and vulnerable, prolific but often empty, the sea inspires hope and despair, enthusiasm and rage, peace and terror. She can be beautiful and inviting, but becomes at once terrifying and cruel. She is loved and hated at the same time by most sailors. An impossible mistress for millennia, the sea became our indispensable spouse.

"*Odi et amo* [I hate and I love] may well be the confession of those who consciously or blindly have surrendered their existence to the fascination of the sea," wrote Joseph Conrad.

The Flying Dutchman

The opera has, on occasion, taken the sea as its dominant image. Bizet's *The Pearl Fishers* is a familiar example. Benjamin Britten's *Peter Grimes* is also about a fisherman and his struggle with the elemental forces in nature and his own soul. Better known is Richard Wagner's *The Flying Dutchman*. This work has kept alive a very old legend of the sea.

It tells of the Dutchman who is doomed to sail forever on his ship, the *Flying Dutchman*, until the love of a faithful woman redeems him for his challenge of heaven and hell. Once every seven years he is permitted to go ashore to find that love. During one of these periods, he is driven by a storm to a Norwegian harbor. Here he meets Senta, the woman of his dreams. Senta rejects her former lover to give herself to the Dutchman. Reasoning that if she could be unfaithful to Erik she could also be unfaithful to him, he suspects that his hopes for redemption are again to be shattered. He returns to his ship and, though a storm is raging, sets sail. Senta climbs to the top of a cliff, shouting to the Dutchman that she has always been faithful to him and will be until death. Then she throws herself into the sea. The *Flying Dutchman* immediately vanishes beneath the waves. The Dutchman has after all been redeemed. Embracing, the forms of Senta and the Dutchman rise heavenward.

The Greatest Whale of All

The sea has always been significant in American life, first with its colonies huddled close to the Atlantic shore, and later when the nation pushed south to the Gulf of Mexico and then to the Pacific. It was inevitable that literature in the United States should reflect that proximity.

James Fenimore Cooper invented the sea novel in the United States and Herman Melville brought it to perfection. Cooper is probably best known as a writer of the forest frontier, as in *The Last of the Mohicans*, but he wrote a dozen novels of the sea as well. In *Red Rover, Afloat and Ashore, The Sea Lions, Ned Myers, The Water-Witch*, and others, one can see the beginnings of an attempt to make the novel of the sea a vehicle for meanings as universal as birth and death. The man at sea begins to become representative of all men.

In the years shortly after Melville returned from his wanderings in the South Seas, he published two novels based on his experiences there—*Typee* and its sequel *Omoo*. In 1851 he published what is perhaps the greatest novel of the sea ever written—*Moby Dick*.

Perhaps the most famous of ill-fated ships is the Flying Dutchman, *still supposed to be seen off the Cape of Good Hope in tempestuous weather.*

289

Moby Dick is a great white whale, and the novel tells how the whale is pursued by the crew of the *Pequod* under Captain Ahab until it destroys the ship and only one man survives.

Of Moby Dick himself, Melville writes that "a gentle joyousness—a mighty mildness of repose in swiftness, invested the gliding whale. Not the white bull Jupiter swimming away with ravished Europa clinging to his graceful horns; his lovely, leering eyes sideways intent upon the maid; with smooth bewitching fleetness, rippling straight for the nuptial bower in Crete; not Jove, not that great majesty supreme! did surpass the glorified White Whale as he so divinely swam."

It is a novel that can be taken simply as the story of a single American whaling voyage, in which there are descriptions of the whale, whale burning, and the whole gory process aboard a ship that reduces once-proud whales to barrels of oil in the hold. But it is also a novel that can be regarded as an evocation of the whole human drama. Perhaps most importantly, the novel is one that can help us to understand that life can only be honestly confronted in the loneliness of each human heart and that it is in a relation to the sea that we might best know that loneliness. It also suggests that there is no better place to search for a meaning to existence than in the inscrutable sea, where everything is in eternal flux.

The Chantey

Workers who labor at a common task often need a song to lighten their burden. There was a day when every lumberjack and railroad worker, every worker on the dock and every roustabout at the circus had his own song to sing. And, of course, the sailors had theirs. While hoisting sail or hauling anchor, the leader, the chanteyman, sang the verse and all the sailors joined in the chorus. The rhythm helped them coordinate their efforts. The songs they sang might be about any of the many thoughts that occupy a sailor's mind, and they are often perceptive and humorous.

Oh, a ship was rigg'd, and ready for sea,
And all of her sailors were fishes to be.

Windy weather! Stormy weather!
When the wind blows we're all together.
Blow, ye winds, westerly, gentle southwesterly,
Blow, ye winds, westerly—steady she goes.

Oh, first came the herring, the king of the sea,
He jumped on the poop, "I'll be captain," said he.

The next was a flatfish, they call him the skate,
"If you be the captain, why, sure, I'm the mate."

The next came the hake, as black as a rook,
Says he, "I'm no sailor, I'll ship as the cook."

The next came the shark, with his two rows of teeth:
"Cook, mind the cabbage, and I'll mind the beef."

And then came the codfish, with his chucklehead.
He jumped in the chains, began heaving the lead.

The next came the flounder, as flat as the ground:
"Chucklehead, damn your eyes, mind how you sound."

The next comes the mack'rel, with his striped back,
He jumped to the waist for to board the main tack.

And then came the sprat, the smallest of all,
He jumped on the poop, and cried, "Main topsail haul."

Windy weather! Stormy weather!
When the wind blows we're all together.
Blow, ye winds, westerly, gentle southwesterly,
Blow, ye winds, westerly—steady she goes.

The Sea in Dance

It should not be surprising that the sea, and especially the rhythmical movement of the sea, has been an inspiration to the choreographer. Mary Anthony's *Threnody*, Frederick Ashton's *Ondine*, and John Cranko's *Pineapple Poll* are a few of the story ballets that derive their principal imagery from the sea.

In 1963 Gerald Arpino gave us his *Sea Shadow*, a dramatic pas de deux in which a sprite from the sea visits a dreaming boy on a beach. Attracted as the boy is to the nymph, he soon understands that she will lead him not only to happiness but also to his death. It is one of the most beautiful duets in all ballet.

Perhaps the perfect example of a presentation of the sea in dance is Doris Humphrey's *Water Study*. In this ballet without music, the dancers—12 girls—imitate the undulation of the waves and the movement of the surf onto and from the shore. To feel the sea with one's whole body, as these dancers must, and as the audience is asked to do, is to understand something very fundamental about it in a way that can mean more than all the books one might read.

Man Fishing

"It is enough that we do not have to try to kill the sun or the moon or the stars. It is enough to live on the sea and kill our true brothers." These are the words of Santiago, the fisherman, in Ernest Hemingway's short novel *The Old Man and the Sea*.

The old man has gone for 84 days without catching a fish. He is unlucky, and he has only the youth, Manolin, a boy lovingly described as "already a man" in his knowledge of things of the sea, to give him encouragement. But Santiago sets out alone in his skiff and sails out far beyond the other fishermen. He is alone on the Gulf Stream, until he sees a flight of wild ducks go over. Then he remembers that "no man was ever alone on the sea." Of the two porpoises that frolic near the skiff he says, "They are good. They play and make jokes and love one another. They are our brothers like the flying fish."

The marlin that he catches is both his "brother" and his "friend," and all the qualities he sees in the fish—beauty, nobility, courage, calmness, and endurance—are the qualities he values most.

But both Santiago and the marlin have enemies in the sea. Tied along the side of the boat, the marlin is attacked first by a mako shark, one of the biggest and most dangerous sharks. Then, as the skiff sails toward home, come the shovel-nosed sharks in packs, and they leave nothing but the skeleton of the great fish for the old man to bring to shore.

But he has not been defeated. The old man's triumph is not so much in his endurance in having put up a good fight or even in his acceptance of the loss of the fish but in his understanding of man's kinship with all the other creatures of the world. Santiago is not simply a fisherman. He is Man Fishing.

THE
ADVENTURE
OF
LIFE

CHAPTER
13

YOUTH
OF LIFE

The development of the theory of evolution began with the question, "Where did I come from?" It is understandable that early man believed that he had been "created." How could he have guessed the age of the earth and understood fossils as the history of life recorded in stone? Many still do believe that the earth and life on it have remained essentially unchanged from the beginning. Change is new to man's consciousness. The collapse of old ideas came when Darwin, a religious man, became converted to evolution by evidence he saw in living plants and animals. The newly accepted concept of change both of land and life provided evidence from which we could search out our past and proceed with the detective story of man's origin.

Having realized that we share ancestors with monkeys and apes and that through our vertebrate relatives we are distantly related to starfish, we were able to delve further into ourselves. The birth of human consciousness is a good example. Evolutionary theory states that alterations in a species are solely the result of blind chance—random mutations that are most often lethal. We are told by psychologists that we use only a small fraction of the total potential of our brain. Is it possible that the evolutionary process had selected the most intelligent man-apes to survive but that random chance produced a mind with capabilities that far exceeded the needs of the animal at the time? Could our mind really be a preadaptation—a quality possessed by an animal not essential now but of vital importance to a future state of development? Or could it be another lethal mutation whose effects will doom the species at a future date? Our mistreatment of other men, animals, and our planet may indicate that the faculty we call intelligence and conscious thought is a curse destined to bring about our downfall as a species. Personally, I feel certain our struggle to achieve, progress, and search the unknown is leading us to actualize hidden dimensions of our potential and will ensure a positive future for mankind.

Regressive evolution is seldom a success; the system perpetuates those who move ahead. This is an alarming consideration. Man has stopped his own evolution! Instead of selecting the most fit to survive, we select the environment to suit our needs. Our medical sciences heal the sick, allowing them to reproduce and pass inborn deficiencies to their offspring. We consider ourselves humane but we cannot supplement the forces of evolution by selecting the best of our species. In fact, our intellect may have rendered us incapable of even judging what is good or beneficial for our species. However, it is our understanding of the basic principles of evolution and that same medical science which will permit us to play sorcerer's apprentice and cure the ills of mankind. Through genetic manipulation we can and will revitalize ourselves physically and, if our intelligence is a positive attribute, we will learn to use our mind and increase its capabilities to make life pleasant and worthwhile for everyone. Progress has already been made—an artificial gene has been synthesized.

In defense of evolution, I do not believe that its doctrine negates the existence of a divine being. It merely states that life arose gradually as a result of physical and chemical interactions in a primordial sea. It was recently proved that organic substances, such as hydrocarbons, are being routinely produced in outer space throughout our galaxy. These substances are turned into more complicated molecules that are the building blocks of proteins and of DNA in a catalytic process occurring during the formation of a star. The hypothesis of universal insemination of the cosmos is very promising.

The story of evolution is probably the most fascinating of all we have to tell because it reaches into our very fiber. It ties all forms of life together and provides us with a common bond to plants and animals. Basically we are composed of the same substances and possess the same basic drives, making us all brothers in a cosmic experiment. Astronomers tell us that the earth should continue to exist for another five billion years. Since life has only been around for three billion years, we are still in our youth. The insights gained through the study of evolution will assure us a future and allow us to determine our destiny.

Beach with red shrimp and seals

EVOLUTION OF EVOLUTION

When in 1530 Copernicus wrote in his *Commentariolus* that the sun and not the earth was the center of our universe, man's ego was badly shaken. But Darwin's *On the Origin of Species* (1859) was infinitely more damaging to man's self-image.

Darwin, a most religious man, explained that his observations of natural life made during his voyages on the H.M.S. *Beagle* reluctantly led him to the unbiblical conclusion that no single flashing moment of creation could account for the great variety of living creatures and plants. Darwin expounded the view that all living organisms come from various lower forms through natural selection and the survival of the fittest.

Darwin was not the first to speculate on evolution. However, it was not until the time of the French Revolution, in the early 1790s, that the French philosopher Jean Baptiste Lamarck studied seriously the acquired characteristics of animals.

Lamarck

Lamarck's theory was first published in his *Philosophie zoologique* in 1809. He held that an animal's adaptation to the environment was based on use and disuse of body structures and organs.

It is very easy to conclude that we have no tail, in contrast to our relative the monkey, because we have lost it through disuse. Or that a giraffe has a long neck because he continually stretched and reached for higher leaves causing it to grow longer. Lamarck's belief was that whenever such adaptations arose they were inherited and carried from one generation to the next. This meant that a change in an organism, induced by the environment, was transferred in the genetic makeup of the individual.

An early theory, disproven in 1830, claimed that wild geese began life in the sea as gooseneck barnacles (seen here) which emerged from the water to become adult geese.

These finches are descendants of those that Charles Darwin observed in unusual diversity in the Galápagos Islands; although isolated, they bore a remarkable resemblance to species on the American continent over 600 miles away.

Although Lamarck's theory seems logical, it has never been substantiated experimentally. However, his views were generally accepted and 70 years later were used to combat Darwin's theories.

Darwin's Finches

The scientific views of creation that prevailed in Charles Darwin's nineteenth-century England were based primarily on a narrow interpretation of the Book of Genesis. Men of knowledge interpreted it to mean a miraculous creation of life and pinpointed the year it occurred—4004 B.C. Darwin, as a young man, was not ready to challenge this concept of creation.

In 1831, on the *Beagle*'s second exploratory-scientific expedition, while crossing the Atlantic from England to South America, Darwin read Volume I of Charles Lyell's *Principles of Geology.* In it, Lyell argued that the earth's continents, plains, and mountains were not shaped by the Flood but rather by the action of wind, rain, earthquakes, volcanoes, and other natural forces and that these remodeling processes were continually at work.

In Charles Darwin's view, life was taking on a new meaning and time new dimensions when, in 1835, the *Beagle* headed westward through the Pacific to the Galápagos Islands. Exploring the dry, volcanic islands, he sought to collect one specimen of each species he observed. What amazed him was that almost every species he found was peculiar to each of the islands. Of the 26 species of birds he captured, 13 were finches; "a most singular group of finches," Darwin called them. They all bore striking resemblances to each other yet each was a distinct species. He wrote of the finches that "seeing the gradation and diversity of structure in one small, intimately related group of birds, one might really fancy that from an original paucity of birds in this archipelago, one species had been taken and modified for different ends." In addition, they bore a great resemblance to mainland finches.

The finches gave Darwin insight into what was becoming more and more apparent to him: species were not instantaneously created, but arose from common ancestors.

Darwin arrived back in England in 1836. He published his ideas in two books: *On the Origin of Species* (1859) and *The Descent of Man* (1871). When he began his career, the doctrine of special creation was doubted only by heretics. When he finished, the idea that evolution occurs was accepted by a majority of scientific people.

Though unexceptional in all other ways, the batfish gets its food with the aid of a piece of natural fishing tackle that sticks out from under its head.

SURVIVAL OF THE FITTEST

Divergence

Animals that live in the ocean have evolved what seems to be an endless array of adaptations, enabling them to take advantage of conditions unique to their environment.

Perhaps the most diverse group of animals in the sea are the crustacea; over 25,000 species have been identified. They come in assorted sizes, shapes, and colors and with a variety of life-styles. Barnacles spend their entire adult lives attached firmly to rocks or pilings; only their nauplius larvae are free-swimming. A relative of barnacles, *Sacculina*, is a parasite that infests crabs, sending rootlike processes throughout the host's body.

Many crustaceans are microscopic in size. Copepods are microscopic crustaceans that make up the bulk of the zooplankton in the oceans of the world. Shrimp have adapted to a wide range of habitats, from sandy tidal flats to coral reefs. Crabs are a specific group of crustaceans that exhibit a wide range of evolutionary divergences. The fifth pair of legs of the American blue-claw crab has been modified into paddlelike structures for swimming. Spider crabs, in turn, have long spindly legs. The giant of all crustaceans is the Japanese spider crab. Found in deep waters offshore, it reaches a length of 11 feet from claw to claw.

Some crabs inhabit the cold waters of Alaska, some the deep-sea floor, and others with modified gills live on land. All crustaceans, however, arose from the same primitive stock many millions of years ago and

One of the most beautiful of the starfish, the basket star is a living bush that grows, wilts, and walks.

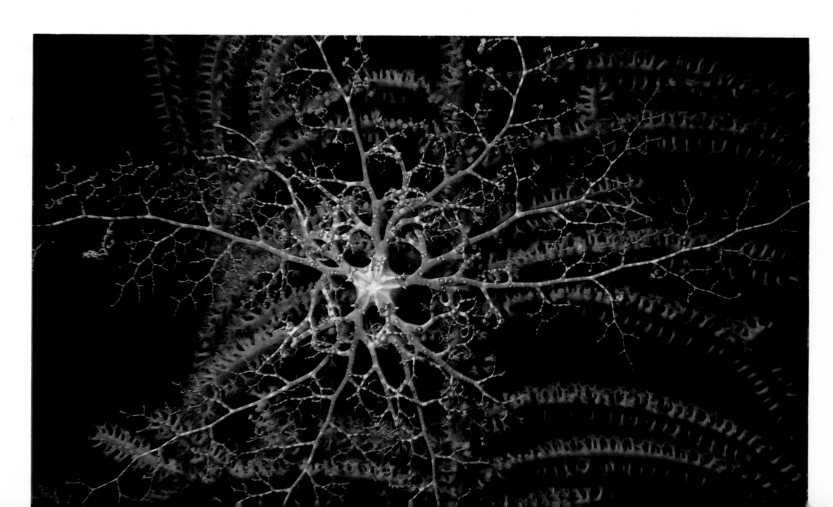

Stalking on spindly legs, the sea spider eats the tissues of small colonial animals, such as certain coelenterates and bryozoans. The starfish nearby is in no danger.

through the phenomenon of divergent evolution have radiated out to fill the innumerable habitats they occupy today.

Convergence

In contrast to the diverse crustacea, we also find cases of animals that are not closely related but that develop under similar conditions and may exhibit superficial similarities. This phenomenon is known as convergent evolution.

A good example of convergence is the similarities between tunas and isurid sharks, which include the makos, mackerel sharks, and great whites. Both have similar color patterns, fin arrangements, and streamlined bodies. Their gills are large; muscles used in locomotion are well developed; and the circulatory system has developed in such a way as to make them warm-blooded. These convergences reflect the way these fish live: both are restless, far-ranging wanderers capable of great bursts of speed.

Evolutionary Experiments

In nature, experiments are constantly carried out, producing a never-ending array of strange, bizarre, and sometimes grotesque creatures. As odd as some animals may seem to us, their features usually represent special capabilities that have enabled them to survive.

A well-documented monstrous invertebrate is the giant squid. One huge sperm whale examined by scientists had clearly identifiable sucker marks on its skin that measured 18 inches in diameter. The researchers estimated the squid that inflicted those wounds must have been about 200 feet long.

Vertebrates in the sea have also been experimented upon by nature, and some strange and odd creatures are the result. Perhaps the largest bony fish in the sea is the ocean sunfish. Covered with a thick leathery skin, it seems to be all head and no body. A specimen caught in the Pacific Ocean off the coast of California measured 11 feet in diameter and weighed one ton. The sunfish occasionally comes to the surface where it basks in the sun.

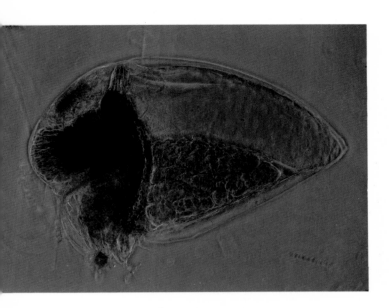

Before a barnacle reaches its mature phase and clings to a rock or piling, it drifts as plankton in this larval form.

The batfish hopping or waddling across the bottom is the ocean's clown. By chance, any structure on the batfish that might look somewhat normal has been altered. The pectoral fins are enlarged to the point where they project forward and outward, looking like a frog's front legs. Underneath the clumsy body, the pelvic fins too serve as walking legs. Still it cannot walk well. When the batfish is hungry, it sends out its tackle, via a sort of pistonlike mechanism, and vibrates it for a while to lure in a naive fish. This vibration is in only one direction; some fish of the same species are "righties," some are "lefties."

The basket star, which resembles a pretty plant until it gets up and walks away on its branches, has a way of feeding that has evolved unlike that of most other starfish. Instead of extending its stomach out over its prey, it carries the food to its stomach with its multibranched arms. Basically it functions as a filter feeder; it extends its arms at night, uncoiling all the tendrils and ensnaring any small animals that can be caught. When food has been trapped, the arm wilts and brings it to the mouth. By day it wilts into a bundle of arms; by night it feeds.

The number of branches on a single basket star can exceed 90,000. They all eventually join toward the central disk into the customary five arms of most starfish. The common Atlantic genus, *Gorgonocephalus*, may be two feet across, from tip to tip; the body is about four inches across.

The body of the pycnogonid, or sea spider, is reduced about as far as it will go—the animal is almost all legs. In fact some of the organs that usually are present in the body reside in the first joints of the legs. The eyes (usually four) are gathered together in one mass on a short stalk. The eggs are carried by the male on two legs especially modified for this purpose.

Not all the sea spiders are as big as the three-foot *Colossendeis giganteis*, but they all seem to share the same general mode of life. Like tall, skinny men, they walk slowly around and now and then stop to suck up a bit of refreshment of small animals with their tubular probosces.

Lingula, which although it resembles a clam with a fleshy tail, is not a mollusc. This small, bivalved mud-flat dweller is the longest lived animal genus, and a modern *Lingula* is virtually identical to those dating from the Cambrian and perhaps the pre-Cambrian periods. This means that *Lingula* is at least 570 million years old! It is a generalized animal that has changed with the changing nature of the sea and has established a life-style that could adapt to almost any event in the earth's dynamic history.

The phytoplankton lithodesmium contributes directly to the chain of food by adding to the surface waters.

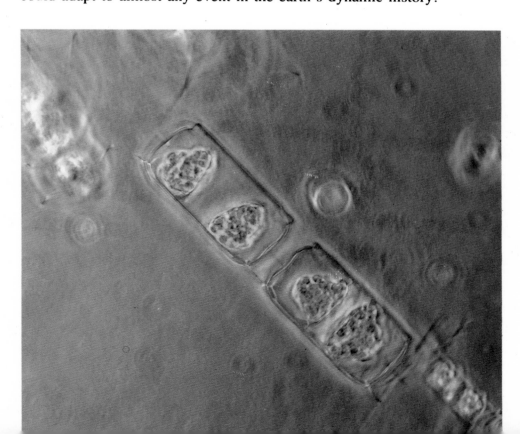

300

A gastropod larva shows traces of its eventual shape.

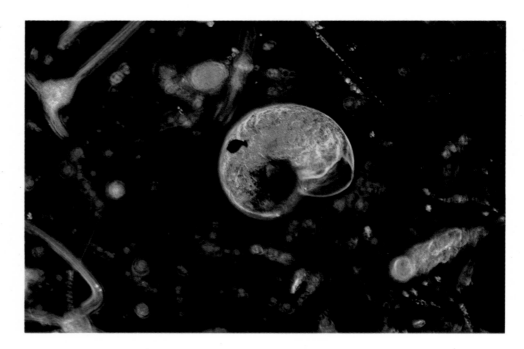

LARVAL LINKS

As animals have developed, adapted, and diverged from one another and have formed new species, they have often retained links with their past. In general, groups of animals can be separated or linked by the characteristics of their larvae, although there are many individual larval adaptations that help to ensure the survival of the individual.

The anemones and jellyfish are a group of simple multicellular animals whose relationship to protozoans can be inferred from their larvae. To make an anemone or jellyfish from a colony of protozoans or planula larva, one side of the ball of cells must simply be pushed inward forming a double layer surrounding a chamber with a single opening. In a jellyfish the outer layer would constitute the ectoderm, and the endoderm would surround the indented digestive cavity, whose opening would be the mouth. The body form of the most simple flatworms (platyhelminthes) suggests that they may share some affinity with this primitive planula larval form.

Another very interesting larval form is the trocophore larva. It is somewhat rounded and in the shape of a compressed diamond. There is a U-shaped gut, which extends from one extreme edge to the most ventral portion of the body. A band of cilia surrounds the larva at the equator, and another region near the anus may also possess cilia. In addition to the gut, trochophores contain a nervous system, sensors, and a kidneylike system. This larval type is seen in a number of the invertebrates, and serves as a basic link between annelids and molluscs.

One of our main reasons for placing echinoderms (starfish and urchins) on the line leading to fish is the similarity of the echinoderm larva to the tornaria larva of the acorn worm, a primitive chordate (an animal with a longitudinal gelatinous chord, called the notochord, which was the predecessor of the vertebral column). The free-swimming larva of echinoderms is very different from the trochophore larva of molluscs and annelids; it is more flattened and has looped bands of cilia for locomotion. Thus the similarity between the starfish and acorn worm larvae leads us to conclude that man is more closely related to starfish than to the highly intelligent octopus.

ORIGIN OF LIFE

When an ancient Chinese poet wrote that Creation was something that came from nothing, he touched on the truth. Who could have dared dream that a loose cloud of stardust and gases could have ultimately produced our earth with its oceans, land, atmosphere, and life? That the whole process

301

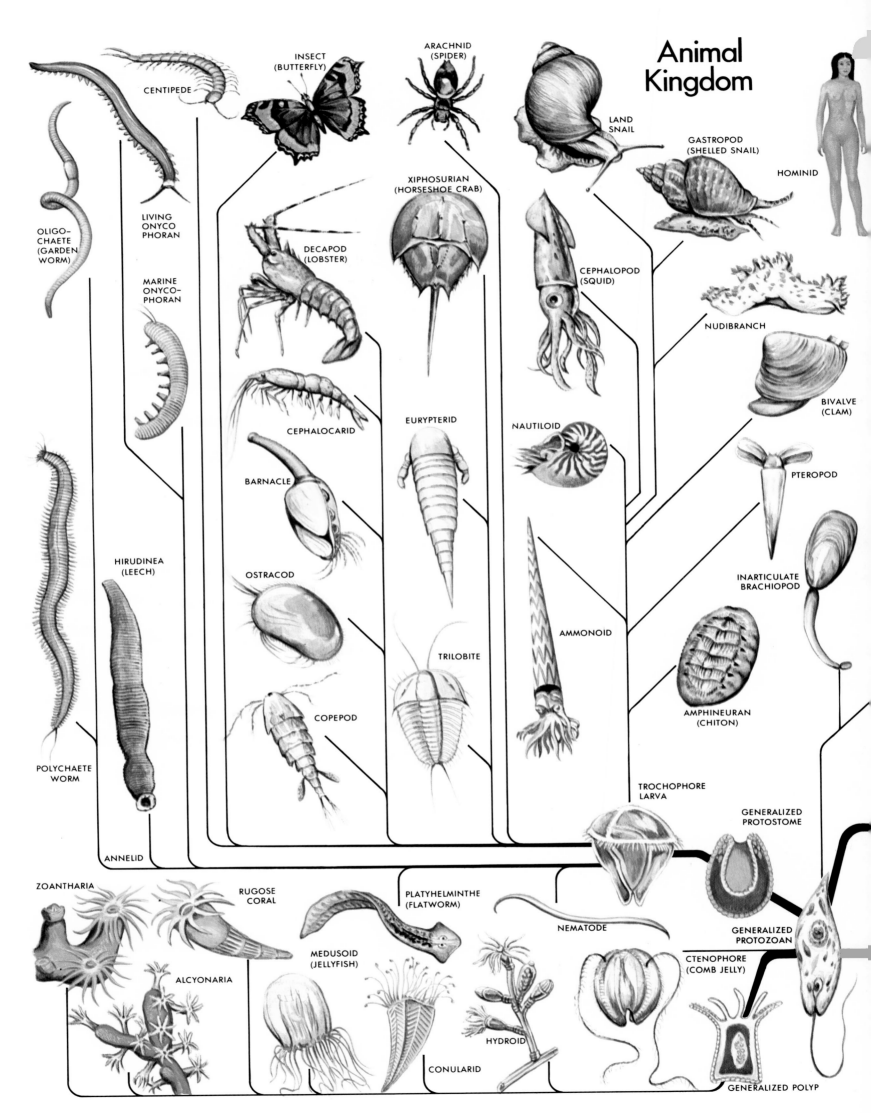

Animal Kingdom

CENTIPEDE

INSECT (BUTTERFLY)

ARACHNID (SPIDER)

LAND SNAIL

GASTROPOD (SHELLED SNAIL)

HOMINID

OLIGO-CHAETE (GARDEN WORM)

LIVING ONYCO PHORAN

MARINE ONYCO-PHORAN

XIPHOSURIAN (HORSESHOE CRAB)

DECAPOD (LOBSTER)

CEPHALOPOD (SQUID)

NUDIBRANCH

BIVALVE (CLAM)

CEPHALOCARID

EURYPTERID

NAUTILOID

PTEROPOD

BARNACLE

HIRUDINEA (LEECH)

OSTRACOD

INARTICULATE BRACHIOPOD

AMMONOID

TRILOBITE

POLYCHAETE WORM

COPEPOD

AMPHINEURAN (CHITON)

TROCHOPHORE LARVA

GENERALIZED PROTOSTOME

ANNELID

ZOANTHARIA

RUGOSE CORAL

PLATYHELMINTHE (FLATWORM)

NEMATODE

GENERALIZED PROTOZOAN

CTENOPHORE (COMB JELLY)

MEDUSOID (JELLYFISH)

ALCYONARIA

HYDROID

CONULARID

GENERALIZED POLYP

302

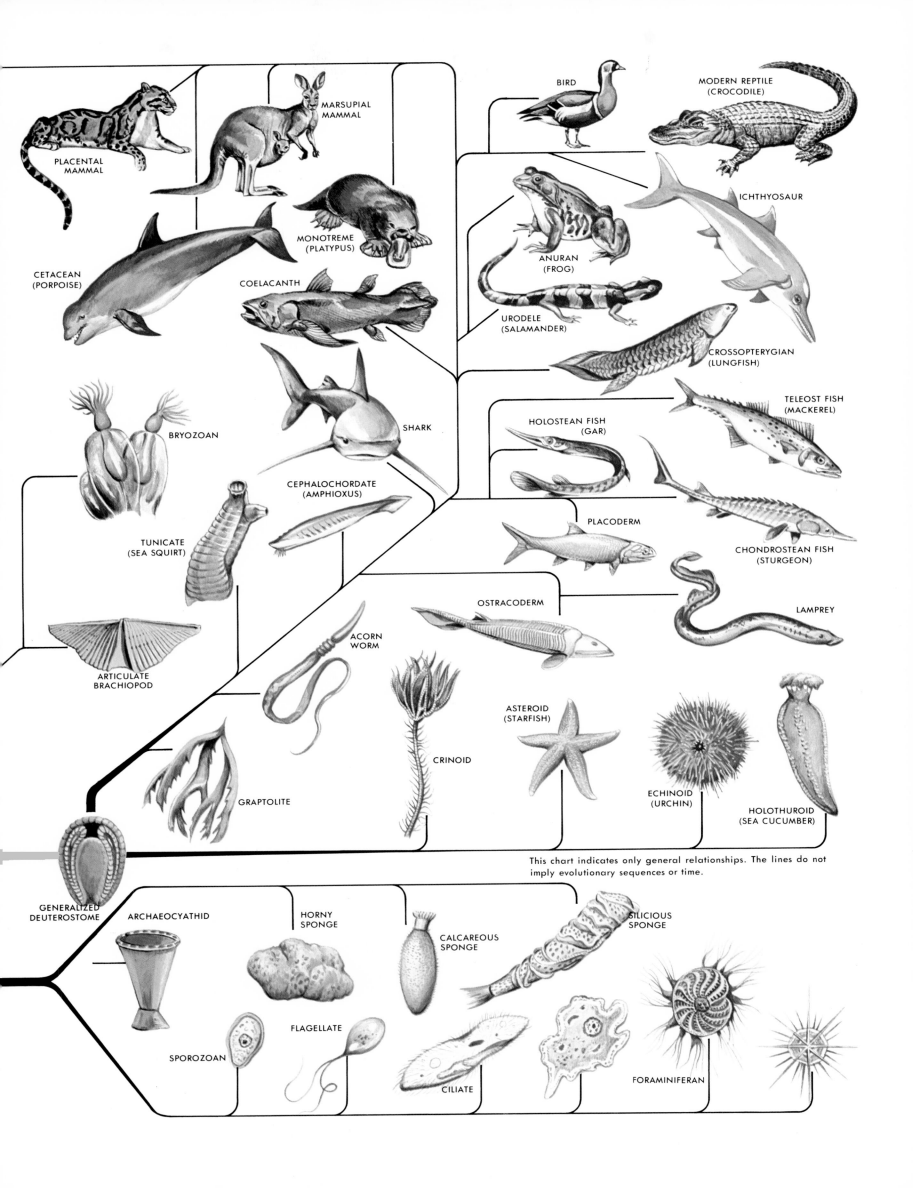

PLACENTAL MAMMAL

MARSUPIAL MAMMAL

MONOTREME (PLATYPUS)

CETACEAN (PORPOISE)

COELACANTH

BIRD

MODERN REPTILE (CROCODILE)

ICHTHYOSAUR

ANURAN (FROG)

URODELE (SALAMANDER)

CROSSOPTERYGIAN (LUNGFISH)

BRYOZOAN

SHARK

TELEOST FISH (MACKEREL)

HOLOSTEAN FISH (GAR)

CEPHALOCHORDATE (AMPHIOXUS)

PLACODERM

CHONDROSTEAN FISH (STURGEON)

TUNICATE (SEA SQUIRT)

OSTRACODERM

LAMPREY

ARTICULATE BRACHIOPOD

ACORN WORM

ASTEROID (STARFISH)

CRINOID

ECHINOID (URCHIN)

HOLOTHUROID (SEA CUCUMBER)

GRAPTOLITE

This chart indicates only general relationships. The lines do not imply evolutionary sequences or time.

GENERALIZED DEUTEROSTOME

ARCHAEOCYATHID

HORNY SPONGE

CALCAREOUS SPONGE

SILICIOUS SPONGE

SPOROZOAN

FLAGELLATE

CILIATE

FORAMINIFERAN

took approximately 4.5 billion years does not diminish the wonder of its happening.

Although every branch of science has its own definitions of life and no one has as yet been able to define precisely its countless functions and varied manifestations, it is generally agreed that life came into existence about 3.5 billion years ago.

Life without water would have been impossible, and it seems that all the essentials for life are universally present in seawater. As the most important component of protoplasm, water accounts for approximately 80 to 90 percent of all living matter. And because water is a universal solvent, it could carry all elements necessary for nurturing life, in suspension or solution, and take them from areas where they were abundant to other areas in the ocean where they were scarce. This gave life a better chance to start on our planet.

Ultraviolet light, if it reached the earth unshielded today, would kill most organisms. Four billion years ago, however, ultraviolet energy established the groundwork for life to begin. These powerful rays split the inorganic molecules that were in the primitive atmosphere. Some of these fragments reunited to form organic (carbon) compounds such as aldehydes and amino acids.

Organic chemicals occurring in the sea had to be presented to each other in the right concentration in order to react. The organic chemicals in the early sea were in low concentrations, and there had to be some mechanism to get them together for life to form.

Certain clay minerals have the unusual property of adsorption—that is, they collect and accumulate substances on their surfaces. This natural phenomenon leads to a second. When oily or fatty substances collect on the right medium, poorly understood ionic forces induce the resulting globs to behave in a rather lifelike manner. These accumulations of organic molecules are called coacervate drops, and they are thought to be an early stage in organic evolution.

The coacervate drops in the primeval sea broth were probably made up of prototype enzymes and complex organic compounds. They contained the basic substances of life. Enzymes are essential to all living things, and they control the chemical reactions taking place within a cell. They are basically proteins to which another substance, such as a vitamin, can attach. As catalysts, they change the rate of chemical reactions without being changed permanently themselves. An enzyme may separate a molecule, or it may bring molecules together by uniting, temporarily, with a specific part of a molecule. After the appropriate chemical reaction has occurred, the unchanged enzyme breaks away, leaving a newly synthesized chemical compound.

A subsequent event vital in the origin of life was the appearance of nucleic acids (DNA and RNA). These chemicals had the ability to reproduce themselves. ATP (adenosine triphosphate), the compound that spreads and stores energy in the cell, was also present. These are the essences of life.

The ingredients for life now pervaded millions of watery places over the primitive earth. Life was finally ready to begin. Or had it already begun?

After millions of years of chemical evolution, the primordial sea soup yielded a living creature—a one-celled entity that could reproduce, take in food, and grow.

The earliest successful forms of life were the autotrophs—organisms that could put together all or some of the organic nutrients they needed from inorganic substances. Modern autotrophs have evolved into forms

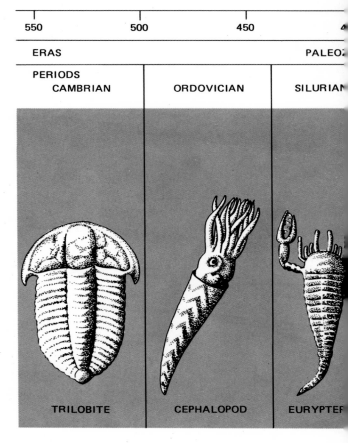

550   500   450

ERAS   PALEOZ

PERIODS
CAMBRIAN   ORDOVICIAN   SILURIAN

TRILOBITE   CEPHALOPOD   EURYPTER

The Cell

304

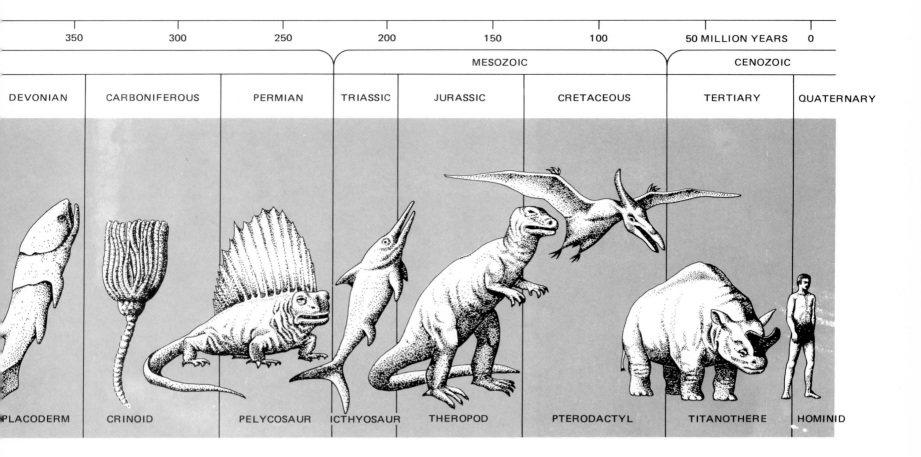

DEVONIAN	CARBONIFEROUS	PERMIAN	TRIASSIC	JURASSIC	CRETACEOUS	TERTIARY	QUATERNARY
PLACODERM	CRINOID	PELYCOSAUR	ICTHYOSAUR	THEROPOD	PTERODACTYL	TITANOTHERE	HOMINID

that can assimilate a whole range of nutrients, including substances that are far from the common notion of food. Still other cells have developed a way to use light from the sun to supply them with the energy to convert inorganic material to organic substances. These cells, called photoautotrophs, all developed a system of light-sensitive pigments with which they could capture solar energy and use it to make food. The process they utilized—photosynthesis, basically the conversion of light energy to chemical energy—was perhaps the most important and crucial step in the continuing evolution of life.

Photosynthesis not only created a new supply of organic nutrients for all heterotrophic organisms, including man, but also liberated free oxygen into the atmosphere. This apparently trivial by-product of photosynthesis, the release of oxygen, changed the evolution of the world and its inhabitants. It not only provided the planet with a protective ozone layer but also led to the evolution of a whole new line of organisms. These creatures had oxidative metabolism; that is, they could utilize oxygen in their energy-producing processes. These organisms had somehow hit upon the food-to-energy conversion system which has become a model for all higher forms of life. It allows a more efficient transformation of food into the free energy needed by the cell to do its mechanical, chemical, and biological work.

In our 4.5 billion-year-old detective story concerning the origin of the world and life on it, we have been left with an incredible number of special clues—fossils, or remains of living organisms preserved in various ways for up to billions of years.

Quick burial is the simplest and most dependable form of fossilization. The "reinforcing" process during which a shell or a bone becomes coated and its pores or canals filled by a solution of calcium carbonate also provides fine fossils. Petrification, one of the most common types of preservation, takes place when part of the shell is replaced by a mineral in solution. An animal's form is preserved when groundwater dissolves an

THE INVERTEBRATE EXPERIMENT

These stony remains are ammonoids, a primitive shellfish form that replaced early cephalopods in the late Paleozoic and Mesozoic eras, then became extinct.

organism's shell but leaves the shell's impression on the enclosing substance. Today tracks, trails, burrows, teeth marks, and excretions of organisms provide evidence of former living things, even when they leave neither impressions nor parts of their actual organisms. Fossils also help date rocks. By knowing the age of one species we can infer the ages of others depending upon whether they lie above or below the known species.

The First Animals

Protozoan means "first animal." Flagellata, one group of protozoans, have varied body forms, but all have flagella—whiplike projections that lash about, providing propulsion for the organism. Most investigators believe the Flagellata were the group from which higher forms of life developed.

The amoebas belong to another group of protozoans—the Sarcodinia. It is believed that the amoeba evolved from flagellates that had abandoned their former mode of life. Sarcodinians have evolved into many specialized types. Foraminifera and radiolarians inhabit the oceans. Their skeletons, made up of calcium and silicon compounds respectively, abound on the ocean floor.

The progression from one-celled to multicellular organisms was a gradual one. The first step upward was probably the aggregation of single cells in a colony of individual cells, but not yet forming a coherent organism.

Sponges are among the most primitive coherent multicellular organisms, but it is universally agreed that sponges are a primitive stage in the evolution of multicellular organisms and that they do not form a part of the main evolutionary line from protozoans to higher forms.

Multicellular Animals

The cells of the simpler metazoans, or multicellular animals, tend to be arranged in three layers: an outer sheet of cells called the ectoderm, a middle layer called the mesoderm, and an inner layer called the endoderm.

In still lower metazoans the mesoderm is absent, and the digestive system has only one opening to the outside. Simple animals that abound in the sea today fit this last description—the Coelenterata or Cnidaria.

Coelenterates are often called flowers of the animal kingdom. They are represented by such common forms as jellyfish, sea anemones, and corals.

A characteristic of the phylum Coelenterata is the presence of nematocysts—stinging cells used in defense and food capture. Because higher metazoan forms lack these structures, some investigators are reluctant to term coelenterates as ancestral metazoans.

Flatworms

Platyhelminthes are soft-bodied, usually much flattened worms. Their organization places them somewhere between the coelenterates and annelids, a highly developed group of worms. The smallest flatworms are microscopic, and the largest are the ribbon tapeworms, which spend most of their time in a vertebrate host and may reach a length of 50 feet or more.

Beginning with the platyhelminthes, most higher groups of organisms are two-sided, in other words bilaterally symmetrical. This led to the differentiation of the front end into a sensory center with concentrated nervous tissue and well-developed sense organs, giving bilateral animals an immediate and apparent advantage over lower forms such as coelenterates.

In place of the jelly that provides much of the coelenterate bulk, flatworms have a solid cellular middle layer—the mesoderm—which includes sets of muscles and a variety of organs. Like the sponges and anemones, flatworms have amazing regeneration capabilities. When cut into a number of pieces, each will develop a head, tail, and full complement of sensory and other organ systems.

Segmented Worms

Annelids are the most highly organized worm forms, and members of this phylum include earthworms and seaworms.

Annelids, in many ways, have expanded and modified the more primitive metazoan body plan as it exists in the coelenterates and platyhelminthes. They have improved the efficiency of the digestive system above that of flatworms by having two openings to the gut: a mouth and anus. And in the marine annelid *Nereis* a set of pincer jaws which can be thrust out aids the

Last of a long lineage that once ruled early seas, the modern nautilus, member of a group that almost disappeared in the Mesozoic era, represents one of the rare reversals in evolution.

307

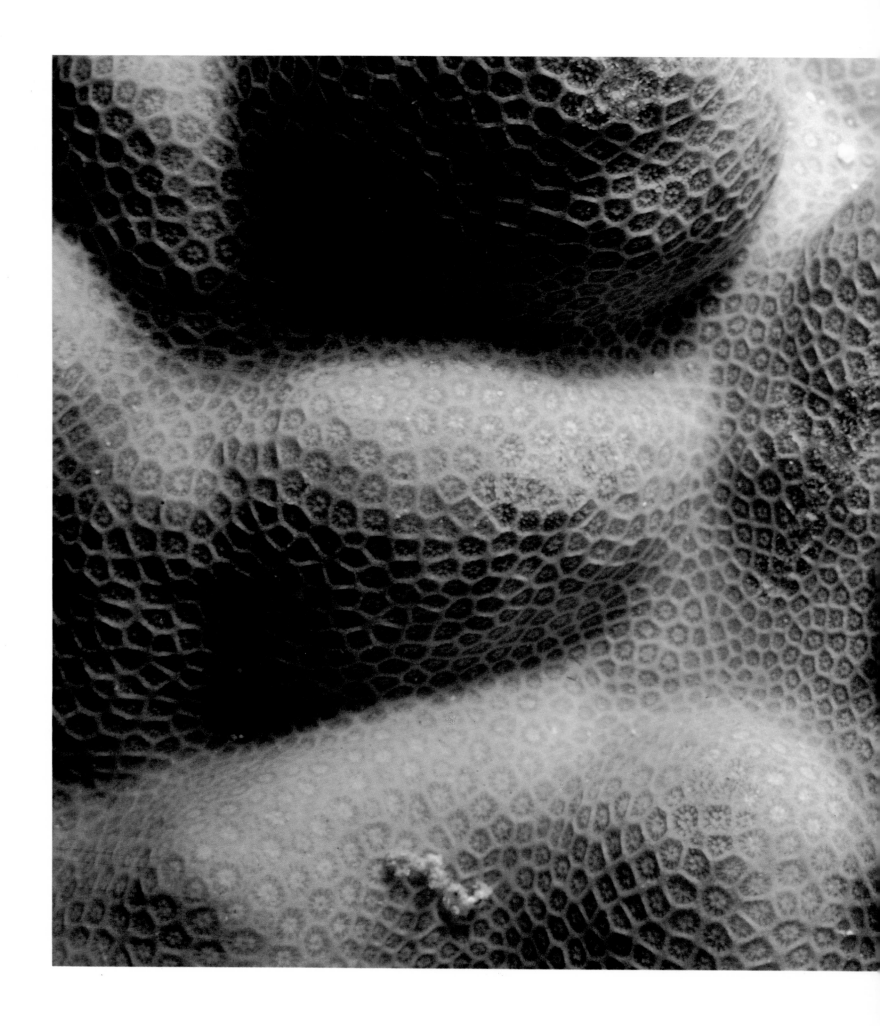

Without obvious defenses this colorful flatworm
seems to advertise its presence to predators. Still, it
has survived a long while, though a poor fossil
record makes tracing its exact history difficult.

worm in capturing food.

In the more advanced annelids the mesoderm is arranged in sheets to form a liquid-filled body cavity called the coelom. The coelom appears in all higher forms and allows for more complexity of muscle development and the evolution of more complex organ systems.

The phylum Annelida derives its name from the Latin "anellus," a ring. Visibile externally are a series of rings encircling the body, each delineating an internal partition dividing the body into segments. *Nereis* has developed, or has begun to develop, almost every type of organ found in more advanced forms: a blood system; kidneylike tubules to eliminate wastes; and a well-defined nervous system with a primitive brain. The annelids are an ancient group, originating about 450 million years ago. Because the fossil record is poor, annelid ancestry is debated. Embryological similarities indicate a strong relationship between flatworms, annelids, and a higher group, the molluscs.

Like some other fossil forms, trilobites have disappeared, leaving no known descendants.

Clams, Snails, and Squid

It is very hard to tell that a mollusc is a mollusc by just looking at it. Molluscs as a phylum include more different-looking creatures than any other group of animals, and any family resemblance between them is often hidden. It is difficult to see that the barely moving blind clam and the graceful squid with its huge eyes and jet-propelled movements are both molluscs.

Every type of mollusc has a two- or three-chambered heart, a kidney, gills, and a well-developed nervous system. Not all of them have complete shells, but most have some form or remnant of shells.

The molluscs are an old family whose history can be traced back more than 500 million years to the Cambrian period. Many of these early cephalopods were often encased in beautiful straight or coil-chambered shells. The first of these known were nautiloids with flat or scooped septa (walls) dividing the individual living chambers of their shells. In later Paleozoic times, the shells developed elaborate dividing walls, perhaps to offer the organism more insulation against the stress of deep dives or protection from predators. Such cephalopods with newly evolved shells were named ammonoids. These organisms were dominant through the end of the age of the dinosaurs (the Mesozoic era) during which there seemed to be very few of the original nautiloids around. For some reason—nobody knows why—all the ammonoids suddenly died out, and only a small group of nautiloids survived, or perhaps became transformed from the complex and seemingly vulnerable ammonoid forms.

Animals with a Past and a Future

If majority rule reigned in the evolutionary world, the arthropods would rule. Arthropod means "joint-legged," and this name well describes the most striking feature of this group of animals. But joint-legged can refer to anything from a lobster's claw to a butterfly's antenna to a spider's leg. Over 80 percent of all living species are arthropods, which should give mankind a gentle hint as to the expendability of our species.

Arthropods are the most ancient well-preserved animals that one finds in the fossil record. And trilobites are the oldest known arthropods. Indeed, the classic definition of the Paleozoic period is "that time interval from the first appearance until the final disappearance of the trilobites."

In spite of their great numbers, this family of organisms has not left a good fossil record except for the trilobites and the sea scorpions. However, there is enough evidence to suspect that most of the varieties of arthropods have been around from the earliest Paleozoic era—about 570 million years ago. This group's additional claim to evolutionary fame rests on the fact that the true scorpions (arachnids) were the first creatures to venture out of the sea on to the land in the Silurian period.

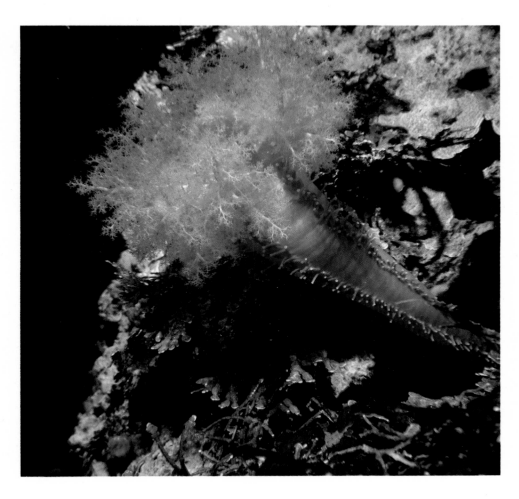

ALMOST A FISH

The routes that evolution takes to accomplish its work are sometimes
devious. Important changes in living creatures, with the most far-reaching
effects, can often begin in some hidden corner of the sea among common,
unspectacular organisms. The development of a cartilaginous structure—
called a notochord—is just such a momentous innovation. It is the most
important evolutionary step in the history of higher animal life.

Fish, birds, whales, and man all have many things in common. But to
a marine biologist, the most striking common feature is that each has a
backbone. Acorn worms, sea lancelets, tunicates, and a few other not
commonly known creatures also have things in common with the vertebrates. These creatures have a notochord present at some stage of their life
cycles, and they are chordates. All vertebrates are chordates, but not all
chordates are vertebrates. For clear communication we lump the
nonvertebrae-bearing forms into a group called protochordates.

Spiny Skin

While being true invertebrates, the echinoderms (meaning "spiny
skin")—the starfish, sea stars, sea cucumbers, sand dollars, crinoids, and
forms that exist only as fossils—hover somewhere along the edges of the
chordate lineage. As we have seen, there are a number of similarities
between the larvae of echinoderms and acorn worms. Starfish and their kin
also have a number of advanced features.

Symmetry means the pattern upon which the body is organized. The
starfish has radial symmetry as an adult, based on multiples of five, but as a
larva it is bilaterally symmetrical.

Echinoderms have a water-vascular system, which is an analog of the
closed chordate internal circulatory system. The other internal structures of
the echinoderms show that these animals are not nearly so primitive as they
seem. They seem to be an odd mixture of primitive and advanced life.

The starfish has almost legendary regenerative powers. Oddly, this is
usually considered a primitive characteristic. But how can one judge?

In an unusual cutaway tank two sea lancelets (Am-
phioxus) are seen protruding from shell gravel.
Amphioxus is an advanced protochordate with typi-
cal chordate gill slits, a notochord, segmented
musculature, and a dorsal hollow nerve end.

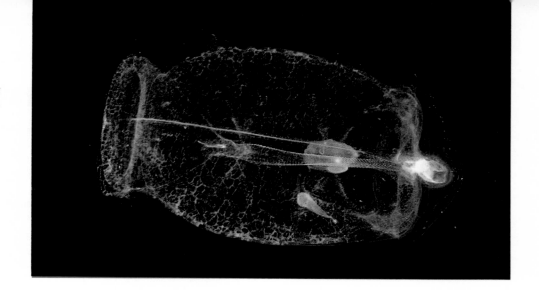

The translucent salp, a primitive chordate, cruises the open sea, forming long chains by budding.

Although the great masses of late Paleozoic-age crinoids, which were at times so numerous that cubic miles of limestone were made up of their skeletons, are gone, the modern crinoids survived. Sea stars and sea cucumbers blanket the deep sea. Our tide pools are full of urchins and starfish. We need not look down on our distant ancestors—they are well suited to their life-style.

Tunicates

In searching out the ancestors of modern vertebrates, we must descend far down the tree of evolution. An important link in the chordate evolutionary chain are the tunicates, or sea squirts, members of the subphylum Urochordata.

Adult tunicates are quite unlike any vertebrates. They are little creatures that sometimes float but most of the time are found attached to rocks or other objects in the shallow water. They look like a lumpless mass of protoplasm held together by a tough, leathery skin, or tunic. They are filter feeders; each has an individual incurrent chamber, and in the colonial

This is the working end of a lamprey, a living Agnatha, showing its jawless, suction-cup mouth full of teeth. Attaching itself to a hapless fish, the lamprey sucks out the body fluids through a hole rasped by its tongue, eventually killing its host before moving on to another.

312

forms they combine their excurrent chambers in a common sewer. Adult tunicates have no brain or spinal cord; in fact they have very little nervous tissue. They also lack a notochord or any skeletal system. How, then, are tunicates related to vertebrates? A gill system is the only chordate characteristic exhibited by the adult tunicate. However, vertebrate relationships become apparent upon examining the tunicate larva.

In many tunicates the larval stage resembles a tadpole in shape, with an enlarged head region and slim tail. A dorsal nerve cord and a well-developed notochord are present, and at this point the free-swimming larva is very similar to *Amphioxus*. Eventually, the head region of the larva attaches to the bottom and the larva gives up its active life. The gill region expands and the nerve cord and notochord disappear, resulting in the adult tunicate body arrangement.

It is not clearly understood which is the more primitive type, the adult tunicate or the larvae. It is reasoned that under favorable conditions some tunicate larvae may have remained in the neotenous state, that is, as juveniles, retaining the tail and motile habits throughout life, instead of settling down to the adult stage. This, some researchers believe, set the stage for the evolution of the chordates, which filled the seas with their kind.

TO HAVE A BACKBONE

In our everyday speech "spineless" denotes a person lacking in strength and resolution. We are vertebrates and can easily recognize the fact that the spine gives shape and support and helps keep us upright.

In most adult vertebrates the notochord is replaced by the bony vertebrae. Some primitive vertebrates retain the notochord (lampreys and hagfish, for example), with only a trace of bony tissue growth. Sharks and their relatives never develop true bony tissue; instead, their skeletons are composed of cartilage. They represent a separate line of evolution from the primitive jawless fish. But all these fish are vertebrates rather than protochordates, because they have a definite development of the vertebral column.

Some of the other characteristics of backboned animals are less obvious. All vertebrates, even men, have internal gills and gill slits. Vertebrates also have a hollow dorsal protective passage made up by the vertebrae; this dorsal passage is topside on a horizontal-swimming form. The position of this cord is very important because it takes many evolutionary steps to move a nerve cord from one side of an animal to the other.

The First Fish

If "fish" means a swimming, cold-blooded creature, then the first vertebrates were fish. Technically they are Agnatha—jawless creatures from the late Ordovician period whose only living direct descendants are the lampreys and hagfish. But in a world where jaws had not evolved, the Agnatha were temporary rulers.

By late Silurian times the armored ostracoderms (the name means "bony skin") were well established and were filling the oceans. Ostracoderms came in many varieties; some had electric cells, some had projecting spines, and some almost looked like modern fish—but none had jaws. They didn't last long, only through the Devonian period, when they were replaced by the placoderms, which had the advantage of jaws.

A Giant Step Forward—Jaws

Fish and animals use jaws to grasp food and other objects, in some cases, as we do hands. It was a great breakthrough when the first rudiments of jaws appeared.

In primitive fish, there are a series of forward-opening V-shaped gill arches, as many as ten, formed by several bones. The first two pairs of V-shaped gill arches were eliminated entirely. The third pair became the

primitive jaws, while the fourth pair in later fish became a supporting structure for the jaws. The remaining gill arches carried on their original function.

A very nice thing happens when one has jaws—teeth have a strong place to attach. The presence of teeth can allow for totally new and more efficient methods of feeding. All fish from the Devonian period onward are jawed, except for the lampreys and hagfish. Today's jawless fishes are parasitic, feeding on the tissues and fluids of living fish.

But with teeth you can attack your prey, and a whole new world of feeding possibilities arose. In the Devonian period, the Age of Fishes, a variety of better-equipped vertebrates evolved, and among them a group of predatory jawed fish arose. They are called arthrodires (part of the placoderms), and king among them was *Dinichthys* (the name means "terrible fish"). *Dinichthys* was armed with strong jaws and bladelike sheets of bone imbedded in those jaws. They may not have been true teeth, but they worked very well. Specimens are known up to 30 feet long, the head making up about a third of the total length. To bring its teeth into best play, *Dinichthys* had a hinged skull; thus the skull moved up as the jaws moved down.

Most placoderms had armor, and they were generally cartilaginous (as opposed to the bony fish, which arose later). The Devonian period was a time of rapid evolution among fish. Forms came and went in relatively short intervals. It was in this period that the lobe-finned crossopterygians dared the land.

The lines of evolution, leading from placoderms to the higher forms of modern fish, are not clear. Somewhere among the primitive jawed fishlike creatures was an ancestor of today's denizens of our oceans. Probably the first thing that comes to the attention of a person who eats a modern fish, such as flounder or cod, is that it has bones.

The most primitive fish managed to get by without jaws but, as this gaping grouper shows, jaws were an important innovation in enabling a fish to capture, kill, and dismember its food.

To Modern Fish

The shark group (Chondrichthyes) has a skeleton of cartilage and is usually thought of as more primitive than the bony fishes (Osteichthyes). Actually both evolved during the Devonian period. The sharks, however, are largely unchanged from the Paleozoic forms. They quickly hit upon some successful formulas for survival: generally live birth; continuous replacement of teeth; swift, aggressive behavior; absence of lungs or swim bladders; and the overall sharklike shape.

The bony fish have a more complicated history. Most remained aquatic, and the group comprising these fish is called the Actinopterygii.

The most primitive forms (Chondrosteans) are characterized by heavy, strong scales and a heterocercal tail (upper lobe longer than the lower). The only living representatives of this ancient group of fish are the sturgeon (a strange living fossil), the birchirs, and a few other rare types.

In the Mesozoic era the holosteans developed from the Chondrosteans. They retained heavy rhombic scales but had a newly evolved swim bladder to replace the protolungs of the Chondrostei. All the members of this group are extinct except for the gars and bowfins.

Finally the Cenozoic era presented us with our whole range of modern fish which belong to the group called teleosts. Teleosts range from the smallest minnows to the giant molas and swordfish. They are probably the most numerous and most successful vertebrates that have ever lived, the culmination of 400 million years of evolutionary history.

Almost fully amphibian, the mudskipper can travel on land and even obtain oxygen from the air.

TO LAND

For 200 million years life flourished in the sea. With the appearance of jaws and fins, primitive vertebrates became active, aggressive animals. The greatest evolutionary challenge still lay ahead—the invasion of land. But before truly land-dwelling animals could survive a life away from water a source of food had to be available—plants. Paleontologists tell us that land plants probably developed from unspecialized forms of green algae. To succeed out of water, plants needed radical alterations. They required a hard outer covering to resist drying; organs to absorb water and nutrients; spores that could be spread by the wind; and a system of support and conducting vessels that could withstand the forces of gravity. A few sea creatures were able to succeed out of water, and once vertebrates became established on land, they adapted to almost every conceivable habitat and swarmed over the face of the earth. Some took to the air and some even returned to the sea.

What evolutionary changes in the past made it possible for animals to leave the water and survive on land? The remains of early amphibians, the first land dwellers, have been found in Devonian and Carboniferous rock deposits 350–400 million years old. The evolutionary process is a slow and gradual one, so surely changes were occurring in fish, making them able to explore land long before the first amphibian appeared.

Lungs were an essential structure for living on land. Early in the evolution of bony fish, the family Dipnoi, or lungfish, emerged. Modern representatives of this group still retain functional lungs; they are truly air-breathing fish.

In order to live on land and to be able to search for food, animals had to have a way of moving about. Obviously, fins were of little use out of water. The family Crossopterygii, or lobe-finned fish, were abundant in the Devonian period and are believed to be the ancestors of the land vertebrates. It was believed crossopterygians became extinct some 70 million years ago, but in 1938 an Indian Ocean fisherman caught a coelacanth, a member of that family, thought to have been extinct since the age of the dinosaurs.

Modern fish such as walking catfish, mudskippers, and desert pupfish can live out of water for stretches at a time and are able to move on dry land. Environmental factors that influenced the evolution of these creatures were probably the same ones that caused the first fish to leave the water millions of years ago: pools of water dry up, food becomes scarce, and the fish must find another place to live or it will die.

Coelacanth

The coelacanths' thick, fleshy fins have a simple skeleton but exhibit the basic pattern from which land limbs were derived. At the insertion of the fin is a single bone, two bones are present in the next segment of the fin; and beyond this the bones branch irregularly. This is structurally similar to a human arm, with one bone from shoulder to elbow, two from elbow to wrist, and then a series of bones in the wrist and hand. Bony patterns in the coelacanth skull and braincase also show close relationships to similar structures found in fossils of primitive amphibians.

A few coelacanths have been caught since 1938, all of them around the Comoro Islands—and we have learned more about them. We know that they are strong, heavy-bodied fish that feed on other animals. They reach a maximum weight of about 160 pounds. Though they were fished at night in depths between 650 and 2000 feet, nobody knows where and how they live. Their front and rear fins have been modified into stalked flippers, which probably allow them to walk or creep over the bottom. They have what can be described as a pseudo-lung—lunglike in structure but lined with a thick layer of fat that makes gas transfer difficult. The heart is a very simple structure; the intestine is very similar to a shark's; and the backbone is not bone at all, but rather a cartilaginous rod. Millions of years ago, an ancestor of the present-day coelacanth moved onto land and gave rise to amphibians and ultimately to all land vertebrates.

The coelacanth was thought to have become extinct 70 million years ago, until, to the astonishment of scientists, a living specimen was found near Madagascar in 1938.

OUTER
AND
INNER
SPACE

CHAPTER
14

OUTER
AND INNER
SPACE

Long before the birth of Christ, astronomers contemplated the sky and began to decipher the mysteries of the universe. But the oceans were virtually ignored. The bulk of what we know today about the sea was accumulated in little more than the last 100 years.

The marine sciences first progressed very slowly, by guesswork. The moving shapes observed through the water, the coincidence between tides and phases of the moon, the animals caught in nets and traps, the whirlpools, the reefs, and the currents—all suggested the fantastic rather than the rational. Some of the myths about the sea, born in the Dark Ages and the Middle Ages, are still alive today and occasionally appear as headlines in our newspapers.

Then came the great navigators. Thanks to them, global maps could be drawn, and the prevailing winds and currents around the planet began to be outlined. With the development of the sextant and chronometer, astronomical navigation became the first link between outer space and the sea. Serious scientific exploration of the sea began with the famed voyage of the *Challenger* in 1872. The main research tool, from the time of the pioneers until today, remains the cable! A line, whether of hemp, steel, nylon, or polypropylene, is reeled and unreeled at each "station," lowering and bringing back thermometers, water and bottom samplers, sediment cores, nets and traps, as well as sensors. The picture of the ocean world that one could conjure up from such pinpoint measurements is often compared with what Martians might learn about the earth if they were to lower grabs from a flying saucer and bring back a snail, a half-burned cigar, and a sample of polluted air from the exhaust of a car.

Echo sounding, submersibles, aqualung divers, and cameras surpassed the cables as research tools, revolutionized undersea exploration. But their measurements and observations are still analytical and prove insufficient to give us an understanding of the general laws governing the oceans and of what the sea means for the planet and its life. Our microinstruments cannot quantify all that goes on in this huge three-dimensional environment.

The first attempt to move from analytical to synthetic knowledge in marine sciences concerned the dynamics of ocean masses. At each "station" of a network extending over large provinces of the ocean, temperature and salinity were measured at many conventional depths, and from this data, computations indicated at what level the surface would be if the ocean were entirely made of homogeneous "standard" seawater. Assuming that such ideal water masses would move from high levels to low levels, it was possible to obtain a simplified idea of important displacements of oceanic masses. Today this method is supplemented by large, costly, accurate models of the oceans, studied on revolving benches simulating the influences of the earth's rotation and of the coastline and bottom topography.

Large-scale investigations of the sea require synchronizing the measurements of many ships in a given area: the Intergovernmental Oceanographic Commission has organized such endeavors in the Indian Ocean and the Caribbean. To study the exchanges of energy between ocean and atmosphere in tropical seas, dozens of ships and aircraft, helped by satellites, make studies of all the factors of evaporation in the areas where hurricanes are formed, hoping to indicate how man can better predict and defuse such disastrous tropical storms.

Daily global measurements are badly needed to assess biological primary production and the pollution of open-sea water. "Remote sensing" from high-flying airplanes has recently been developed. But it is inconceivable to maintain aircraft surveys in all oceans. Efforts are underway to develop instruments, of the same nature as those tested in airplanes, that could be used by satellites or skylabs of tomorrow. Such satellites would also gather the conventional deep-water data collected by thousands of instrumental buoys anchored in deep water all over the oceans. In the near future, satellites will carry the bulk of oceanographic study; outer space technology is essential to our comprehension of inner space.

H.M.S. Challenger, *a sailing vessel used for ocean study*

PIONEERS IN OCEAN SCIENCE

Most peoples of the ancient world left little record of what they saw or knew, and we can only guess at their awareness of geography and of the seas around them. Only from the Greeks and the Egyptians, and only from about 1000 B.C., do we have indications of the origins of the marine sciences. We must assume, however, that these records are very incomplete and may be misleading.

At the beginning of the Christian era, the geographer Strabo is said to have measured the Mediterranean to a depth of more than a mile at one point, but no one knows how he did it. In the second century A.D. Ptolemy made a map of the world that gave the circumference of the earth at 18,000 miles. His map was the standard for a very long time. In Rome the naturalist Pliny studied the sea and its life, through secondhand sources, and compiled his great work *Historia naturalis*.

Then darkness fell. For a thousand years classical learning was ignored, and the geography of the world was interpreted by reference to the scripture. The earth was flat again. Yet through these long centuries of blindness there continued to be daring voyages on the Atlantic, most notably by the Vikings, who traveled to the Faroes, Iceland, Greenland, and America. Their discoveries remained largely unknown to the rest of the world.

Laying the Groundwork

By the end of the seventeenth century, the groundwork had begun to be laid for a truly scientific study of the sea. The English physicist and philosopher Sir Isaac Newton had advanced an explanation of the tides. In 1776 the French chemist Antoine Laurent Lavoisier published an analysis of seawater. In England Robert Hooke, who discovered that living creatures were made of cells, delivered a number of lectures on methods that might be used for deep-sea research.

The pioneering scientists generally concerned themselves with one characteristic of the sea, and it was most often an aspect, such as salinity, that could be explored in the laboratory. But these scientists laid a necessary foundation for those who would later seek to understand the interdependences of these discoveries and to make new ones through close observation of the sea itself.

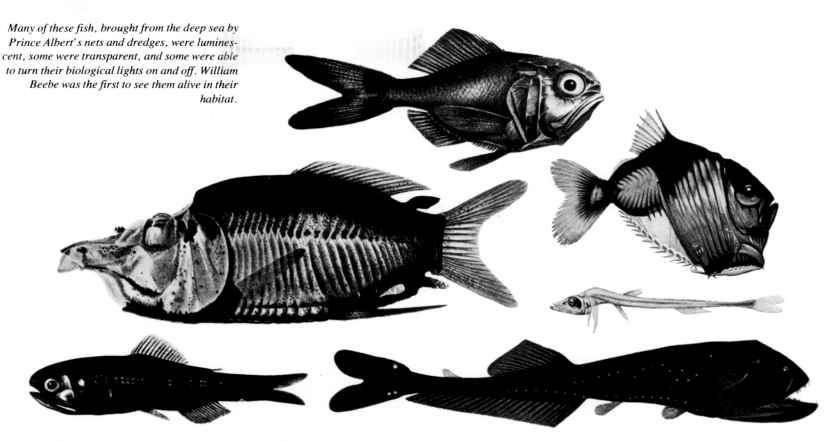

Challenger

In 1871, Wyville Thomson, a Scottish naturalist, was asked to make a trip around the world as director of an expedition to study the deep seas. An old sailing vessel with the appropriate name of *Challenger* was bought for the expedition. Thomson was ordered to find out "everything about the sea."

In its 68,890 miles *Challenger's* team established the main contour lines of the ocean basins and the first systematic plot of currents and temperatures in the sea, and demonstrated that the temperature of deep water in each zone was fairly constant in all seasons. In all, they discovered 715 new genera and 4417 species of living things, demonstrating that the ocean was teeming with life of all kinds. They proved, as well, that life existed at great depths. The findings of the expedition were so extensive that it took 50 volumes to describe them. The science of oceanography had been born.

A Royal Patron

A man who devoted the greater part of his fortune and energies to the study of the oceans and who furthered and encouraged the work of others was Prince Albert I of Monaco.

In 1885 Albert acquired a ship called the *Hirondelle* and began a series

In the natural history workroom aboard H.M.S. Challenger, biological specimens could be examined in detail.

of cruises in the Mediterranean and the North Atlantic that were to make a great contribution to ocean studies. He was early concerned with the question of life at the middle depths, and to this problem he brought new and unprecedented equipment and techniques. Among them were the invention of a high-speed trawl and the examination of the regurgitation of dead whales.

On the *Hirondelle,* Prince Albert made studies of the Gulf Stream. He released a cargo of floats, each of which carried a message in ten languages. Of the 1675 bottles set afloat, 227 were recovered, and their position seemed to indicate the existence both of a clockwise gyre in the North Atlantic and of a current (the North Atlantic Drift) that branched northward from it. The prince's experiment had given considerable weight to those who had argued for a transatlantic Gulf Stream.

Prince Albert's energies were by no means confined to the study of currents. Within the next 25 years he had devoted his attention to biological oceanography, bathymetry, marine meteorology, and education. In 1910 the Oceanographic Museum planned and built by Prince Albert was dedicated at Monaco and in 1911 he established the Oceanographic Institute in Paris, which had professorships in marine biology, physiology of marine life, and physical oceanography.

The conning tower of a nuclear-powered submarine pierces the ice pack like a giant monument on the vast emptiness of the frozen polar sea.

Beebe

The development of ocean sciences in the United States was greatly stimulated between the two world wars by the work of William Beebe. According to one writer, Beebe "was responsible, through his explorations, writings, and lectures, for creating public awareness of the existence of a field of scientific endeavor called oceanography." Beebe's celebrated bathysphere, probably more of an engineering feat than a scientific triumph, focused attention on exploration of the deep sea and encouraged the development of nonmilitary minisubmarines dedicated to underwater research.

The bathysphere made its first descent in 1930 with Beebe and Otis Barton, who had designed the chamber, aboard. *Half-Mile Down* is the record of the work of the bathysphere and its two explorers. One thing that interested Beebe about the descents were the lights and the colors. In a chapter called "A Descent into Perpetual Night," he describes the first time the bathysphere touched bottom: "At 11:12 A.M. we came to rest gently at 3000 feet, and I knew that this was my ultimate floor, the cable on the winch was very near its end. A few days ago the water had appeared blacker at 2500 feet than could be imagined, yet now to this same imagination it seemed to show as blacker than black. It seemed as if all future nights in the upper world must be considered only relative degrees of twilight. I could never again use the word BLACK with any conviction."

But that black world was not without its light. At 2000 feet there were consistently a number of bioluminescent organisms within Beebe's view. A number of previously unidentified fish swam by with large cheek lights. There were hatchetfish, anglerfish, jellyfish, eels, and even snail-like animals, with fleshy extensions called pteropods, used to fly through the water—and they all glowed with their own colored lights. He was the first man to witness the deep-sea shrimps which, instead of discharging ink like the octopus to confuse a predator, extrude liquid fire, a luminescent material. He was overwhelmed by the living fireworks display to the point of using up all the dramatic adjectives he knew, leaving him, after a time, unable to express adequately what he was seeing. The record of his narrative at times becomes dull and lifeless; he would return to the deck of the ship speechless, as though in a trance somewhere between fancy and reality. He concludes that the only other place comparable to these marvelous regions must surely be naked space itself, between the stars far beyond the atmosphere.

After World War II technological advances gave a tremendous boost to oceanography. Since the war there has been an increasing trend toward international cooperation in large-scale oceanographic efforts. No single country has monopolized study of the sea, and in recent years even the most important explorations have not been the private preserve of the larger nations. Furthermore, the great expeditions have not concentrated on any one aspect of the sea, but have usually concerned themselves with the acquisition of data in all areas of oceanography.

The International Geophysical Year of 1957–58 added considerably to our understanding of all the earth sciences. In those two years, some 5000 scientists from more than 40 countries made a concerted, worldwide study of the environment, including some significant studies in oceanography and polar sciences.

Robert E. Peary had gone over the ice to the North Pole in 1902. In 1958 man went beneath the ice to the pole. On August 3 of that year, after two failures, the nuclear submarine *Nautilus* of the U.S. Navy passed directly beneath the ice at the North Pole.

This historic voyage began in the Hawaiian Islands. The *Nautilus,* named after Jules Verne's fictional ship in *Twenty Thousand Leagues Under the Sea,* then went through the Bering Sea and Strait into the Chukchi Sea. While going through the narrow Bering Strait, the *Nautilus* was sandwiched between ice to the extent that 43 feet of water lay beneath her keel and the ocean bottom, while only 25 feet separated the ship's conning tower from the ice sheet above. Soon after the submarine entered the Barrow Sea Valley, near the northernmost point of Alaska, it passed under the polar

RECENT VOYAGES OF DISCOVERY

IGY—Cooperation for the Sake of Science

The New Northwest Passage

pack ice. It was 1100 miles under the ice to the pole—which took four days—and another 800 miles to the open sea near Spitsbergen.

The objectives of this epic cruise to the North Pole were many and varied. They were to take depth soundings and get water samples whenever possible. They were to observe arctic currents, measure air and water temperatures, attempt radio communications with the outside world, study ice pack formations, measure light penetration through the ice, and study the physiological effect of the arctic expedition upon the crew of the submarine.

Shortly before reaching the pole, they passed over the Lomonosov Ridge, a 9000-foot mountain range whose existence the Russian scientist Mikhail Lomonosov had first predicted on the basis of geophysical studies of the earth's crust. The temperature at the pole was 32.4°F., and the depth was 2235 fathoms, deeper than had been reported by Ivan Papanin in 1937 and by Admiral Peary in 1909. The ice beneath the pole extended 25 feet beneath the surface. The undersea mountains were described by Commander Anderson as "phenomenally rugged and as grotesque as the craters of the moon."

In 1959, another U.S. Navy nuclear submarine, the *Skate,* became the first ship in history to break through the polar ice cap to the surface.

One of the most ambitious projects in the history of ocean research involved a ship named after the *Challenger.* The *Glomar Challenger* was designed by the Global Marine Company and its activities are supported by the National Science Foundation. It began its work in 1968 as part of the Deep Sea Drilling Project (DSDP).

The *Glomar Challenger* is a scientific tool especially designed for deep-sea drilling. The vessel is 400 feet long and has a 142-foot drilling tower. Among its more interesting innovations is a system of dynamic positioning, which permits the ship to remain practically motionless over a

Bubbling white foam fore and aft mark the locations of side thrusters, part of the Glomar Challenger's *dynamic positioning system that helps keep the ship in place over a drilling site.*

The Drilling Ship

drill site thousands of feet below. A sonar beacon is placed on the ocean floor; its signals are received on board and analyzed by a computer which controls the main propellers and the side thrusters. The side thrusters consist of propellers in circular tunnels that lead from one side of the ship to the other, enabling the ship to move in any direction from forward to sideways. The ship is stabilized to lessen the roll, further reducing the strain on the drill pipe and its operation. The drilling is done through a large hole in the center of the ship. The *Glomar Challenger* is outfitted with the most complete geological laboratory that has ever gone to sea.

One of the remarkable discoveries made by the DSDP concerned continental drift. The theory of continental drift maintains that a single supercontinent broke up about 200 million years ago. Drilling by the *Glomar Challenger* has revealed shallow-water-deposited sediments and even indications of dry-land conditions that are overlaid with typical deep-sea sediments indicating that when the supercontinent broke up, fragments of continental material that were in the breaking zone slowly sank beneath the sea. The oldest ocean sediments recovered date back 140 million years.

Drilling results have also confirmed a general increase in the age of the ocean floor from the area of crustal generation at the midocean ridges to the destruction zone in the deep-sea trenches, strongly supporting other geophysical data interpreted in terms of sea-floor spreading and continental drift.

In the Mediterranean, drillings indicated that that sea may have completely dried up about 12 million years ago. This was suggested by the mass of salt deposits and related sediments as well as by the various species of marine animals and plants that were found. Scientists have long puzzled over features on the cliffs descending to the bottom of the Mediterranean that suggest erosion by water. It is now conceivable that great waterfalls cascaded from Europe down to the Mediterranean basin thousands of feet below. Imagine the scene created by the rising sea level at the end of an ice age when water began to pour over the Strait of Gibraltar and refill the Mediterranean.

In the *Glomar Challenger*'s first voyage to the Antarctic Sea it was discovered that Australia did indeed break away from the polar continent some 50 million years ago and has been drifting northward at a rate of a few inches every year.

Violent storms that originate at sea can cause great damage to coastal structures such as this cliffside swimming pool, and massive erosion of the shoreline itself.

Recently the United States National Park Service gave up a 40-year effort to maintain artificial barriers against waves and storms. The agency had tried to prevent beach erosion by building giant dunes at the Cape Hatteras National Seashore in North Carolina. They finally discovered that their efforts were doing more harm than good. The artificial dunes had resulted in considerable damage to beaches and had caused an ecological disruption because the force of waves was abruptly blocked by the artificial structures rather than being allowed to dissipate itself in a long sweep across the beaches. It was just one more example of what man has been learning about the sea since he first tried to build on its edge—the sea will sooner or later have its way.

Waves can be extremely destructive. One of their most spectacular targets has always been the lighthouse. At the mouth of the Columbia River there is a lighthouse that stands several miles out at sea, atop a rock whose nearly vertical walls rise to a ragged surface about 90 feet above the mean low water mark. The entire rock shudders whenever there is a severe storm, and fragments that are torn from the base of the cliff are tossed on top of the rock. Once during an especially severe storm a rock weighing 135 pounds was thrown higher than the light, which is 139 feet above the sea, breaking a hole 20 feet square in the roof of the lightkeeper's house. Another time there was trouble with the foghorn, which is 95 feet above water. When the lightkeeper went to investigate, he found the foghorn was filled with small rocks.

The Aleutian earthquake of 1946 (many seismic sea waves originate off Alaska) created waves 55 feet high on the shores of Hawaii, where 150 persons lost their lives and great damage was done to property.

Now that we have a seismic-wave-warning system for the Pacific, accurate predictions of the arrival time of such waves can be made and much loss of life and property prevented. An underwater earthquake occurring anywhere in the Pacific will now set off alarms in the Hawaiian Islands, the west coast of North America, the Fiji Islands, New Zealand, and every other nation, island, and territory using the warning system. The earthquake and the resultant waves themselves are felt by underwater pressure sensors, and the first sea wave arrival is recorded on coastal tide gauges, as sea level recedes many feet before the first wave hits.

There are four major kinds of structures designed to combat waves: jetties, breakwaters, seawalls, and dikes. Jetties, usually in pairs, extend into the ocean to confine the flow of water in a narrow zone. A breakwater is built well out from shore to provide a substantial area of quiet water. A seawall separates land from water at the shoreline. A dike is essentially an impermeable breakwater that functions like a dam.

Each year, in the United States alone, tiny animals eating their way into ships and waterfront structures cause about half a billion dollars' worth of damage. They are the borers and are found everywhere, from Spitsbergen to Tierra del Fuego, and have been a menace to man ever since he had anything to do with the sea.

There are two main types of borers—shipworms and gribbles. The shipworm is not a worm at all, but a relative of the clam. The gribble is a crustacean, and it causes the greatest damage to waterfront structures. Unlike the shipworm, which excavates a tunnel in which to make its home, the gribble does not remain in its tunnel but comes and goes. A heavy infestation of them can eat away more than half an inch of wood in a year.

Chemical preservatives are used to treat wood that might be exposed to the borers. Since different borers react differently to different preservatives, considerable research is being done to improve chemicals. Synthetic resins may render the wood less easily penetrable; metal and sheathing is only efficient if it has no breach, no discontinuity. Still no methods are foolproof.

THE DESTRUCTIVE FORCES OF THE SEA

The Power of Waves

Like a piece of abstract art, this unusual photograph shows driftwood that has been bored by shipworms.

The Battle Against the Borers

CLIMATE AND THE OCEAN

Probably the most unusual looking vessel for oceanographic research is FLIP, a 355-foot-long Floating Instrument Platform operated by the Scripps Institution of Oceanography. In its working attitude, most of the vessel is submerged, leaving the bow (shown here) high and dry; the bow contains research laboratories, living quarters and, of course, the engine room. Because its long stern section is flooded, in its vertical position FLIP remains unaffected by motion at the sea's surface, making it an excellent platform for carrying out undersea experiments.

In all parts of the world, climates are shaped by ocean currents and other great movements of water. Water has an exceptional capacity both to store heat and to transport it from one place to another. Some of the principal currents that influence climate in various parts of the world are the Benguela Current, bringing cold water up the west side of Africa and keeping a long strip of that coast relatively cool and foggy; the Peru Current, bringing water of antarctic origin almost to the equator; and the Gulf Stream, with its tempering effect on the climate of northwest Europe.

Our knowledge of such currents comes to us not so much through the use of current meters and drift bottles as through the study of salinities, oxygen content, plankton, and temperatures. From such physical observations we can learn where a particular current comes from and where it is going.

Some of the instruments for measuring temperature can be overwhelming: the "thermistor chain" used by Woods Hole Oceanographic Institution was an enormous reel on a ship's fantail, storing hundreds of feet of streamlined chain, housing hundreds of thermistor sensors, all connected electrically to recorders on deck. Once the chain was unreeled and hanging under the hull, the ship traveling at speeds up to eight knots recorded simultaneously in great detail and with great accuracy the water temperature at all levels for thousands of miles and as deep as 600 feet and even deeper. Such an abundance of data can only be processed by computers.

NIMBUS A giant step in applied meteorology came with NIMBUS, the weather satellite program of the National Aeronautics and Space Administration. NIMBUS has already had a wide variety of assignments—making the first vertical temperature readings from space through clouds, monitoring a mysterious disappearing current off the west coast of South America, and thermally mapping the earth's surface.

With the instruments that NIMBUS carries, it can penetrate cloud cover, measure temperature precisely, identify heavy-rain clouds, and also differentiate between old and new ice in the arctic and antarctic regions, another big help to weather experts. NIMBUS has revealed, in addition, that the polar ice caps have boundaries that are not accurately drawn on world atlases. The pack lines at both poles are not smooth around the ice edge as shown in the atlases, but consist of a great many indentations.

The weather satellite has also been measuring rainfall over the oceans on a daily basis. Before NIMBUS, weathermen had no adequate way to monitor ocean rainfall on a global scale. Knowledge of its rate and extent can give us a good idea of how much heat energy is being released into the atmosphere. In turn, knowing rainfall rate and the amount of heat released will greatly aid in reaching the goal of long-range weather forecasting and improving short-term forecasts of hurricanes so that lives and property along coasts can be protected from the onslaught of storms.

VIBRATIONS FOR SCIENCE

A data buoy that drifts in constantly moving arctic ice can operate effectively for about 15 months, transmitting information about the harsh environment.

The story of underwater sound had a humble beginning in a freshwater lake. In the early nineteenth century two scientists submerged a church bell in Lake Geneva and measured the time it took for the sound to cross the lake.

Acoustic techniques have become one of the most efficient and versatile tools for scientific study of the oceans. For accurate measurements, the speed of sound, as well as the physical characteristics of seawater, must be known with precision. Echo sounding was a revolution. It did away with the time-consuming inconvenience of cable sounding. With sonic depth finders the ship need never stop.

The speed of sound in seawater is about four-and-one-half times its speed in the air. It is dependent on the temperature, salinity, and pressure of the water and increases with increased salinity or increased depth (pressure).

A refraction or a reflection of sound leads to the creation of shadow zones—where sound is greatly reduced and which can provide a protective shield for submarines that want to avoid detection—and sound channels (or rather, layers) where sound is concentrated and travels extremely long distances. There are often two of these layers, the lower one being the most efficient. An explosive charge detonated in the SOFAR (Sound Fixing and Ranging) channel, through which sound can be detected from anywhere in the world, has traveled as far as 15,500 miles.

Other distinctive layers are the deep scattering layers (DSL), sound-reflecting layers caused by the presence of marine organisms rising toward the surface at night and descending at sunrise.

Measuring of the ocean depths by dropping a line to the bottom and hauling it back up began as early as 2000 B.C. with the Egyptians. Today, it is possible to make acoustic soundings for detailed global maps of the sea floor with an accuracy of a few inches, if the velocity of sound has been acoustically determined and the tides are well known. The main inaccuracies in such maps originate in the fact that the ship's position is not as accurately measured. Satellite navigation systems do not provide sufficient accuracy. Very expensive radio-navigation systems are today capable of locating the ship within yards.

Echo sounding is also used to help marine geologists identify undersea rocks and help locate important mineral deposits. An underwater explosion is detonated to generate a series of sound waves that fan out to the ocean bottom and to the rocks beneath the sediment layer. The echoes vary as the rocks that reflect them vary. Sound waves travel faster through some rocks than through others. Thus, by comparing the acoustic properties of known rocks with those of the sea bottom, geologists can identify the rocks that make up the sea floor.

The echo sounder has proved of considerable value, too, in locating schools of fish, especially herring, at depths to more than 500 feet. The echo sounder not only tells the depth at which the school is traveling but also gives information on its size, density, and upper and lower limits.

Probably the most valuable technological advances in oceanography that have come in the recent years have been in the development of various remote sensors, by which we can study certain phenomena without needing a human presence on the spot. The greatest advantage of remote sensing is the fact that man is expensive to place on site and maintain, and he is not expendable. Equipment can be lost with no great consequence. A satellite that fails to achieve orbit and burns up in the atmosphere can be replaced. An astronaut who died in the accident would be irreplaceable.

Acquisition of data by instruments dangled from a ship is an expensive procedure largely because of the cost in ship time. If the instruments are suspended from buoys moored at the surface of the sea, data collection is cheaper and much long-term data can be acquired from an array of them.

REMOTE SENSING

Buoys

A newly installed rubber sonar dome bulges from the bow of a U.S. Navy frigate. Sonar is a great aid in the detection of submarines and other objects below the surface.

The ERTS satellite makes about 14 revolutions around the earth each day and covers the whole globe every 18 days as it travels 570 miles above the earth in an orbit near to the poles and synchronous with the sun.

Sensors attached to surface buoys record weather conditions, surface currents, and waves among other things. Subsurface buoys permit accurate measurements of currents, temperature and salinity, sound velocity, and so forth.

The Naval Oceanographic-Meteorological Automatic Device (NOMAD) can remain unmanned on station for months at a time, relaying data to shore facilities. The 11-ton, aluminum-hulled buoy is equipped with a complex of electronic instruments in its floating framework.

Buoys will probably play an increasingly important role in ocean research. There is too much variability in the sea to generalize from a few isolated spots about what is happening on a global scale. However, problems arise when instruments are left unguarded. They become prime objects for scavengers, or they are mistaken for food by predators, or they are considered valuable salvage or dangerous weapons by humans who come across them at sea.

In 1966 a tethered unmanned submersible called CURV (Cable Controlled Underwater Recovery Vehicle) helped in the search off Palomares, Spain, for a hydrogen bomb lost at sea after a refueling accident involving two United States Air Force planes. It was CURV that managed to grab the bomb with its mechanical arm and bring it to the surface. This was just one of the more spectacular jobs that this kind of submersible has been doing lately. Its normal mission is to recover spent torpedoes and missiles. CURV sees with television and sonar, moves with propellers, and wields a mechanical arm that it can shed if it must.

Robots on Leashes

Photographs from Space

Of our many devices for remote sensing, photographic equipment developed to give us pictures from satellites orbiting the earth may prove one of the most valuable.

Photographs taken in space have already revealed where sediments and pollution discharged by rivers drift into the sea. They have been used in locating shoals and reefs, studying geological features, measuring heat flow from the sea surface, and interpreting surface detail to reveal the local effect of wind and tide and the transportation of sediments. They show the daily meandering of the Gulf Stream, its beautiful blue being distinguishable from the neighboring, more productive green waters.

Not long ago a disastrous drought hit parts of Africa. From a satellite scanning the earth, chromatographic analysis of "photographs" revealed the possibility that there were hitherto unknown sources of underground water in the area.

The satellite that made the discovery was ERTS—the Earth Resources Technology Satellite. Of the countless discoveries that ERTS has made since it was launched in July 1972 this is one of the more spectacular.

Among other functions, ERTS is providing valuable information about ocean currents, sediment distribution near shoreline areas, and large upwelling areas. These will be mapped and monitored to determine their relationship to fisheries. ERTS is also monitoring pollution of coastal areas of the oceans as well as inland seas and waterways. Parts of the world that have never been mapped are being accurately described for the first time. Ocean dumping and surface pollutant films are being plotted around New York, including acid-iron wastes, sewage sludge, and suspended solids. It has monitored ice floes in the arctic and detected plankton in the Atlantic.

A major objective of the program is to outline the first comprehensive inventory of the earth for future reference. The *Calypso* and its crew were able to participate in this important project by collecting oceanographic data in the antarctic when our course crossed the path of the satellite. Among other things, we measured the chlorophyll content of the summer antarctic waters and then beamed our information to a satellite above, which relayed it to NASA headquarters in the United States.

OVERLEAF: Different wavelengths reveal different characteristics of sea and land as shown in this infrared photograph of San Francisco Bay taken by the ERTS satellite from 50,000 feet.

(Below) A NASA photograph shows a section of the New England coast with Block Island and the Connecticut coast at bottom and the Boston area at the upper right.

(Below right) This is a color composite photograph of the same area from the ERTS-1 spacecraft. Healthy crops, trees, and other green plants are shown as bright red; very young or diseased vegetation appears as light pink; barren land, cities, and industrial areas show as green or gray, and clear water is completely black.

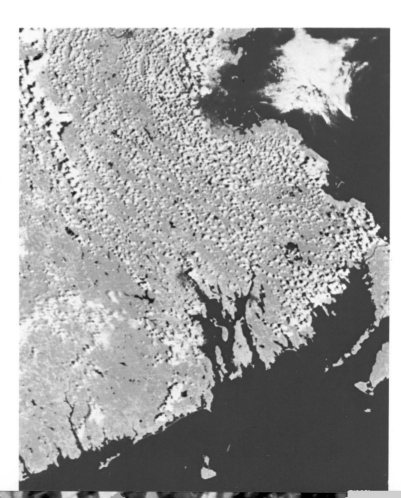

NO SUBSTITUTE FOR MAN

Man cannot leave the study of the sea to automated mechanical or electronic equipment. Dials, computer readouts, and cathode-ray tubes can tell the scientist much of what conditions lie beyond his range, but personal experiences often lead to intuitive insights that may solve perplexing questions or open up totally new fields of thought about the sea. In addition, man in the sea can detect subtleties and make quick decisions that no technology will ever surpass.

Tools of the Archaeologist

Ships have sunk every year since man first floated down the river on a raft. Entire cities have vanished beneath the waves. The possibilities of undersea archaeology are enormous, and some remarkable discoveries have been made as new tools have been developed. Sonar, both side-scanning and sediment-penetrating, underwater television, core samplers, magnetometers, metal detectors, and submersibles—all are valuable, but none so much as the relatively simple development of free-diving itself, allowing the archaeological worker unencumbered access to the object of investigation.

A land archaeologist needs a shovel, but for the worker underwater it is the air lift, first used in our 1953–54 excavation of the Greek wreck at Grand Congloué Island near Marseilles. Since then it has been used on virtually every major undersea excavation. It is a suction hose, a vertical pipe or tube that acts like a vacuum cleaner. As air, pumped through a hose from a compressor on the surface, enters the tube near the bottom, it rises toward the surface and expands as the water pressure decreases, causing suction at the

Man's ability to perceive, decipher, and understand is an indispensable partner with technology. In underwater exploration, the most important factor is human curiosity about the ocean world.

Submersibles have increased human ability to study the soil of the ocean floor and marine life, and even to reclaim wrecks from the ocean depths. The diving Saucer, the two-man submersible used in Cousteau Society expeditions, aids the divers in an archaeological "dig" off the coast of Greece.

mouth of the tube. This suction pulls in water, as well as sediments and other material small enough to enter the tube, and clears the wreck site.

Our work on the Grand Congloué wreck demonstrated that it was physically possible for free divers, if the time they spent underwater was strictly controlled, to work thousands of man-hours with little or no danger of decompression illnesses.

Probably most important for the future of underwater archaeology is the use of saturation diving and undersea habitats. The day is not far off when archaeologists will live for weeks at a time beneath the sea. But whatever technological advances are made, archaeological digging operations are not aimed so much at bringing back objects from the bottom as they are at investigating the relative position of each component to allow a reconstruction of the site or the ship. Reconstructions can lead us to fascinating discoveries about past cultures.

Tektite II   William High has as part of his responsibilities the well-being of the sablefish fishery in the northwest United States. He was part of the Tektite II mission in Lameshur Bay of the Virgin Islands, and his job was to discover what he could about the behavior of reef fish in relation to traps and other fishing gear. He made some remarkable discoveries.

Most astounding, probably, was that bait in a trap meant little or nothing to a fish. For centuries the fishermen of the Virgin Islands, like fishermen everywhere, have been baiting their traps. High's observations showed that the fish came in out of curiosity or out of apparent desire to occupy a new space, whether or not there was bait in the trap.

High also noticed that a fish caught in a trap seemed to act as a visual stimulus to other fish to come in. Eventually, when a critical point of crowding was reached (the experiment was done with squirrelfish), they began to make frantic movements in a rapid darting fashion, which High believes was to warn others to stay away.

Some of the observations made by the scientists of the Tektite II mission, of whom High was just one, were made while sitting inside the habitat and letting television do the work. They soon found, however, that television can do nowhere near what the diver can do. For instance, the diver was able to track the fish over long distances—something television could not do. Furthermore, television has night blindness. A diver can see far better in the dark. Most important, however, are the myriad of incidental observations that the man on the spot can make and that television ignores.

The two-man submersible *Nekton* recently went oil prospecting in Glover's and the Barrier reefs off the coast of British Honduras. Samplings from these areas have proved positive—oil must be there.

Nekton's scientists spent 120 hours underwater to detail the profile of the reefs and measure the depth of reef-building organisms.

The thousand pounds of rock samples they brought up proved relatively youthful—aged from 5000 to 10,000 years. They all came from reefs that at one time had been living coral. As the seas rose at the end of the last ice age, the corals died. Their stony skeletons were smashed by storms and tumbled down the reef front where they were impacted with calcium from marine plants and other sediments to form a solid limestone. The seaward reefs hardened to nonporous stone rapidly, while their counterparts in lagoons remained porous.

This exploration of living reefs has given oil geologists a better understanding of the ancient reefs now found on land in such places as the Swiss Alps and Texas. Today about half the petroleum found on land is located in porous limestone. Drilling is generally abandoned when nonporous rock is struck. The findings of the *Nekton* scientists may lead to the resumption of oil prospecting in locations that were previously thought to be worthless.

FOR THE SAKE OF KNOWLEDGE

Man as a species has progressed to this point only because of his ability to keep written records. The wheel does not have to be reinvented every few generations. A young scientist can rely on the work of the past, on basic principles that need not be proved again; he can pick up where his predecessors left off. Today science proceeds at a rate limited only by the number of men and computers that can be put to work.

As basic research advances, applied scientists are a few steps behind, trying to do the job of converting the data supplied by pure researchers into practical applications for the advance of civilization. It is our duty to future generations of mankind to continue faithfully to nurture and replenish this storehouse of knowledge, never letting it run dry.

More and more, pure research is contributing to the practical aspects of scientific development. Only by experimenting with extremely cold liquid gases have we found that as some substances approach absolute zero they offer no resistance to an electrical current—they become superconductors. When applied to its fullest, this principle may revolutionize the field of power transmission and electronics, on which we are increasingly dependent. Experimenting with quantum mechanics and substances that can emit light led to the development of lasers, which are becoming more and more useful daily. Studies on the nervous system of a large nudibranch are providing us with insights into the basic functioning of brain cells. Routine measurements of water conditions in the Red Sea led to the discovery of one of the world's greatest concentration of minerals. Biochemists working on obscure marine animals are daily finding sources for potentially miraculous drugs. These are examples of pure research that have found practical applications.

Understanding the spaceship-planet we live on should be as much a goal of mankind as living in space and traveling to distant stars. And the oceans are the bulk of the living part of our earth. There are no greater mystery tales than the accounts of scientists in pursuit of elusive explanations for what occurs in the world around us. Moments of intellectual insight that result in exciting discoveries are to be cherished.

The ocean world does not relinquish its secrets with ease but demands diligence and daring, a sincere dedication to science, and an unflagging fascination for this last frontier.

BALANCE
AND
THE WORLD

Life in nature is, in essence, a struggle. A struggle for food, for space, for safety, for perpetuation of the species. But also, a struggle against adverse conditions—heat, cold, salinity, drought, wind, mud, and dust. The plants and trees in a forest compete for air, water, and soil as fiercely as wolves, hawks, or sharks fight to maintain the pattern of their existence. There are two sets of conditions—the scarcity of water and the rigor of cold—that, where they hold sway, have almost obliterated all forms of life. These "desert" realms may be yellow with sand and rock as the Mojave, Sahara, or Gobi, or white with ice as the polar caps of the arctic and the antarctic. In either case, they constitute the forbidding limits beyond which the provinces of death begin.

The North and South poles, the top and bottom of the earth, are as different as an ocean and a continent can be, but the waters bordering the lands are surprisingly similar. Because water needs a lot of calories to warm up, a lot of cold to cool off; because it is a much better heat conductor than air or earth; because water requires such huge quantities of heat to freeze or to evaporate, the ocean is the great climatic moderator of our planet. While the air temperature may reach $+136°F$. in the Sahara and drop to $-126°F$. in Antarctica, the temperature of the seas rarely exceeds $+80°F$. or drops to $+30°F$. Paradoxically, warm-blooded creatures, such as mammals and birds, that must maintain a fixed central body temperature, are better equipped to resist the extremes of temperature—thanks to fat, fur, and vascular controls—than cold-blooded animals, which would literally freeze if water temperature were to drop to as little as, say, $+25°F$. In both arctic and antarctic oceans the water temperature remains fairly close to life's edge, and some of the antarctic fish are believed to discharge a sort of "antifreeze" protein into their blood to avoid being turned into blocks of ice.

Another major paradox of the polar seas is that, so close to universal death, marine life is several times more plentiful than in any of the temperate or tropical areas of the ocean! This is because nutrients are carried to the polar zones by deep-ocean bottom currents and back to the surface by upwellings. In these frigid but nutrient-rich waters, all the sun's energy during the uninterrupted summer day is turned into vegetable plankton; crustaceans thrive on these glacial meadows and become abundant food for larger animals. Unfortunately, the prodigious tonnage of living organisms produced in both frigid seas is distributed among very few species, perhaps because the variety of ecological niches is so limited. The polar pyramids of life are high and thin, extremely vulnerable, ready to collapse if seriously disturbed.

As a matter of fact, the delicate web of life in the boreal and austral seas has already been endangered, partly because in such remote reaches the crimes of man remained without witness. In the arctic, the last true penguin (Auk) was killed in 1948; the polar bear, the narwhal, and the sea otter are endangered; the populations of belugas, walrus, arctic fox, and wolves have been decimated. In the antarctic, only six percent of the whale population is left; the Ross seal is on the verge of extinction. Knowing the destruction man has wrought in the past few decades, I was unable to reconstruct what the first polar explorers must have witnessed. I sat in an Eskimo umiak, in the middle of the Bering Strait, imagining the mass northward migrations of walrus 300 years ago, before they fell prey to frustrated whalers. And in the antarctic, I often flew in *Calypso*'s helicopter in search of the few remaining whales and thinly scattered seals, while, below, millions of the smaller penguins feasted on the unused whale food. The splendor of the poles today is but a dim image of what they were before human intrusion. In the north as in the south, the dazzling landscapes are already stained by the sinister silhouettes of oil prospecting derricks.

We have done so much harm to our planet already . . . now we want to extend our destruction to those areas that are least capable of accommodating change. How much longer before we make a truce with our world, so that the ends of the earth will remain what they are?

Crabeater seal

THE POLAR SETTINGS

North of the Lands

Six million square miles of ice and cold. The ice is bright, but not crystal clear. Unending opaque whiteness curves off into the horizon in the polar regions of the earth. The similarities between the northern and southern polar areas are obvious, but so are their differences.

The arctic world is a sea world—the north end of the globe is covered by an ocean dotted with masses of floating sea ice. There is no land to mark the spot of the North Pole. But surprisingly this region is not as cold as the interior of landmasses to the south (Alaska and Siberia)—even in the arctic the ocean tempers the weather.

Traditionally the arctic has been defined as that part of the globe above 66°33′03″ north latitude. But perhaps a better definition of arctic would be that northern area where no trees grow. In some places the timberline moves well above the Arctic Circle and skirts 75° north latitude, while in other places it reaches as far as 56°, the latitude of Copenhagen, Denmark.

The arctic is hard to describe in other than general terms. It is marked by frequent high winds; long, cold winters and short, cool summers; low precipitation (lower than many desert areas of the world); and permafrost, the layers of subsoil that remain frozen throughout the year.

Water is the primary feature of the arctic landscape. Flying toward the North Pole in summer, one first sees frozen land and then the sea ice on its border. A few miles out, the dark ocean waters make their first appearance, gradually becoming lighter green studded with pieces of ice called brash, which are splinters of the mammoth floes. Beneath the sea is an ocean basin as warped and rifted as any on the face of the earth. The major feature is the submerged Lomonosov Ridge, which divides the arctic floor into the Amerasia Basin on the Pacific side and the Eurasia Basin on the Atlantic side.

The water flow of the Arctic Ocean is part of the worldwide oceanic system, with about 60 percent of the outflow spilling into the Atlantic between Spitsbergen and Greenland. Because of land barriers and the earth's rotation, there is almost no flow from the Arctic into the Pacific. However, about 35 percent of the Arctic's inflow comes from the Pacific. Most of the other inflow into the Arctic Ocean comes through the Norwegian Sea.

Unlike the antarctic region, the arctic has been settled. The peoples that populated the subarctic region were primarily of Mongoloid stock—North American Eskimos, various Siberian tribesmen, and the European Lapps, who are believed to be partially Mongoloid.

South of the Oceans

The Arctic is an ocean surrounded by continents, but Antarctica is a continent surrounded by ocean. Therefore the antarctic is subject to the extremes of continental weather. The lowest temperatures on earth are recorded in Antarctica—including one of -126.5°F.

The southernmost continent is roughly circular in shape. Palmer Peninsula extends up toward South America, well out of the Antarctic Circle, and is punctuated with mountains, which are an extension of the Andes. Though ice and snow are the dominant features of the landscape, Antarctica has coastal cliffs that are scoured by wind and are too steep to retain snow. In some places, the sun of summer—November to March in the Southern Hemisphere—raises temperatures above freezing, and areas with thin snow cover become exposed. Like its northern counterpart it has alternating six-month days and nights.

Just as the borders of the arctic region are difficult to pinpoint, so are the limits of the antarctic. Astronomers have designated a theoretical line at 66°33'03" south latitude as the Antarctic Circle. A better boundary might be the Antarctic Convergence, the point where the colder and saltier waters flowing north drop below the warmer waters of the Atlantic, Pacific, and Indian oceans. This boundary, which runs roughly between 60° and 50°

Surrounded by the only waters that completely encircle the earth lies the continent of Antarctica. Scientists disagree about the name for the ocean that is formed by the circling waters—some prefer Southern Ocean, some Antarctic Ocean—but except for the tip of South America, there are nearly 1000 miles of open water separating Antarctica from the nearest landmass. The Antarctic Circle is a theoretical boundary for the area, while the Convergence and Divergence represent physical and biological limits for the waters. The cutaway shows the incredible thickness of antarctic ice and the effect this enormous weight has had in pressing some land surfaces of Antarctica well below sea level.

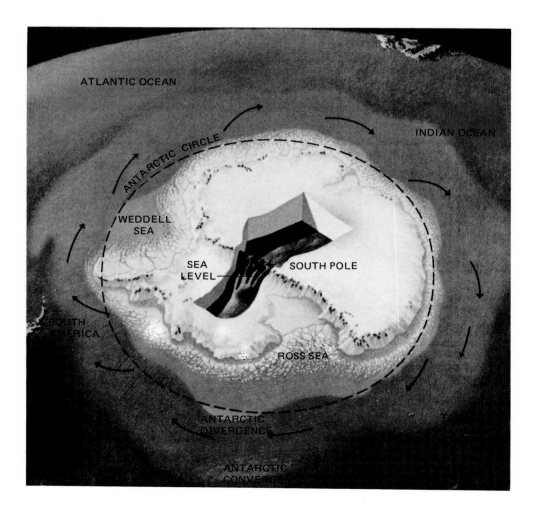

According to most scientists, the rounded, pancake shape of some types of young sea ice results from jostling in the sea. Pancake ice also shows turned-up edges that come from impact among pieces. Calypso crew members observed pancake ice forming within a half-hour after the water's surface began to freeze.

south latitude, comes within 300 miles of Tierra del Fuego. It is an irregularly shaped circle that marks a sharp distinction in surface temperatures of the water and in the life-forms found in the upper reaches of the sea.

Perhaps the most spectacular frontier of the antarctic region is the almost impenetrable ring of ice, hundreds of miles wide, that girds the continent tightly in winter.

Much is still unknown about Antarctica. That it influences weather throughout the world is accepted, but there is still speculation on how much and why. We know there is land beneath most of that ice, but its extent and topography are still being determined. It appears that there may be a land area, East Antarctica, and a series of islands in the west, similar to the arctic islands of Canada.

The southernmost continent has not always been ice-covered. A bone of a now extinct amphibian called a labyrinthodont more than 200 million years old was found in Antarctica, indicating that at least semitropical animal life was abundant there during that period. The other continents and islands, of course, have living relatives of the labyrinthodont.

Ice from the Sea

The most extensive ice area in the northern hemisphere is the arctic ice pack, a floating slab of frozen seawater. In winter it extends downward and averages between 8 and 12 feet in thickness. Further growth is inhibited because the covering ice, being a poor conductor, prevents the heat of the water from being absorbed by the atmosphere. From August to March the sea ice doubles in size. The new ice is called annual, or winter, ice and is distinguished from the polar ice that lasts from season to season.

The surface of the pack ice is often distorted as a result of ridges formed by grinding pressure. Hummocks and cracks appear as a result of the pushing and pulling of winds, swell, currents, and tides. With the formation of hummocks, the action is accelerated since they present a sail-like surface to catch the wind and are moved about more rapidly.

When the sea ice breaks into pieces, the smaller units—which may measure miles across—are called the floes. Cracks large enough to steer a ship through are called leads, and those openings that are completely surrounded by ice are called polynyas (from the Russian for ''hollow'' or ''open''). The amount of open water in the arctic ranges from five to eight percent in the winter to as much as 15 percent in the summer.

Sea ice is a relatively complex material, and its physical properties are dependent on temperature, salt content, crystal structure, and air bubbles. Salt has almost no solubility in solid ice and most of the salts are trapped in pockets of liquid brine. As the temperature of the ice increases, and especially as it approaches the melting point, the brine escapes from the ice, greatly reducing the salinity. This migration of salt out of the sea ice is so effective that Eskimos have used sea ice as a source of drinking water.

Freshwater Ice

It has been calculated that the ice mass covering Antarctica contains nearly 90 percent of all the snow and ice in the world. This freshwater mass is the result of 20,000 years of snow accumulation. The transformation of snow to ice requires the interaction of time, wind, and sun and the pressure of layers of snow, which forces out the air between the snow particles and compacts the crystals.

This mass of antarctic ice is a gigantic glacier that averages more than a mile in thickness and measures 2500 miles in breadth. The glacier is not a flat tabletop; in addition to the sculpting action of the wind and the irregular melting by sun's rays, the ice has a tendency to creep. Solid matter in large amounts, whether it is ice in a glacier or rocks beneath the earth's crust, displays a plasticity usually associated with fluids. The substance remains hard and brittle if subjected to sudden force; but if constant, yet not extreme, pressure is exerted, it will flow, however slowly. For this reason, the glacier

Aurora borealis, the spectacular atmospheric color and lighting of the north polar region, inspires photographers, yet eludes accurate capture on film.

covering Antarctica and also those of Greenland flow toward the edges of the continent, along the contours determined by the topography and the pull of gravity.

Floating Ice Mountains

Icebergs come in two styles—northern and southern. The northern, or arctic, iceberg is a jagged peak of irregularly shaped ice broken from a land-based glacier; the southern, or antarctic, iceberg is a flat-topped chunk calved (spawned) by the massive ice shelf surrounding Antarctica.

As many as 10,000 northern icebergs may be calved in a year's time, some of which can tower 500 feet above the surface, with 85 percent of the mass still submerged. Currents and wind push the icebergs into the North Atlantic where they can drift into the Labrador Current toward the Gulf Stream and present a real hazard to oceanic transportation. In 1912 the supposedly unsinkable *Titanic* struck an arctic iceberg and went down, taking the lives of 1500 people.

As these icebergs drift into warmer waters, as far south as Bermuda and the Azores, they begin to melt and deposit silt, pebbles, rocks, and large boulders on the ocean floor. This terrestrial material was picked up during the thousands of years it took the glacier to move across the face of the land, scraping and scouring the earth.

The tabular icebergs that are found in waters of the southern hemisphere have no counterpart in arctic waters. These table-topped formations measure up to 100 miles across and have faces like sheer cliffs. Antarctic icebergs may rise as much as 250 feet above sea level and have been sighted as far north as 26° south latitude, nearly the latitude of São Paulo, Brazil.

Visual Riches of the Poles

Visually, the polar regions are storehouses of treasure, with the wind and water combining to form stark kinetic sculptures. The most spectacular phenomenon, though, is not in the sea or on the land but in the air. Auroras—borealis in the north, australis in the south—provide dancing, swirling, undulating sheets of colored light. The physics and chemistry of the forces involved render photography incapable of capturing the richness of color and fullness of beauty.

But perhaps we can explain it. The aurora occurs, not because of the cold in the polar region, but because of an opening, a gap, in the earth's magnetic shield that allows cosmic radiation to penetrate. Solar energy in the form of highly charged ions and particles bombards the earth in the area of these holes and reacts with thin gases in the upper atmosphere, somewhat like neon tubes are activated by electricity, to produce the light show shifting back and forth across the sky during the six-month-long nights.

LIFE IN THE ARCTIC

Life in the arctic is the story of predators and prey, yet the limited number of species in the arctic indicates that the territory—however large it may be—is not big enough for vast numbers of creatures competing for the same small amount of food. Polar bears stay away from the inland areas populated by the caribou and musk ox. The large herbivores, in turn, developed separate habits, with the caribou migrating seasonally in search of food, and the musk ox developing special defenses against wolves and cold that allow it to live a more sedentary life.

Among the smaller animals, wolves and foxes divide the territory so as to reduce the competition; and the hares and lemmings, though both rodents, do not often impinge upon each other's existence.

But the edge of the land presents a different story. Life in the sea is abundant; vast numbers of seals and walruses survive offshore, while mammoth whales roam about in its depths.

Great White Hunter

The polar bear is really the great hunter of the north, for it ranges on land, ice, and sea seeking food. It will do battle against walruses, small whales, and, of course, its favorite, the seal. Bears are reluctant to enter water but once in it are so mobile that some scientists consider them aquatic mammals. On land the bear may range inland eating vegetation and allowing its claws to grow longer for another season on the ice. Although small foxes and hares prove to be elusive, lemmings—especially when they appear in great numbers—are taken by bears, as are bird eggs and an occasional salmon during spawning season.

The bear is certainly omnivorous. But it is primarily a blubber eater, and seals are the tastiest, most numerous, and easiest-to-catch source of blubber a bear can find. The polar bear is well fitted for hunting aquatic animals from a base on the ice, since it has an exceptionally well-developed neck, shoulder, and forearm. Generally rangier and leaner than its more southerly relatives, the polar bear is capable of delivering a killing blow with its left paw and hauling out the dead seal in almost one motion.

Hunting Parties

The polar bear almost always hunts alone but is so closely attended by the arctic fox that this little creature, no bigger than a common house cat, has been called the jackal of the north. The fox does nothing to aid the bear in its work and may actually rest while waiting to pick up the leftovers. Once the kill has been made, the bear will often throw pieces of meat toward it to keep it quiet. An oddity of this throwing ability is that the bear always uses its left paw.

The fox is a good hunter in its own right. It has been reported to kill a seal in her calving chamber by surprising her and striking directly at the fleshy area around her mouth. Another prey of the fox is the lemming; once the fox has detected the rodent moving under the snow, it springs into the air and, with its paws and nose close together, dives into the snow. If its aim is true, as it usually is, the lemming is a goner.

Another predator of the north is the polar or arctic wolf, with its massive head and jaws. It can grow as long as six feet and weigh more than 150 pounds. Its main prey is caribou, and so the wolf is found on the grazing lands of the tundra and especially in the large islands of the Canadian Archipelago. When caribou is scarce, wolves live on a diet of carrion and lemmings.

The Victims

One of the most sought-after prey of the arctic world is the little lemming, barely five inches long. The lemming breeds very quickly; its usual litter of eight is born after a gestation period of just 21 days—the shortest gestation among mammals. And the offspring are ready to breed just 20 days after birth. Within a year a pair of lemmings and the generations of their offspring would number 170 million. Such overpopulation obviously works against the lemmings because the more lemmings there are, the less food and shelter there is for all of them. When things get out of hand and lemmings overpopulate an area, they commit what is virtually mass suicide.

The arctic hare has no problem with overbreeding. Its major predator is the fox, and one of the few times the hare is relatively safe is in the heavy snows of early spring, where its speed and elusiveness are particularly effective. The hare itself is a vegetarian, feeding on roots and shoots that it can extract with its tweezerlike incisor teeth. But in extreme cases, such as after waiting out a storm by crouching in a rocky crevice, a hare may be driven to eating meat from a fox trap or even a human storehouse.

Wanderer of the North

One of the most spectacular natural sights of North America is the migration of the caribou. Thousands upon thousands of caribou can be seen moving along with a gentle lope in a stream that appears a half mile wide and can run on for over an hour.

Like most large migratory herbivores, caribou move about on a seasonal basis in search of food. The sedges, willows, birches, and grasses that provide food during the summer are not sufficient for the cold, hard winter, and the vast herds move farther south to higher ground where lichens are exposed on rocks or can readily be found beneath the snow.

Wolves, waiting to single out and attack a lagging member of the herd, are an ever-present danger to the migrating caribou. But especially in the spring, unexpected water is another hazard. There is a report that 500 caribou were drowned in a flooding river.

The musk ox relies on a thick undercoat and a long overcoat of fur for survival in the cold. To weather the worst storms, or when facing danger, the herd forms a ring or semicircle, with the young and females at the center, protected by the males with their sharp pointed horns.

Displaying impressive antlers and deep body, the caribou is an herbivore, wandering from range to range in search of food.

LIFE ON SOUTHERN ICE

As forbidding a continent as Antarctica is, it is not totally devoid of life. Although no large animals live there permanently, both penguins and seals frequent the shores at various times of the year. The vegetation is scarce, but there are various species of mosses and lichens that grow on the bits of exposed rock in the interior. And the tiny pink mite *Nanorchestes antarcticus* was found only 300 miles from the South Pole.

Flightless Birds

Penguins can be called the most primitive of birds because they have changed the least of any bird since branching off from some reptilian ancestor. The fact that they can't fly is curious, since it is almost certain that they evolved from flying stock. Some of the penguin's ancestors may have been taller and perhaps bigger than the modern four-foot emperor (the largest of the penguins), for five-and-a-half-foot-long fossil skeletons have been found that date back 60 million years.

The penguins inhabit only the southern hemisphere, but their range is surprisingly wide for flightless creatures. The majority of the species are found in and around Antarctica and the nearby subantarctic islands, but their relatives have reached Chile and Peru in South America, the Galápagos Islands, South Africa, Australia, and New Zealand. The cold ocean currents along the coastlines of these places provide the kind of environment the penguins can live in.

In most species the females lay two eggs, but among the emperors there is only one egg laid. This is most likely because the emperor chooses to breed in winter on a bleak nesting site—a vast gray sheet of ice floating above the dark sea. Like its parents, the chick-to-be will never know land; its life will be spent on ice and in the water.

Once the egg is dropped, the female positions it on her feet and covers it with folds of skin hanging from her breast. Within three hours after the egg has been laid, the egg is transferred from the female's feet to the male's, where the incubation continues. The hen, which has been without food for

OVERLEAF:
Dwellers in the tundra, caribou wander in herds looking for vegetation to feed on. The caribou migrate each season from mountain valleys to the tundra, creating what observers have called "rivers of brown."

two months, waddles off for a while to feed in the sea—which may be some miles away by this time, depending on the growth of sea ice. The chicks are hatched by the males in the dead of winter when the winds may reach gale force and the temperatures may dip as low as $-70°$ and $-80°$ F. The watchful parents provide for the young during this harsh period, and by the time the next winter rolls around, the chicks are big enough to take care of themselves.

The king penguin, slightly smaller but otherwise almost indistinguishable from the emperor, also lays but one egg, but has a more temperate breeding site in the mud flats of the subantarctic islands. The Adélies, the gentoos, and the chinstraps are the only other penguins to breed on Antarctica.

The penguins are not particularly defensive birds, and even the Adélie tolerates the presence of man after a brief get-acquainted period. During the breeding season, skuas may snatch an egg or hatchling that is left alone too long, but otherwise the penguins have little to fear on land or from the air. It is on the edge of the ice and in the water that the only danger lies—the leopard seal. When penguins reach the water's edge, they take a close look before plunging, since a hungry seal may be lurking about hidden in the murky shore waters. The entry is the only dangerous moment in the penguin's life, when it could be caught by surprise; the seal can barely out-swim a penguin, and in open water the penguin easily outmaneuvers the seal.

As graceful in water as they are clumsy on land, penguins use only their feet to steer. They have the ability to leap out of the water, sometimes as high as ten feet; in exiting from the ocean the birds rocket toward shore and pop up onto a rock or ice floe and land feet first. Groups of penguins are frequently met at sea, hundreds of miles away from any shore, voyaging at a very constant pace and swimming very much like dolphins, clearing the surface every time they take a breath and opening their mouths wide for a fraction of a second. This "porpoising" type of progression is only used when penguins "make way." When they are hunting, their group scatters temporarily and they dive individually, often to great depths, to find the shrimp, little fish, and larvae they feed upon. The emperor penguin has been recorded diving to 800 feet and staying 15 minutes underwater.

A most remarkable bird, the albatross features in many sailors' legends and superstitions. With its huge wingspan and ability to soar for days on end without landing, the albatross is a familiar and magnificent sight over the seas.

Mammals of the Antarctic

Only about ten species of seals and sea lions survived the slaughter of the nineteenth-century sealers who butchered true seals for their fat and fur seals for their pelts.

In the antarctic the rarest of the earless seals is the Ross seal, named for James Clark Ross. It is a rotund creature with bulging eyes that lives on the pack ice and emits not only the usual grunts of many pinnipeds, but also some trilling "coos" that have earned it the name of the singing seal.

The most numerous of the southern true seals are the misnamed crabeaters, which use their interlocking teeth to strain small crustaceans from the water in the same way whales do. Crabeater pups are born with a full set of teeth and nurse for a number of weeks before taking off on their own. The solitary leopard seal is widely distributed in the southern hemisphere, reaching as far as Australia and New Zealand, although more common in and around the subantarctic islands.

The Weddell seal stays in the higher latitudes for the whole year, spending most of the winter in the water, which is much warmer than the air. The Weddell has the ability to stay under water for close to an hour and can reach depths as great as 2000 feet while diving.

The southern elephant seal is the largest pinniped, ranging up to 20 feet in length and 8000 pounds in weight. Its most distinctive feature, apart from its size, is its large proboscis, which gives it its name. The seal uses this nose as a resonating chamber to produce a variety of sounds, especially during the breeding season before the cows arrive and while the bulls are still establishing their territorial rights.

These breeding bulls stake out their territory in midwinter and wait for the cows and mating season to come with spring. After establishing territories, defending them, servicing the harem, and protecting the cows, the bulls may then finally return to the water to feed themselves. It may have been as long as three months between meals.

Such a long period without food requires a vast amount of blubber, which is what led to the exploitation of the elephant seals and very nearly resulted in their extinction in the last century.

A Bird by Any Other Name

Seamen were ready observers, quick to name the birds they saw, sometimes pinning two or three appellations on the same bird. In other instances, many different species were lumped together under one name for the sake of quick identification.

Common names, which differ from island to island and country to country, may be based on something as concrete as the color of a beak or may arise from a sailor's imagination—one man's petrel could be another man's shearwater. Many seabirds fly very close to the surface of the ocean, seemingly skimming on top of it. Hence the name shearwater is given to many birds. But other sailors, observing the same flying maneuver, were reminded of Saint Peter walking on the water, and named the bird after him—Little Saint Peter. Over the years this was corrupted into petrel. Albatross is supposedly derived from *alcatraz,* Spanish for "pelican," while mallemuck, small albatross, was formed from words meaning "foolish" and "gull." Scientists, however, group all petrels, shearwaters, and albatrosses in the order *Procellariiformes,* tube-nose birds.

The albatrosses (which have been known to live for 70 years), with their long, narrow wings, are gliders that use the force of the wind itself to aid them in remaining aloft. First they bank one way, then the other,

Of all different kinds of penguins, only four species —including the Adélie seen here— are known to breed on Antarctica. Though they appear dignified while standing still on land or ice, penguins look comic and clumsy when they begin to waddle toward the sea. Once in the water, however, they display speed and grace.

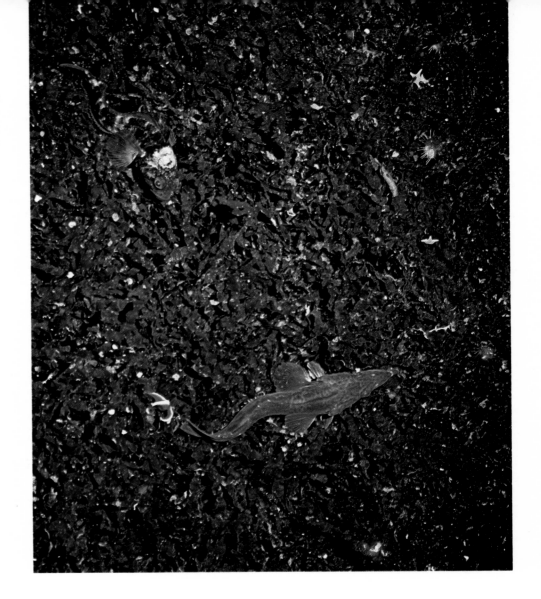

alternating in a gentle roll. There is only an occasional beat of the wings. The albatross not only uses the wind as lifting power, but it also manages to remain master of its own course, either being carried all the way around the globe or, on the contrary, making way against the strong winds without a beat of its wings.

The dependency on the wind is so great that albatrosses find it difficult to become airborne in areas of calm; they are, accordingly, absent in the low-wind areas of the equatorial zone known as the doldrums. Only rarely have albatrosses been seen north of the equator or inside the Antarctic Circle; they choose to remain closer to the "roaring 40s," the latitudes of strong and constant winds.

The skuas are distinctive scavengers, preying on isolated smaller birds and abandoned eggs, and especially on the Adélie and gentoo penguin eggs and chicks. With its strong beak and curiously clawed—yet webbed—feet, the skua has been improperly dubbed the eagle of the antarctic. In fact, a skua is easily chased away from a colony by the smallest adult penguin and can make a living only by patiently waiting for the penguin parents to make a mistake.

BELOW THE ICE

Life in the water world above the Arctic Circle and below the Antarctic Circle would seem to be much easier than life in the polar land or polar skies. And, judging by krill alone, this is at least partially true. These shrimplike creatures provide food for the largest animals on earth—the whales—as well as for large fish, seals, penguins, and seabirds. The krill itself feeds on the abundance of diatoms. These and other species of phytoplankton abound in the polar waters because of the wealth of nutrients constantly brought up in surface waters from the deep "upwelling currents." There are certainly many invertebrates in polar waters, both pelagic and benthic, but there is a relative paucity of fish.

One trait that appears common to most polar fish regardless of species is that they all lay their eggs on the ocean floor, apparently to avoid contact with the ice cover or with the low-salinity surface layer in the summer. The eggs are relatively large, with massive yolks for their size. When the larvae hatch, they are also fairly large and they can immediately cope with the rigorous environment.

The flow of currents and the temperature distribution in the sea tend to isolate the antarctic waters—in terms of marine life—from other parts of the ocean. Thus it is not surprising that when whalers plying the southern waters returned home they told tales of bloodless fish having white gills instead of red. This unusual family, the icefish, do indeed have blood in their semitransparent bodies. It is just that their blood appears pale white because it lacks red cells, while there are some nucleated cells that are similar to white blood cells. Ordinarily, red blood cells, with their hemoglobin, are used by vertebrates to carry oxygen through their system. Icefish accomplish this in some other manner, for despite the lack of red cells, their blood is about 0.7 percent oxygen by volume, about one-tenth that of the blood of other fish. A puzzle to be solved, then, is why and how did the icefish lose their red cells, and if these fish can get along without them, why do other fish have them?

The icefish are not the only fish in the antarctic waters. In fact, the most numerous species belong to the order *Nototheniidae*, and include antarctic cods, plunderfish, and dragonfish. The cods, probably the most numerous, are bottom feeders and live on small invertebrates and algae. They are also quite well adapted to living under sea ice. The plunderfish look somewhat like antarctic cods, but have larger heads and, like the icefish, lack scales. The dragonfish are the most distinctive, having an elongated shape and no spiny anterior dorsal fin. Some dragonfish have pointed noses and large canine teeth, and some are almost scaleless.

There are others, including eelpouts, some true cods roaming from

Unique Differences

This menacing looking isopod is one of the marine creatures of the polar regions.

357

milder waters, flatfish, flounders, and snailfish. These latter, though they are certainly not the most common fish, are widely distributed in the area, in shallow as well as deep water. Snailfish, which are characterized by heavy, gelatinous bodies, are sometimes quite colorful, ranging from an almost transparent white to purple-brown, red-orange, or pink.

Life on the bottom may be more abundant and more exuberant in the antarctic than anywhere else in the world. Not far offshore—in water 50 to 100 feet deep—there is a bare patch caused by large ice crystals, called anchor ice, which form in the bottom during the winter and rise to the surface when the temperatures rise, carrying with them any creatures which may have taken up residence.

Slightly deeper, from 100 to 140 feet, there are many sessile coelenterates, while the next zone contains large starfish, bryozoans, ascidians, isopods, non-reef-building corals, and sponges. There are more sponges found here, displaying their sharp-pointed spicules, than can be found on any tropical seabed.

There is a greater abundance of molluscs in these waters than was known even recently, and those that are found are generally not much different from those of milder regions slightly to the north. What differences exist are usually due to the fact that the antarctic species live in much deeper waters than those living along South America's coast. Other abyssal organisms found in antarctic waters—although only rarely—are crinoids, sea cucumbers, sea urchins, and sand dollars, and myriads of tiny sea snails.

One of the more unusual findings about life in the Arctic Ocean, the shallowest of all the world's oceans, is that though the fish life is sparse in terms of number of species, there is an abundance of marine invertebrates. The most common fish in the arctic are those that are bottom dwellers, including eelpouts, cod, snailfish, bullheads, sculpins, and flatfish. The bullheads and armed bullheads have developed perchlike features that make them resemble the polar fish of the antarctic. They have oversized, frog-shaped heads and large gill chambers, as well as broad pectoral fins and nonforked tails, and they have poorly developed muscles for lateral movement.

The severity of the environment has an inhibiting effect on fish. Limiting factors include the low temperatures, the destructive ice action near

Under the Arctic

the inshore waters where it is broken up by tidal action, fluctuating salinity due to the freezing and melting of ice, and a lack of large plant life.

An expedition organized under the direction of Dr. Joseph MacInnis of Canada had an observation sphere called Sub-Igloo, for divers working in Resolute Bay, 600 miles north of the Arctic Circle. Their findings confirmed the disparity between fish species and other marine life—biologists collected fewer than a dozen species of fish and almost 100 different species of invertebrates. One of the biologists, Dr. Alan Emery, described what he observed: "I found plants and animals more abundant than I ever expected, though compared to the tropics, these waters have much less variety. Without actually diving into the cold depths, we could never have realized how plentiful arctic marine life really is, and yet how painfully slow it grows, moves, and reproduces." Among the more frequent sightings the Sub-Igloo divers made were lion's-mane jellyfish, sometimes with tiny sea fleas, called amphipods, hitching a ride on them. There were also numerous ctenophores and nudibranchs, many with partially transparent bodies, as well as many tiny sea snails, which dotted the seabed or were feeding on the algae that coated kelp fronds. Flattened armored isopods walked along the bottom, looking very much like extinct trilobites. There were also tiny planktonic crustaceans as well as squids, octopuses, and cetaceans.

THE POLAR LABORATORY

Emperor penguins may stand as tall as four feet and weigh up to 90 pounds after a good summer of feeding on plentiful antarctic krill, shrimplike creatures that also nourish whales and seals.

The polar climates, which are as intense as they are harsh, offer examples in the adaptive capabilities of living organisms. Biological laws—actually general statements describing how phenomena naturally occur—have been formulated that deal directly with polar areas. One of these—Gloger's law—states that in the northern hemisphere birds and mammals with a north-south distribution tend to display lighter colors in the northern part of their range. Since darker colors are associated with higher humidities, it is not certain whether the humidity or temperature is the determining factor.

In the nineteenth century German biologist Carl Bergmann formulated a law that held that warm-blooded animals tend to be larger in the cool parts

The Eskimo way of life was developed over centuries of constant contact with the environment. Traditionally Eskimo hunters waste nothing. It is said that with the large animals they kill, "they make use of everything except the roar."

of their ranges. The logic behind this, of course, is that the larger the total bulk of the body, the smaller the proportion of surface to volume. This is important in considering that body heat is dissipated most readily through the surface area of an animal, generally the skin.

Bergmann's law helps explain, in part at least, the giant size of whales, which feed mainly in colder waters or in the colder currents of milder environments. The walrus and the elephant seal are huge, and the polar bear, though it lives in the most inhospitable of regions, is one of the largest of the mainly meat-eating ursines. In the southern hemisphere, the emperor penguins and others ranging the subantarctic are larger than the Magellan penguins, which live in Patagonia and the Falkland Islands. Both are larger than the Humboldt, or Peruvian, penguin living farther north.

Bergmann's rule is by no means absolute, for certain temperate-zone animals like raccoons and moles show a decrease in body size as the climate of their range gets cooler.

The American zoologist Joel A. Allen emphasized the tendency for the appendages of polar inhabitants to become shorter, as compared with races of the same species living in warmer climes. The effect of this is a lessening of the surfaces from which body heat may be radiated away. Allen's law is evident in a number of polar creatures. The ears of arctic foxes are smaller than the ears of foxes living in milder climates. The arctic hare has ears much shorter than the snowshoe hare, which lives farther south. The snowshoe hare, in turn, has much shorter ears than the antelope jackrabbit, which lives in Arizona.

Seals, whales, and penguins all have extremely short appendages in relation to their body size. Among penguins, there is an even finer differentiation. The emperor and king penguins are almost indistinguishable in coloration and markings. Yet the emperor, which lives year round on

Antarctica or the ice surrounding it, has proportionally shorter flippers, feet, and bill than the king penguin, which lives only a few hundred miles farther north.

IN HARMONY WITH THE ENVIRONMENT

To survive in the arctic means making judicious use of every possible natural resource. The legends and tales of Eskimos are filled not with the exploits of great warriors, but rather with the lessons of learning to cope with the environment and the feats of successful hunters. In the polar regions one must live with nature, not fight it.

It is interesting to note that Eskimos have no one word for snow. Rather, they describe it with a variety of terms which refer to its moisture content, texture, or any other physical property that has a practical meaning.

An important principle in making the most of polar life is conservation of material. Even human waste products are put to use. Urine, for example, was used to ice the runners of a sled—wax was unheard of and the friction of the snow on a cold day could tire the dog team needlessly.

Eskimo, or Chukchi, arctic residents were ingenious in their use of animal products. A walrus, for example, provided meat for both people and dogs. Blubber burned in lamps and stoves; hides became boats (the smaller, covered kayak and the larger, open umiak) and floor coverings, or were cut into thongs for fishing or harpoon lines. Tusks were carved into hunting implements, the keels of umiak boats, household tools, and religious or decorative figurines. A needle, awl, comb, scoop, knife, buckle, or pair of sun goggles could be made from walrus ivory. The intestines were converted into waterproof rain gear or translucent window coverings, or they were stretched over wooden frames and decorated colorfully to become ceremonial drums.

Eskimos had to make a life out of the materials the environment offered. There is little vegetation in the arctic, so the Eskimos did without it. They lived—before the coming of the white sailors and traders—almost entirely on the flesh and fat of animals, whether mammal or fish. They regard the fishheads and head meat of mammals as not only the most nutritious of foods, but also the tastiest.

TO LEARN OF THE WORLD

One of the first systematic attempts to study the polar regions was made by the International Polar Conference, which, during the winter of 1882–83, established a series of scientific stations, eleven in the arctic and four in the antarctic.

Early antarctic study included many unscientific attempts by adventurers and explorers who collected odd specimens and natural rarities, such as penguins and their eggs. For the most part, scientific endeavor at the South Pole was dominated by publicity and commercialism, overlaid with the land claims and counterclaims made on the territory by sovereign nations.

Following a fruitful "Polar Year" in 1932–33, the International Geophysical Year of 1957–58, involving 30,000 scientists and technicians from 66 nations, brought an increase in scientific activity. Many of these were stationed in the polar regions and their efforts not only capped, but also initiated, intense research efforts.

A major effort of the IGY was devoted to the study of interactions of the Southern Ocean and the atmosphere above. Before the IGY it had not been clear whether the Circumpolar Current involved the entire water body from the surface to the bottom, or whether a deep countercurrent existed below the eastward surface flow. Observations of the distribution of temperature, salinity, oxygen, and other substances proved that the eastward motion persists throughout the entire water column. The total flow, however, seems

to consist of a complex of separate streams with fast-moving cores and even some subordinate countercurrents.

From outposts in the far north, task groups set out by ship, dog sled, tractor train, and aircraft to fill in some of the big gaps in our knowledge of the Greenland Icecap and polar Canada. The most spectacular polar projects, however, were in the all-out multination effort in Antarctica, with its 16,000 miles of relatively unknown coastline. Eleven nations—Argentina, Australia, Chile, France, Great Britain, Japan, New Zealand, Norway, the Union of South Africa, the Soviet Union, and the United States—established bases on Antarctica or its offshore islands. Magnetic observations and gravity measurements were made in dozens of locations. A network of meteorological stations made surface weather observations and upper-air observations to 100,000 feet twice daily by balloon with radio transmitters that sent back reports on temperature pressure, moisture, and wind. Glaciologists drilled holes through the ice to a depth of 1000 feet and more to obtain ice cores and to measure temperature gradations.

These were only a part of the scientific investigations of the IGY—the most ambitious cooperative study of our environment ever carried out.

TO PLUNDER OR TO MANAGE THE POLES

Man, at least so it seems, is incapable of leaving a place the way he finds it, whether it is outer space, the moon, or Antarctica. The mark of man's presence is often an absence—the disappearance of blue whales in familiar waters or the absence of fur seals on certain islands. Elsewhere man leaves something behind, a physical reminder.

Antarctica is a continent recently dedicated to science, for the 500 winter residents and the 5000 summertime guests are engaged in learned work or aid it indirectly by providing support services. Yet even with such noble purpose, Antarctica is being ruined. The most staggering problem is waste disposal. Garbage dumps are difficult to dig in land that is permanently frozen. Petroleum-based plastics are virtually indestructible. Even incineration is no great solution; not only does burning pollute the air with foreign chemicals, but there is always residue.

Power plants, no matter what kind of fuels they use, are another disaster at the poles. Fossil fuels pollute the air, and nuclear plants can superheat the water and cause a problem of radioactive waste disposal.

Man is not always conscious of his own plundering. But ships bring rats; sailors bring dogs and cats; settlers bring sheep and goats and rabbits. Overnight an environment has changed. Birds—like the blue petrel—that through generations of evolution developed the ability to nest in burrows for protection from flying predators suddenly become acquainted with a rat's maze-ranging talents. Strains of grass and plants hardy enough to withstand polar climates begin to yield and finally die out under the persistent feeding of herbivorous hares and rabbits.

Even in the arctic, where man has had a much longer association, assaults are being made. The polar routes used by jet airplanes are paths of pollutants. Oil exploration on the continental shelf of North America promises the wellhead leaks, seepage, and tanker accidents familiar to the North Sea and Gulf of Mexico.

What to do, what to do? The hope is that man will somehow be able not only to define the problems, but also divine the solutions.

What to Expect

Hope is an ephemeral thing. It can easily be shattered by the unscrupulous, the ignorant, the distrusting, the economically deprived. The very nations that found it profitable to exploit the polar regions are the countries that must be relied upon to protect these areas.

Perhaps the wasteful killing will stop. There may even be a time when no animals will need be killed, for fashions change, substitute products can be found, zoos may learn to breed rare animals in captivity.

Slowly, as man contaminates the water, poisons the air, he is killing himself. To reverse the trend before it is too late, all human beings have not only to change their approach to nature but probably their very way of life.

An era of hope came with the signing of the Antarctic Treaty in 1959. The signatories were Argentina, Australia, Belgium, Chile, France, Great Britain, New Zealand, Norway, South Africa, the Soviet Union, and the United States. They agreed to waive any territorial claim for 30 years, to ban nuclear explosions from the continent, provide for inspection of each other's scientific stations, and to take unresolvable disputes to the International Court of Justice. But even more in line with the spirit of cooperation that prevailed, the treaty stipulated that "it is in the interest of all mankind that Antarctica shall continue forever to be used exclusively for peaceful purposes and shall not become the scene or object of international discord." Thus, a continent became dedicated to science and the peaceful purposes for all peoples based on the open exchange of information between nations.

The Alaska pipeline brings oil from the North Slope to southern Alaskan ports for shipment to the lower states. Numerous ecological problems have arisen in the course of building and using the pipeline and its long term effects remain in question.

MANNA FROM THE SEA?

When mathematicians entered a new field of speculation just by deciding that an imaginary number would have a negative square, it sounded like a harmless exercise. But while the improbable hypothesis was extending the scope of abstract thinking, it was also set to work in electronics, helping to develop transistor radios—and devastating guided missiles. Advances in science, however pure in their intentions, are always available for possible uncontrolled applications.

Aesop's "tongue parable" should help remind us that men can use anything for the best or for the worst. Science originally was intended to increase our understanding of the universe, but it opened doors inadvertently to deadly, as well as helpful, inventions. Technology was meant to increase the goods available in order to satisfy essential human needs, but, instead, it was used to create artificial "needs." Diverted from its objective—the quality of life for all—technology has given rise to a dangerous myth: quantity for the sake of quantity; more goods, even if they satisfy only social status, even if they are unevenly distributed; more energy, even if most of it is wasted; more money, even if it is devalued in a mad inflationary race. Today a nation's importance is measured by its rate of production rather than by its intellectual contribution, by its gross national product rather than by its artists and composers. With increased production the only goal, the responsibility of nations to the environment and to future generations is abandoned. The emphasis is on "more" rather than "better." We have ignored the wisdom of Mies van der Rohe's statement: "Less is more." Today the word "progress" is used as a synonym for "growth." And growth grows out of hand, unchecked, like a tumor on mankind.

Our very minds are so contaminated that when explorers open the gates of the ocean or of outer space for mankind, we ask: What resources do the moon or the sea have to offer? But it is essential to discuss what a reasonable approach should allow us to expect.

Mass slaughter of whales, incessant scraping of the North Sea's bottom with heavy trawlnets, killing of porpoises and dolphins in huge tuna purse nets, ravages of coral reefs by spearfishermen, hasty oil drillings in unsafe offshore areas—these are examples of how a distorted image of progress can lead to a shameless rape of the sea. We must not wait for obvious warnings—like the imminent collapse of industrial fishing—to switch to a rational, internationally controlled management of marine resources. These resources could be more than sufficient to allow us the time needed to check the world's population, reconsider the priority of needs to be satisfied, and better allocate the planet's wealth.

Within the next fifty years, fishing will be progressively replaced by mariculture. With offshore oil reserves almost exhausted, mining the ocean floor will provide a bounty of ore for several useful metals. New drugs such as antibiotics will be extracted from marine creatures. But the greatest material contribution of the sea to man's welfare will be in the field of energy. Our present Western civilization depends on coal, oil, and natural gas—three fossil fuels that are nothing more than energy from the sun, transformed in plants and plankton in an extremely low-efficiency process, stored up through many millions of years. It is high time to find more direct access to the sun's power; the major problem is that even in deserts, solar energy is scattered over huge areas. In the sea, on the contrary, it is *naturally* concentrated in ocean currents, and even more inexhaustibly in the temperature differences between the surface and the deep waters in tropical zones. Marine thermal plants of gigantic proportions could provide massive quantities of electricity for the production of liquid hydrogen, the clean fuel of the future. Such plants will also generate artificial upwelling currents and thus fertilize the surface of the oceans.

Having acknowledged our past mistakes, we are being handed a world in which to demonstrate new-found abilities to exploit without greed and without pollution; to colonize without conflicts; simply to contemplate and create. The greatest riches are those of the heart, and the sea is capable of literally flooding us with aesthetic and intellectual joys.

SUNKEN TREASURE

Locating sunken treasure is hard work, a lot more difficult than the child's dream of finding a lost map marked with an X. Sometimes years of research precede a search. Manuscripts written in ancient languages must be deciphered; expensive reconnaissance missions must be mounted. Even if the treasure is found, the lucky finder is rarely the keeper. For instance, a French citizen, no matter from what area he salvages a wreck, must turn over 100 percent of the booty to his government and the government alone decides upon the compensation. Spanish and Portuguese government regulations state that ancient ships that once flew the flags of these countries still belong to them. If a wreck is located off the coast of Florida within territorial seas, the state demands a fourth of the treasure. But even against these odds, some men still get an itch in their palm and a "Long John Silver" glint in their eye at the mere mention of the magic words "sunken treasure."

Silver Bank   In the summer of 1643 the flagship of the Spanish Silver Fleet, a galleon called *Nuestra Señora de la Concepción,* crashed on the treacherous reefs of the Silver Bank north of Hispaniola, the island in the West Indies now divided into Haiti and the Dominican Republic. With the ship went a treasure trove of pearls from Venezuela, gold and silver from Mexico and Peru, and various jewels valued at over $3 million in today's currency that had been destined for the coffers of the Spanish king.

William Phips was a Bostonian ship's carpenter with treasure fever. He persuaded two kings of England, Charles II and James II, to back his efforts to find the ship. In 1686 he was successful and the final take was measured to be 27,586 pounds of silver, 347 pounds of plate in precious metals, 25 pounds of gold, and great quantities of pieces of eight and jewels. Phips was given $75,000 and made governor of Massachusetts as his reward. However, like many treasure seekers, he found that his wealth could not buy happiness. He died penniless before his forty-fifth birthday.

Treasure Today

The war of the Spanish Succession delayed the sailing of the Silver Fleet for over two years, from 1713 to 1715. When it finally set sail from Havana, laden with two years' collection of riches, the combined armada of 1715 carried $14 million in treasure. The traditional June departure was delayed. When the fleet finally set out on July 24, it was the hurricane season. At 2 A.M. a hurricane roared down, smashing the ships one by one on the ragged reefs off what is now Cape Canaveral.

Kip Wagner, a Florida beachcomber, who in 1955 had found a coin stamped with the Spanish seal, formed the Real Eight Company, a treasure-seeking corporation. He began his search for the remains of the Silver Fleet of 1715 by searching for the base camp along the Florida coast, as a hobby. Finally, after three years of picking at the sandy coasts, he found at his feet a large, crudely made gold ring set with an enormous diamond.

The next step was a reconnaissance mission over the Florida bay in a rented plane. Wagner was hoping to see ballast stones, all that usually remains of a ship long at the mercy of the sea. A suspicious dark blotch on the reef was finally spotted from the plane and the Real Eight Company began in earnest to find the armada of 1715.

The first four dives yielded pottery, of interest only to museums. On the fifth dive, pieces of eight were found. A dredge and water jet were moved in to remove sand. As clouds of sediments and shells moved to one side, the divers were amazed to see thousands of golden doubloons exposed on the ocean floor. The Real Eight Company located all eleven of the fleet's vessels and has brought up over a million dollars in gold, silver, and jewelry.

MINING THE SEA

The ocean is said to be a vast storehouse of natural resources. Successful exploitation of the ores and minerals from the sea depends upon three factors: expansion of geological knowledge to facilitate the location of resources; technological advances to enable pollution-free extraction and mining; and definition of international law regarding marine mineral rights.

Dredging for Gold and Gravel

In the United States gravel and sand, which are needed to make cement, constitute a billion-dollar-a-year industry. Growth rates projected to the year 2000 estimate that the demand will increase fourfold. Unfortunately, dredging the continental shelf turns entire marine provinces into disaster areas.

Several other products are mined by dredging the ocean floor. Seabeds all over the world are littered with manganese nodules ranging in size from cannonballs to grapes. Some areas look like cobblestone streets, so heavily are they paved with the nodules. Calcareous shells, whose calcium carbonate content is used for cement and fertilizer, are mined from deposits in the Gulf of Mexico and off Iceland. Tin is taken from ancient riverbeds that lie under the sea in Indonesia, Thailand, and Malaysia. Gold deposits have been found off the coast of Nova Scotia and the vestiges of the great Yukon strike are still there—where the Yukon feeds into the sea. Around 12,000 carats of diamonds are taken each month off a fearsome shore in South Africa called the Forbidden Coast, and extensive new diamond fields have recently been found off the coast of Africa near a group of small islands with the unlikely names of Roast Beef, Plumpudding, and Guano.

Geological Movements

Deciphering the origin of the earth and the great geological events that followed is a quest that is difficult but not impossible since both on land and under the sea clues to the puzzle are planted. Some of these clues have given

369

rise to the theory of continental drift. Corollaries of this theory are giving ocean prospectors some solid guidelines to finding mineral deposits in the ocean.

At the places where one plate dives under another, minerals emanate from the converging plate as it melts into the earth. These molten minerals then combine with sulfur to form metallic sulfide deposits. The Kuroko deposits of Japan, the sulfide ore veins of the Philippines, and the rich metal deposits in former convergence plate areas, such as are found in the Rocky Mountains and the Andes, are examples of this phenomenon. Localities where plates diverge, commonly found in the midocean regions, are also important areas of mineral formation.

The Red Sea is considered young when we speak of oceans. A divergent area is spreading the sea floor between the continents of Africa and Asia. Discovery along the center of the Red Sea of rich metallic sulfide deposits in sediments and in solution in hot brine pools directly above the sediments has sent scientists scurrying to see if similar concentrations of metallic sulfides appear in older ocean basins. Veins of copper located by deep-sea drilling at a depth of 7380 feet beneath the Indian Ocean indicate that other "old" oceans may hold similar deposits.

Black Gold

The greatest mineral resource of the sea is black gold—oil. The world's use of oil has increased from 11 million barrels daily in 1950 to 46 million today. Seventy percent of the oil and natural gas is consumed by the United States, Canada, Western Europe, and Japan and the majority of it is used for transportation. In most cases, land areas have been tapped. Only the offshore and deep-sea oil sites remain to be exploited. More than a million dollars *per day* is being spent along the United States Gulf Coast alone in the search for and development of offshore sources of oil.

The process by which oil is formed is still not clear. The generally accepted theory is that marine organisms are trapped in areas where oxygen levels are low and decay bacteria cannot break down the organic structure of the plants and animals. The hydrocarbons produced by this partial decay collect in layers of porous limestone or sandstone. The oil may float above a layer of water that prevents it from migrating downward. It must be sealed from upward movement by an additional layer of impermeable rock that overlies the deposit. Oil-yielding rock is often associated with large deposits of salt, called salt domes, that can be located by seismic studies. The nature of the bottom and subbottom topography can be determined by the use of sonar. Profiling aids, such as explosions that further expand the sonar readings, can reveal the geological structure of the ocean bed to at least 2000 feet below the ocean floor. Eventually, after an area has been identified as having possibilities as an oil field, samples are taken by core drills.

POLLUTION-FREE ENERGY

The use of fossil fuels (gasoline, coal, oil, and the like) as a major energy source is far from an ideal method of obtaining power. The supply is finite and the environment is endangered both from accidental oil spills and from the hydrocarbons emitted when the substances are burned. In the face of the power-pollution problem and increased needs for pollution-free electricity, we can once again look to the sun and to the sea for answers.

Tidal Power

By the late Middle Ages tidal mills (harnessing the gravitational energy of the moon as a power source) for grinding wheat were in existence all along the coasts of England, Holland, and Wales. The first tidal mill in the United States was built in 1635 at Salem, Massachusetts. It, like all the mills previously mentioned, used waterwheels, low in energy production compared to the only modern contrivance that harnesses the power of the tides to generate electricity on the Rance River in France.

The essence of the Rance project is 24 novel turbines that catch and heighten the energy of flowing water. As the tide moves in, blades on the turbine are turned by the water's motion and electricity is produced. At the highest tidal point, the turbines become part of a dam that blocks the water in a basin. As the tide ebbs, a head is created. Then the turbine gates are opened. The water rushing back to sea again generates electricity.

Such generators can be constructed wherever great extremes exist between the level of high and low tides. As early as 1919, engineers proposed designs for a joint United States–Canadian power plant for the Bay of Fundy in Nova Scotia, where the tidal range is 50 feet. Franklin D. Roosevelt finally gave his approval to the plan, but it was halted by engineering and political difficulties. Revived in the 1960s by John F. Kennedy, it again failed to receive congressional approval.

Windmill in the Gulf

One of the most imaginative methods of capturing the power of the seas uses the antique principle of the windmill. Within the core of the Gulf Stream, in an area 10 miles wide and 450 feet deep, running for 350 miles off the United States coast, the velocity of the water is strong enough to drive rotor-type machines. It is proposed that a row of 12 turbines abreast be spaced a mile apart for the 350-mile stretch. Over 100,000 megawatts of pollution-free power would be produced.

Offshore oil drilling facilities like this North Sea platform may be exploiting the last regions where oil is to be found in quantity.

Beginning in the sixteenth century, Mediterranean men went out in boats in search of the precious red coral that is really the nonliving skeleton of coral animals. Much of the coral is worked into jewelry but sometimes large branches of coral are left intact, and simply polished and mounted. Depending on their shape and size, these coral "trees" may sell for thousands of dollars.

FOR THE CONNOISSEUR

The unique and the rare as well as oddities and curiosities have always captured the eye, and the purse, of the wealthy. From the sea come some of the world's most prized collectables, giving pleasure and profit to the connoisseur. Hardly a culture exists that doesn't cherish some item from the sea.

The legendary cloth-of-gold from antiquity came not from underground mines but was spun by marine bivalves known as pen shells. These molluscs secrete byssus fiber, a milky substance that hardens into bronze threads that the animal uses to anchor itself. Byssus was first woven into cloth in the Kingdom of Colchis on the Black Sea. Jason and the Argonauts called the elusive golden fleece "Colchis," giving rise to the modern theory that the fleece was made from byssus.

Perhaps the first nation to grow into world fame and financial power because of a luxury item from the sea was Phoenicia. The Phoenicians discovered that the murex snail harbored a dye that would turn silk a delicate shade of purple. They developed the art of dyeing cloth and carried their purple product to Greece and Rome. Thus arose "royal purple," since only the highest members of ancient communities could afford robes made from the cloth. Purple robes today remain a symbol of royalty in Europe.

The modern connoisseur seeks fabulous treasures in the sea but does not look for murex or cloth-of-gold. Instead he searches for novel foods, beautiful and rare shells, and "red gold"—coral. As general affluence grows, all these natural refinements are seriously endangered. We urge the hobbyist to halt collection of these items.

The Gift of the Oyster

In France, as late as 1720, the wearing of pearls was forbidden to any but royalty as a measure of the esteem in which pearls were held. However, in 1891 a young Japanese noodle salesman conceived the idea of pearl culture. By 1920 the process had put an end to most searches for natural pearls. The jewel was now within the financial reach of the middle class. Pearls are induced to grow in an oyster by the insertion of a piece of shell or plastic into the mantle of the oyster. If the insertion is successful, the tissue forms a pearl sack around the irritant.

Today most pearl divers seek the oyster for the mother-of-pearl, or nacre, the iridescent substance that lines the shells of many oysters and abalone. The demand for this product has caused some oyster culturists to have as their prime business mother-of-pearl rather than pearls themselves.

Skeletal Jewels

Red coral has always been valued for its beauty. Coral pieces from the sixteenth century were so highly prized that they became a medium of exchange in financial dealings between Europeans and Asians.

Coral today is increasingly scarce, and scarcity means high price. In a bracelet with alternate, entwining circles of coral and gold, the price of adding a link of coral is almost as expensive as a gold one. Pinkish white coral, the most sought-after variety, sold for $300 per pound in the uncut state just a few years ago. Today a New York jeweler states, "We used to buy coral by the kilo; now we take what we can get—by the carat."

Marine Architects

Man's attachment to shells, and his fascination with their beauty, surpasses his involvement with almost any other natural object. American Indians wove cowrie shells into money chains they referred to as wampum. Shell motifs are found in art throughout the world. Religions and folktales abound with stories of the power of the shell. Aphrodite, the Greek goddess of love and beauty, arose from a scallop shell. The same fable is related about the birth of Quetzalcoatl, the Aztec plumed serpent god.

The popularity of shell collecting has stimulated the opening of hundreds of shell shops around the world, especially concentrated in tourist and resort areas near the sea. Fifteen such shops operate in Honolulu alone. Collecting marine life for their shells alone must surely be condemned. A diving biologist recently visited some South Pacific islands, intending to photograph some of the more interesting shelled animals. He found none. Islanders explained that he would find good shelling areas no closer than a day's boat ride from any local airport. All other areas had been scoured by shell collectors, who had taken living animals as well.

(Left) Holothurin, a useful drug, is isolated from the mucous covering of the organs of the sea cucumber, released during evisceration and seen here being nibbled on by a goby.

Two Indians fishing alone with simple gear can take enough fish for only a few days. For centuries the sea's living resources were not overly exploited and the sea did not feel the stress of man's predation.

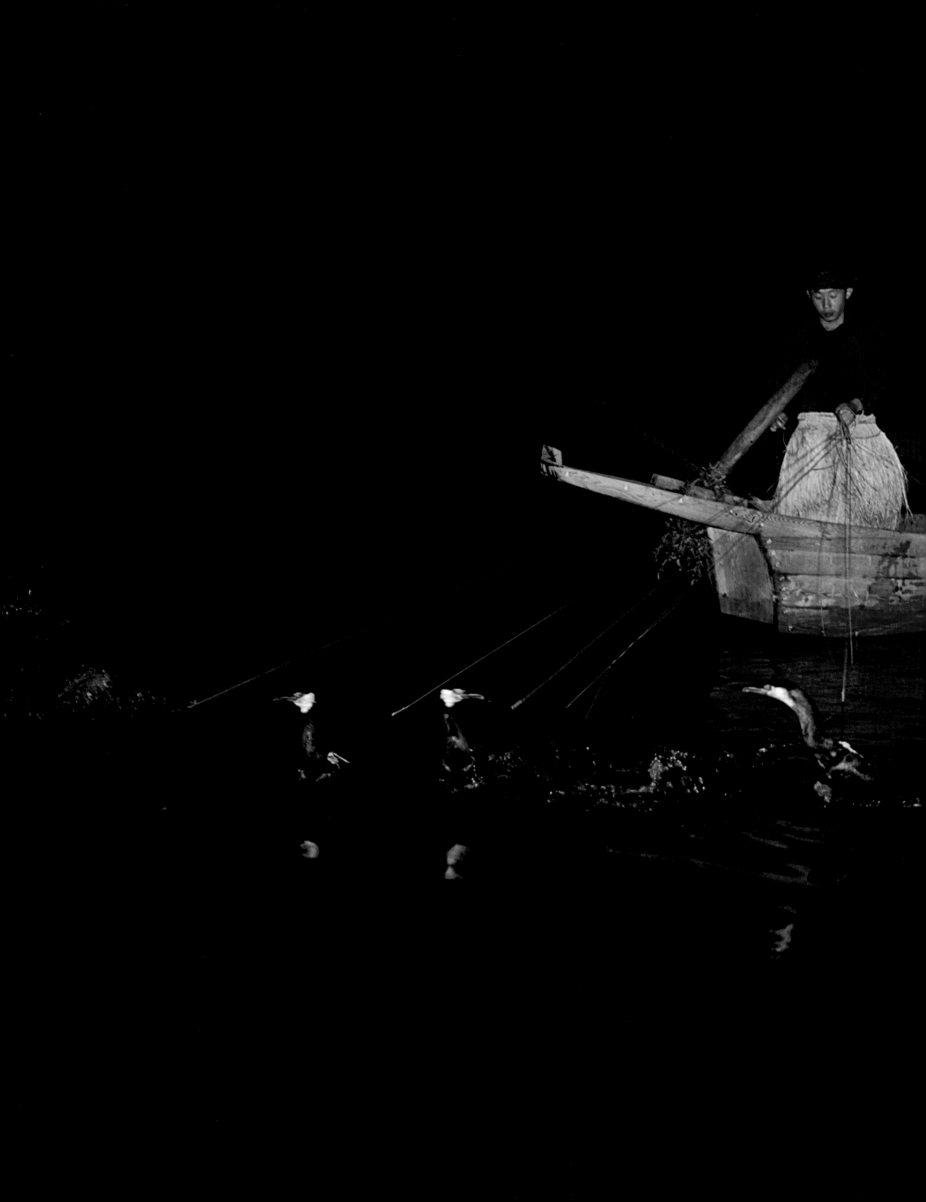

A cormorant fishes for its own prey underwater (top). A sleek, fast-moving bird, the cormorant is well adapted to underwater life.

With a bright fire hung off the side of his boat to attract fish, a Japanese fisherman uses tethered cormorants to retrieve his catch. A ring around the bird's neck prevents it from swallowing the fish before the fisherman can extract it.

Anchovy fishermen using a purse seine net.

THE CLEVER PREDATOR

Societies that depended on fish as food and whose fishermen worked singly could not seriously deplete fish stocks in their waters. Even so, they recognized that overfishing could eventually affect their food supply and had to be avoided. Understanding that preserving the source of food was good strategy, they very spontaneously invented the first laws of ecology.

Godly Quotas

South American Indian religions still prohibit the taking of fish during the spawning season and dictate that young fish be tossed back. The number of fish taken must not exceed the amount that can be eaten in one day. Fish left on the beach to rot are an insult to the ghosts. Poor fishing, famine, illness, and death will soon follow.

Modern fishing gear has contributed to the loss of old ways and traditions that respected the creatures of the sea. A purchaser of native-made nets in Vietnam, for example, must pay a tax to buy a sacrifice to the sea goddess. The purveyors of imported, machine-made nets command no such homage. The implication is that the modern equipment is so efficient that it does not require outside, godly assistance. Thus allegiance to the old rules has been diminished and irreverence has led to a loss of appreciation for the fragile bounty of the sea.

Old-Fashioned Methods

Fish are highly sensitive to odors. Some strong-smelling substances have been used to attract fish since earliest times. Favorites are anise, musk, heron's oil, and castor. Fishermen on the Upper Rhine River still put female salmon in their traps during spawning season to lure the male to the trap.

The fishermen of Oceania know that sharks are as curious as cats. Vibrations in the water will draw sharks from great distances. The natives beat the surface of the water with coconut shell rattles. The noise and disruption in the water draws sharks to the surface where they are caught with rope loops.

376

Some marine animals ignore exotic odors and members of the opposite sex most of the year. Shy animals try to hide: lobster traps and octopus jars are based on this trait.

Animal Assistants

Any animal whose natural habitat is the water is candidate to be a companion to the fisherman. The otter is such an animal. Marco Polo brought back stories of Chinese fishermen using freshwater otters as assistants in the Yangtze River. The animals were muzzled so that they wouldn't eat the fish as they chased them into nets, like sheep dogs herding their flocks up a hillside. The otters remained with their masters for their lifetime—up to 15 years.

Cormorants have been used in river fishing in Japan since 813 A.D. They are taken captive when only a few months old and undergo training sessions four or five times a day until they are accustomed to their masters. Just like the falcon, they dive for their prey on command. A string around the throat prevents them from swallowing all but the smallest fish, which serve as their food. Efficient birds can catch up to a hundred fish in an hour.

Remoras have a large sucker that has evolved as a spectacular restructuring of the dorsal fin. Fishermen in diverse parts of the world string the remora on a line to use as a living fish hook. Amazingly, the remoras are strong enough to withstand the tug-of-war between fisherman and fish.

Modern Fish Factories

Practically every creature that creeps, swims, or crawls in the sea is relished by the Japanese, and they have traditionally been the leaders in the fish business. The Japanese have recently slipped behind, under the massive onslaught on the sea by the Russians, but they still hold their own as one of the world's most powerful fishing nations.

Japanese fishermen take over nine million metric tons of fish each year, including 60 percent of the world's catch in tuna. Their boats dominate the Pacific and are often in the Caribbean. In 1957 the waters of the Atlantic were entered by Japanese boats for the first time. The tuna catch is already down, indicating overexploitation. An identical situation has developed in the Indian Ocean, where over 200 Japanese boats continually fish for a diminishing number of tuna.

The Russians also have a far-flung fishing operation. Administrators direct fishing activities in each of the major seas like fleet commanders maneuvering their forces in battle. As many as 300 ships will travel in massive flotillas accompanied by factory ships and transport vessels.

Huge factory ships begin the processing just as soon as a catch is taken onboard. One such ship, the *Professor Baranov*, is 543 feet long. It is able to salt 200 tons of herring, process 150 tons of fish into meal, filet and freeze 100 tons of bottom fish, and manufacture five tons of fish oil, 20 tons of ice, and 100 tons of distilled water—all in one day. The cost of a Soviet invasion of the sea is absorbed by the government. A fishing fleet with such a high subsidy from national funds certainly puts other fishing nations—and the fish—at a distinct disadvantage.

El Niño and the Anchovy

Peru's fishermen concentrate on one fish—the anchovy—and fish only within 200 miles of shore. Even so, the world's largest commercial catch, by weight, is Peruvian.

The anchovy industry did not begin in earnest until 1957. Easy success led hundreds of boats into the field. Subsequently, dozens of fish meal factories were constructed to process the catch. In 1970, $340 million in foreign money entered Peru as a direct result of the anchovy industry.

Coastal Peruvian waters are abundant in food for the anchovy. Upwelling, wherein deep water is drawn to the surface, supplies nutrients to support plentiful plankton growth in the cold waters pulled northward by the offshore

Humboldt Current. Every few years a warm current dips down from the equator to displace the Humboldt. It is called El Niño ("The Child") because it often makes its appearance near the Christmas season. The warm current interrupts upwelling, plankton growth decreases, and the life of the anchovy becomes a chaotic search for diminishing food supplies. Their schools disperse as they are further reduced by an influx of predators—yellowfin tuna and hammerhead sharks—that move in on the warm stream. The air becomes foul with the smell of dying fish and "guano birds" that depend on the anchovy for survival. Fishermen come home with empty boats, tarred black from the decay in the water.

The past two El Niño situations, in 1965 and 1972, had particularly extreme effects on the anchovy. Usually the fish population quickly recovers from the devastation. But by the end of the 1972 season, only one-quarter of the number of fish taken the previous season had been landed. El Niño had come but the fishermen had to take their share of the blame too: they had landed in excess of two million tons of anchovies the previous year over what had been recommended by fisheries biologists. The overall population of the anchovy is now estimated at two million metric tons, a reduction from 20 million tons in 1971.

The guano bird population has been similarly affected. Their return to normal population levels after El Niño ordinarily takes only a short time. However, after El Niño in 1965 the birds did not return, their population now stands at 4.3 million, down from a 1957 high of 27 million.

Both the birds, whose droppings are sold for fertilizer, and the fish are important economic factors in Peru. Their disappearance would be a financial disaster.

Peak production of this oyster farm on Oahu, Hawaii, will be six million oysters a month. Here, a worker collects sample spats of the maturing oyster.

Why do we not harvest great quantities of plants from ocean fields as we do from those on land? Primarily because the majority of the ocean's plants are one-celled. At the present stage of technology, it is not economically possible for our machines to imitate the baleen whale and filter tons of water to extract microscopic organisms. Even if it were, the majority of plants in the "algae soup" are encapsulated by indigestible coverings of cellulose, silica (the main ingredient of glass), or calcareous plates.

The only sea plants that are of use as food to man are those that grow in fixed locations—seaweeds. Forests of the giant kelp *Macrocystis* grow along the coasts of California, Brittany, and the southern coasts of South America. Harvested and processed, they prevent crystals from forming in ice cream and give toothpaste and paint a creamy texture. Algin, from kelp, maintains the foamy head on beer. Extracts from brown kelp are found in over 300 products from chemicals to fertilizers.

Commercial kelp operations rely on natural beds for raw material. The giant kelp recovers rapidly from harvesting; it grows at the rate of one to two feet in a single day. Strands and streamers several hundred feet long are cut to about three feet beneath the water by equipment that simultaneously carries the harvest on board barges. Since the reproductive parts of the plants are not destroyed, in some areas the same field can be cut again in four or five months. Cutting the crop may even contribute to kelp growth by allowing greater amounts of sunlight to reach new growth.

Seaweed Succulents

In the past 10,000 years we have learned to irrigate, fertilize, and develop hardy breeds of grain and stock. An acre of land, scientifically farmed, is far more useful in human terms than an agriculturally idle one. Yet thousands of years after we abandoned hunting on land as an efficient method of obtaining food, we continue to pursue the creatures of the sea with the attitudes of cavemen.

FARMING THE SEA

Ocean farming—mariculture—can protect the natural stock in the sea as well as vastly supplement our food supply.

Grow Your Own

Legend says that the 300-year-old Japanese practice of oyster farming was started by a lord who moved from Wakayama to Hiroshima. He carried oysters from his former home to set in a bay near his new lands. His neighbors there had placed bamboo fences in the water to protect their clam beds from roving bat rays. Soon the fences were speckled with tiny oysters, called spat because it was thought that young oysters were blown out of the mouths of adults.

Fishermen in the bay tried to move the attached spat to other waters. When the larvae didn't survive, it was concluded that certain conditions had to exist for successful oyster growth. The water temperature can range only between 59° and 86° F. The tides must frequently change the water but must not be so swift as to tear the animals from their moorings. A supply of phytoplankton has to be available as food. To this day, oyster farmers rely on these determinations to guide the management of their crops.

Oysters are ideal domesticated animals. They are easy to collect, grow rapidly, and require small living areas. Their culture may soon be even easier with the recognition that sewage can be diverted into special ponds to support algal growth. Oysters, seeded into the pond, graze on the algae. Marine biologists theorize that a 50-acre pond could provide the final stage of sewage treatment for a city of 11,000 and an annual crop of one million pounds of shellfish.

Highly Cultured Shrimp

Egg-carrying females are captured and placed in saltwater tanks that imitate the natural shallow bays where the animals breed. Within 24 hours the water in the tanks usually becames pink and foam appears on the surface as an indication that the two or three females in the tank have each produced their expected 300,000 to 1,000,000 eggs. Initially fertilizers are added to the rearing tanks to stimulate the growth of algae—shrimp baby food. Later brine shrimp and then clams and fish supplement their diet. At the postlarval

In this Japanese oyster farm, strings of maturing young oysters are suspended from horizontal poles at the water's surface. The vertical growing of oysters allows more shellfish to be raised than the old way, which limited culture to the bottom.

In Puget Sound, Washington, fenced-in areas of open water are used to farm salmon.

stage, the young take up a bottom existence, burrowing in the sand at the bottom of their tanks. The water circulation is kept swift to prevent cannibalism and to keep oxygen levels high. Between 60 and 100 percent of the infant shrimp survive their stay down on the farm and find their way to a kitchen. Shrimp are among the very few cultivated marine foods that are considered tastier than the wild variety.

Hot Fish

Anyone who has tried to keep a saltwater aquarium thriving has had a glimpse into the difficulties facing farmers of marine fin fish.

The Japanese, whose island life-style traditionally sent them to the sea for food, have achieved success in raising a true ocean fish—the yellowtail. One-third of the yellowtails marketed in Japan are cultured fish. The farms' annual yield is as much as 126 tons of fish per acre. An advanced pig farm produces only 25 tons of live pigs per acre in the same amount of time.

One fish that is a million-dollar-a-year commercial success in Florida is the pompano. Pompano have great potential because they grow rapidly, don't have a highly specialized diet, and can tolerate the abrupt environmental changes that often occur in artificial ponds.

British researchers have coupled sole and plaice farms with warm water discharges from coastal power plants. The heated water allows sole to be raised farther north than they usually range and speeds their growth to the degree that they are ready for market a full year sooner than if they were raised in their natural cool water habitat, although this is still in the research stage.

Tasty Cannibal

Lobsters are ill-tempered, solitary beasts. A tank containing several will soon hold only one—the biggest. The major frustration facing lobster

380

farmers is that their charges persist in eating each other before they can be sent to market. One solution to the cannibal problem may be in an anti-aggression pheromone secreted by the female during mating.

Cultivation begins with the capturing of egg-bearing females. Once the eggs are hatched, the young are collected on screens that filter water from the mother's tank. The larvae are placed in fiberglass rearing tanks where a steady diet of minced clams inhibits cannibalism. Over 30 percent will survive to the fourth molt. Usually the fourth-molt youngsters are released into the wild to supplement natural populations.

Lobster growers are not true farmers because they do not have control over their herds. They are more like ranchers since they release their stock to fend for themselves.

Turtle Soup

Turtle meat and eggs are a favorite meal for many people of the southwest Pacific or the Indian Ocean. Turtle soup is traditionally the opening course at British royal banquets and at the meals offered by the Lord Mayor of London.

Dr. Robert Schroeder, director of the world's first turtle ranch, is attempting to save the green turtle from extinction and to salvage a source of food for man by raising turtles on one of their last remaining rookeries on Grand Cayman Island, British West Indies. Eggs are taken from native breeding grounds in Costa Rica and Ascension Island and flown to the Grand Cayman ranch. Since the natural populations are diminishing so quickly and natural supplies of eggs are small, Dr. Schroeder is expending most of his efforts in encouraging his own turtles to breed.

The Grand Cayman experiment was not started exclusively for turtle conservation. The ultimate aim is to make a profit by obtaining food from the animals so the wild populations will be left alone.

Sea Steak

Abalone steak has been called the filet mignon of the sea, not only because it is delicious but also because, in most parts of the world, its wholesale price is higher than that of prime cuts of beef.

A man who once wrote a paper demonstrating the impossibility of farming abalone is now eating his words along with a steady diet of the shellfish. He and two other researchers oversee a $25' \times 45'$ structure that shelters a million abalone and insulates them from fluctuations in their diet of algae.

Super Salmon

Scientists at the University of Washington are bringing man's knowledge of genetics to the fish farm by using the unique breeding behavior of the salmon. Since these migratory fish mate only at the site where they were hatched, creating a new hatching sea would ensure that only certain fish mated.

In 1947 university scientists released fingerling salmon into an artificial pond that led into a network of streams to the Pacific. The earliest returning, fastest growing, and healthiest females were isolated from the remainder. They were bred with the most desirable males. No other salmon were allowed to reproduce. After 18 years of selective breeding, the salmon have increased in average weight from 10.8 to 12 pounds and are an inch longer than their ancestors. Over half the salmon now mature within three years. Egg production is up 10 percent.

The amount of time it took to achieve the super salmon is a handicap to the project. If mariculture efforts were given time and support equivalent to agricultural research, more practical sea-farming operations could be expected. But mariculture is new and investments in research small. Since most fish take two to three years to mature, a 20-generation experiment would take 60 years to complete.

Mariculture experiments have not been carried out in the open sea, primarily because of its lack of natural productivity. Now, however, in the Virgin Islands where deep water is only a mile from shore, a system has been developed to create artificial upwelling. Pumps bring bottom water to surface tanks where the nutrients activate algae growth. Shellfish then feed on the plant life.

The major problem of creating an open-sea oasis of plant and animal life is containment. No farmer wants to watch his market-ready crops swim away or be taken on another's lines. Some suggest that bottom water could be pumped into a coral atoll. Nutrients in the water would make the basis for a lush community into which food fish could be introduced. The abundance of food in the lagoon and the natural barrier presented by the atoll would discourage wandering.

Another method of fencing in fish is the use of "bubble" fences. A mechanism is placed on the bottom to encircle the open-sea farm. It releases air, creating a wall of bubbles. This may be able to contain some species of fish, although sharks have been known to cross such bubble barriers.

The earliest record of the effect of algae toxins was written by Moses. He described the waters turning to blood and stinking from dying fish. We now know that dinoflagellates cause red tides, killing fish by depriving them of oxygen, and by a toxin they produce. These poisons, when diluted, stop the growth of most types of bacteria and may, in the future, give relief to sufferers of bacterial diseases. The contents of penguins' stomachs support this hunch.

Scientists on an antarctic expedition found that the intestines of penguins were remarkably free of bacteria that usually inhabit animal digestive organs. Clearly, an antibiotic substance was present. Its source was traced through the birds' main diet of krill to the krill's diet of green algae. From the plant was isolated halosphaerin, a strong antibiotic.

Alginic acid from kelp has a unique characteristic. It rids the body of radioactive strontium. Since this isotope of strontium is the most dangerous to human life of all the components in fallout from atomic explosions, this could be an important lifesaving discovery.

The cone snail ejects its venomous darts to penetrate the skin of a suitable prey that falls within its range. We could almost speak of the cone snail as biting its prey since the dart is a highly evolved modification of a tooth. The toxin contained in *Conus geographus* is a muscle relaxant so strong that animals injected with it relax, stop breathing, and die. The toxins may be helpful to individuals whose muscles are in a state of convulsion.

Like the cone snail, clams, oysters, squid, octopuses, and abalone are molluscs. Extracts from particular members of these groups are effective as antiviral drugs. Called paolin, Chinese for abalone extract, components of these materials have been shown to protect laboratory mice infected with influenza and polio virus.

Suicide in Japan is an honorable way out of a dishonorable situation. An ocean fish that is a prized delicacy is also an often-used means to a quick release from life. The skin, gonads, or viscera, depending upon the season of the year, of the pufferfish are so toxic to human beings that death can occur within 15 minutes after they've been eaten. There is no antidote.

A mixture of puffer gonads and sake, a Japanese liquor, is said to increase virility. Drinking the concoction is like playing Russian roulette. There is no way of assessing the degree of toxin present or the chance of dying, without sampling it first.

The essence of puffer poison is tetrodotoxin—a powerful blocking

MARINE MEDICINE CHEST

Herbal Cures

Venomous Cone Snails

Fishy Treatments

agent that acts on muscles and the nerves that govern movement and receive pain. Tetrodotoxin is commercially available for use as an antispasm treatment for epileptics and to relieve the agony of terminal cancer.

The Heart of a Hag

Our heartbeat is regulated by nervous impulses and an intriguing tissue within the heart itself called a pacemaker. If either the nerves to the heart or the pacemaker cease to function, heartbeat stops or becomes so erratic as to be useless. An electronic, transistorized pacemaker must be implanted. A chemical from the hagfish may make electronic pacemakers obsolete.

Only one of the hagfish's three hearts is controlled by direct nerve connections to the brain. A chemical, eptatretin, stimulates and coordinates the beat of the other two. When mice with damaged cardiac nerves are given eptatretin, normal heartbeat is restored.

A SEA OF RECREATION

One of the greatest potentials the ocean has for mankind has little to do with oil or diamonds, pharmaceuticals, or even food. It is the opportunity to contemplate, enjoy, and explore this new realm.

Skimming the Surface

Water skiing was invented in the French Alps. That is not as improbable as it sounds. The Chasseurs Alpins, a select group of soldiers skilled in skiing and alpine warfare, were the first to try skiing on water. After a bout of drinking, one group of "chasseurs" challenged another to try their skill behind a boat. The long, narrow snow skis failed to be practical, but the daredevils soon tried again with wider skis similar to those used today.

The first water ski was patented in 1924 by an American, Fred Waller. In his design, the tow rope was attached to the skis and the skier had to hold on to another rope that was tied to the tips of the skis. Today six million water skiers glide over the surface of lakes and the ocean on fiberglass and laminated-wood skis that give them stability.

Quieter and cleaner than the powercraft pulling the skiers are the sailing boats. They are too slow to pull water skiers, and they are unsuited to so-called sport and "game" fishing. Nevertheless, sailing is the sport of the "true" sailors; it offers to its fans the rare opportunity to forget the pressure of modern life and to struggle, with bare hands, against the natural forces of the winds and the sea.

Fishing for Fun

To the commercial fisherman, the living resources of the sea represent his paycheck. To others, fishing offers a recreational outlet from the daily routine of life, a reason to be out-of-doors, to breathe clean, salt air, and to feel the spray of ocean breezes. One of the other reasons for the popularity of fishing is that the mute animals have a silent agony. To reassure one's conscience, it is said that they are cold-blooded and don't feel pain. Of course such beliefs are totally without foundation.

We look forward to the day when a politician will realize that he will get more votes by his support of a fish sanctuary than by posing in a photograph with a giant dead fish that he and his buddies have caught just to prove that they are "good old boys."

Surfers

Today's surfing enthusiasts are not part of a new wave as many of them think, but are practicing a sport of kings, 1000 years old.

It was on the islands of Oahu, Hawaii, and Maui that the sport first took hold. The early boards were called olos and were made from a balsalike wood, wiliwili. The olos could only be used by royalty. The less desirable beaches and heavier, shorter boards (koas) were used by commoners. The long royal boards were up to 16 feet long and weighed over 100 pounds.

Surfing continued as a popular pastime in the islands until 1821, when

the first of the Calvinist missionaries arrived from Boston to put an end to what they considered the sinful, pagan activities of the Hawaiians. The mumu was designed to cover their bodies and surfing was forbidden to all Hawaiians.

Not until the sport was recognized as an asset to tourism did it regain popularity. After World War II surfing really caught on, spreading from Hawaii to California and wherever else a ride could be coaxed from a wave. Soon a $12-million-a-year business centered in California was in full swing.

Millions of Divers

Diving equipment has opened up the earth's last frontier for exploration by adventurers, photographers, businessmen, and housewives. The world of fish is now our world too. Since the introduction of the aqualung, the sport of underwater exploration has been taken up by a million enthusiasts in the United States alone. Diving is a challenging sport, unique in that it combines the use of skill and strength with self-control and aesthetic appreciation—exercise with educational opportunities. Hardly anyone diving in an area rich in marine life can resist the sudden need to learn about the new world that he sees before him.

Converting Spears to Cameras

In 1936, when the first undersea explorers made their debut in the sea, crowds of fish came to welcome them; they soon understood that the intruder was there to spread destruction.

Many divers take advantage of their time of freedom in the sea to hunt down and kill almost any ocean creature they find there. Over 850,000 spearfishing kits are sold each year in the world. There are known to be over three million spearguns presently owned. It may take as little as one year for one single spearfisherman operating every day to practically annihilate the bustling fish life of a one-mile stretch of coral reef.

The sport of stalking and killing animals is just as odious as the so-called big game fishing, since neither is motivated by the desire to catch choice food, and both end up by wasting large amounts of fish. The game, for them, is only killing in a fictional demonstration of virility.

We encourage spearfishermen to consider the new hunting technique that many land sportsmen have adopted. The prey is stalked; the hunt is on. But the animal is captured on film rather than killed. Underwater photography presents a challenge in itself. Furthermore, the hunter can share his discoveries with nondiving friends and family.

SEEKING
A MORE
ABUNDANT LIFE

Ever since the first living cell divided in two, all the creatures of ocean and earth thrived—thanks to, and at the same time in spite of, their environment. Nature was lavish with both opportunities and obstacles, and to live meant simply to take full advantage of all opportunities and to cope with or, better yet, overcome the obstacles. Survival was a permanent challenge for both the individual and the species. For any individual, the bare essentials were to grow and to last as long as possible, and for any species, to ensure reproduction. But beyond survival, all living things strive for a better life. The awesome migrations of some birds, fish, or mammals are the result of millions of daring attempts by their ancestors to seek for the best, the most secure, the most comfortable conditions the planet could offer.

For three to three-and-a-half billion years, the physical, instinctive, and preintelligent forms of life have met the challenges of the surrounding world; those that were successful have achieved a certain degree of security and access to as large a span as they could conquer. This pattern of existence, at the same time generous and rigorous, was born in the sea long before it was extended on land, and it is not surprising that it is in the sea that the most formidable challenges of nature have been successfully overcome.

Changes in the conditions of habitats have always presented an almost infinite variety of conveniences and of sudden threats. Most probably the diversity of natural challenges triggered the incredible flexibility of the evolutionary processes and helped generate the hundreds of thousands of different plants and animals. Before we extend our remarks to human beings, we have to realize the built-in fecundity of hardships. In a handful of soil, there are billions of microbes, constantly struggling against adversity—too much rain, heat, cold, or drought. Given the innumerable ways in which life is challenged, it is amazing to think of the number of creatures that win the battle for life. The whole earth and all its creatures are intimately involved in a dynamic adventure in which every happening has an effect, however remote, on everything else; and, of course, man is included in this universal dynamic interrelationship. It is overwhelming—or consoling—to realize that we are all dependent on one another in the struggle to survive.

The advent of civilized man modified the pace and the nature of his own evolution. It is highly probable that the success of *Homo sapiens* is not the direct result of the evolution of his brain, but as we have already realized, is due to the simultaneous development of brain, hands, articulated voice, and longevity. These combined abilities gave birth to civilization, which is the storehouse of accumulated experience. Men could not develop as human creatures, independently from their cultural environment. Modern civilization created the means to fulfill easily the basic needs of all men: food, shelter, clothing, health, and education. Although these essentials are not shared by all, their universal availability remains a major goal of civilized men.

Having—at least theoretically—eliminated the natural challenges of physical life, we have turned to spiritual or intellectual challenges in order to increase the scope and the quality of our life. Reaching for the moon or for the bottom of the ocean is a form of the universal quest for a wider frame of life—the same drive that started the ancestors of the migrating birds. Why would the eels of Europe and of North America travel all the way to the Sargasso Sea to lay their eggs? Why would the salmon hurdle rapids to get at its freshwater spawning ground? Why would a sane man choose to go around the world in a rowboat? Most of the challenges we face today are man-made: overpopulation, waste of resources, and destruction of our environment. Our deep motivation is to overcome our problems, even if they originate in ourselves. It is the search, the fight, not the achievement or the victory that provides us with the closest thing to happiness. We cannot do without challenges. If there were none, the world would become meaningless—it would simply obey the second law of thermodynamics and slowly grind to a halt.

Archaeologists working underwater

Elusive green sea turtles find their way 1500 miles across the Atlantic to Ascension Island, the main rookery in the South Atlantic, once every three or four years.

AEONS BEFORE MAN

Aeons before man ventured out on the open sea, thousands of other creatures were striking out over the ocean on journeys thousands of miles long. Some early sailors were aware of the migratory flights of birds and depended upon these creatures as their only method of navigating the seas. Much later, when man finally learned the basic principles of navigation, he still had to struggle for centuries to untangle the patterns of the stars—a process that birds, we know now, understood instinctively.

It is remarkable that so many different groups of animals have developed migration patterns and the ability to navigate. These innate behavioral traits are seen in animals as distantly related as the butterfly and the eel. The same behavior has arisen independently in many animals, and a number of navigational techniques serve various organisms equally well.

Mysterious Pathfinders

There are seasons in the sea just as on land. Some fish respond to these changes as regularly as do migrating birds. The shad, largest fish in the herring family, times its annual spawning runs into freshwater streams to seasonal changes in water temperatures—when the water there is between 13° and 18° C. (55.4° to 64.4° F.). Since all the members of the breeding population in one area answer the same temperature-regulated spawning call, most arrive at the breeding grounds at the same time—an obvious help in ensuring the continuance of generations.

Green sea turtles appear on island breeding grounds at certain times each year in the Caribbean and South Atlantic. Experimental evidence indicates that seasonal variances in water temperature are not the stimulus that triggers the turtles' migratory urge.

Both male and female turtles push their heavy bodies 1500 miles across the Atlantic to Ascension Island once every three or four years throughout their lifetime. Navigational methods used to find their tiny island in the midst of thousands of miles of open sea remain a mystery. The current theory is that the turtles travel up and down the coast of Brazil until they locate (by smell and visual clues) the place where they made their first landfalls in youth. Then a compass sense, combined with an ability to orient from the position of the sun as it moves across the sky, brings them close to Ascension Island; then olfactory and visual messages guide them in.

It is the belief of some scientists that eels, who spawn in the Sargasso Sea southeast of Bermuda, as well as some salmon and sharks, may navigate by use of the weak electric fields generated in the ocean by the movement of currents through the earth's magnetic field. Laboratory experiments indicate that eels are sensitive to currents that are within the voltage generated by the movement of the major ocean currents. Not only can they sense the direction of the current, they may also be able to register, electronically, an upstream or downstream flow.

390

Solar and Celestial Clues

The first clues to avian navigational methods were found in 1949 by Gustav Kramer, a German ornithologist. He observed that during the day in the spring caged starlings faced toward the northeast, the direction that starlings migrate in the wild. Kramer's birds could see only the sky. Apparently the sun was guiding them. But this exciting discovery did not explain how night-flying migrants find their way without cueing on the sun.

An incredible explanation for nocturnal navigation was given in 1955. Franz and Eleanore Sauer observed that warblers, given an unobscured view of the nighttime sky, oriented in the direction of their regular migratory paths. When the sky was heavily overcast, the birds ceased to orient. To prove that the birds used the stars as guiding lights, the Sauers took their birds into a planetarium where the features of the sky at any season in any locality could be projected. When a springtime sky was presented to a warbler he faced northeast. A bird that had spent his life in a cage instinctively oriented toward his species' migratory route.

Living Clocks

Navigation by sun and stars requires a built-in time sense. The animal must be able to adjust the angle of its direction relative to a daily rhythm. It must know the local time in order to make navigational corrections for the local angle of the sun. Celestial navigation must also be based on time: the stars change their positions with the hours and the seasons.

The lives of many animals are regulated by internal clocks. The urge to migrate at a certain time of the year is apparently a response to length of day. Called photoperiodism, it was demonstrated by inducing migration out of season in birds that were subjected to artificially lengthened periods of daylight.

Migratory journeys are not the only seasonal changes in the living world that are responses to a time sense. Each year during the first days of the full moon in October and again in November vast swarms of palolo worms surface above their Pacific reef habitats at dawn. Only the hindpart of the adult worm rises to reproduce. This portion disengages from the head, which stays alive in the reef, and comes to the surface to shed eggs and sperm and then disintegrates.

The cycles of the moon control the lives of all tidal organisms since the tides themselves are regulated by lunar cycles. The reproductive organs of sea urchins and oysters enlarge and mature in response to lunar periods. The very color of fiddler crabs is timed to a 24-hour clock. Their bodies darken in

These acetabulariae, looking like inverted parasols, are large single-celled algae. Both the cytoplasm and the nucleus have biological clocks that regulate photosynthetic cycles.

French painter Théodore Géricault's The Raft of the ''Medusa'' *(1819) dramatically depicts a tragic shipwreck in which hundreds of men were lost and only a handful survived on a makeshift raft.*

the morning and lighten in the evening. Scientists were astounded to discover that levels of activity, as well as the color change, of fiddler crabs that were kept in the laboratory far from their home maintained the tidal times. Compensations were even made for the fact that the tide is 50.5 minutes later every day.

Since the time sense of many organisms is maintained when they are removed from their natural environment and from solar, lunar, and tidal clues, some scientists believe that the electromagnetic fields of the earth and the influences of the celestial bodies may be the mainspring of living clocks.

MEN AGAINST THE SEA

Voyage of Horror

The wreck of the *Medusa*, with its grisly aftermath, was perhaps the most horrible shipwreck in history. On July 2, 1816, the 40-gun French frigate ran aground on a shoal about 60 miles off the west coast of Africa and had to be abandoned.

The six lifeboats could hold only about 250 of her company of 400, and so a great raft was built to hold 200 persons and provisions. One hundred and forty-five men and one woman started out on a sunlit sea. They had no mast, no anchor, no cable, no lines, no chart. They were completely at the mercy of the sea. They did have a 25-pound bag of biscuits soaked with salt water, a few barrels of wine, and several casks of water. But that was about all.

On the first night some of the men were swept away and drowned. On the second night the survivors opened a keg and drank its wine. Madness followed. Crazed by suffering, the men started a frenzied revolt against the handful of officers. Many died beneath saber blows in the darkness—60 in all. One had been hacked to death with an ax, some had been held underwater until they were dead. At dawn 67 remained.

Sharks nosed about the raft now as it drove back and forth before the wind. The water was gone and so was most of the food. It was not long before one man began to hack away at a dead body. A moment later, dozens of them fell upon the corpse like a pack of wolves. One who lived to tell about it wrote: "Seeing that this horrid nourishment had given strength to those who had made use of it, it was proposed to dry it in order to render it a little less disgusting." In the night 12 more died. There were 48 left on the fourth day. That night mutiny raged again on the foam-covered raft. Both sides fought desperately, and in the morning there were only 30 survivors, all of them wounded.

On the sixth day a consultation was held among the healthier survivors. They decided to throw the dying to the sea and the sharks. The woman was among them. It allowed those remaining to sustain themselves for six additional days before they were finally picked up. In the short space of 13 days they "had seen and taken part in such horrors as happily fall seldom to the lot of man."

"Iceberg! Right Ahead!"

The *Titanic* was once the largest ship the world had ever known. She was also believed to be the safest. Her builder had given her double bottoms and had divided her hull into 16 watertight compartments. The world's largest ship was thought to be unsinkable.

When she set out on her maiden voyage from Southampton to New York in 1912, there were over 2000 men, women, and children aboard, including some very prominent persons. At 9 A.M. Sunday, the third day out, a message was received in the wireless shack: CAPTAIN, *Titanic*— WESTBOUND STEAMERS REPORT BERGS GROWLERS AND FIELD ICE IN 42 DEGREES N. FROM 49 DEGREES TO 51 DEGREES W. 12th APRIL. The floating hotel steamed on toward its destination.

At 1:42 P.M. the *Baltic* called the *Titanic* to warn her of ice on the steamer track. The operator sent the message up to the bridge. The officer on the bridge sent it on to the captain, who passed it on to one of the passengers, the director of the White Star Line to whom the *Titanic* belonged. He read it, stuffed it into his pocket, mentioned it to a couple of ladies, and resumed his promenade.

At 7:15 the captain asked that the message be returned to him so he could post it for the information of the officers. And the great ship, its speed unslackened, plowed on through the night. It was exactly 11:40 P.M. when it happened. The lookout in the crow's nest could not believe his eyes. But only for an instant, and then he knew it was a deadly reality. An enormous white shape floated directly in the *Titanic*'s path. Frantically he struck three bells. He grabbed the telephone and shouted into it: "Iceberg! Right ahead!" In the engine room the indicators on the dial faces swung around to "Stop" and then to "Full Speed Astern." But it was too late.

Curiously, the collision did not even wake many of the passengers. To others there seemed to be only a slight jar and a crunching sound. Few could have guessed that the berg had torn a 300-foot gash in the bottom of the ship and that the sea was already surging into the hold. Orders were given that the band take their places on the deck and play—morale had to be kept up. Some of the musicians were to continue to play until the ship was almost gone. Rockets were sent up in the hope that a nearby ship might see them. Lifeboats began to go over the side—only half-filled. Finally the *Titanic* stood on end and began the plunge to the depths, slowly at first, then quicker and quicker. The forward funnel snapped and killed swimmers beneath it. At 2:20 A.M., just a little less than two and a half hours after the collision, the *Titanic* was gone.

Not all accidents end in tragedy; sometimes luck is on the side of the sailors. In 1973 two Britons were rescued from a submersible trapped on the floor of the Atlantic about 100 miles off the coast of Ireland.

The men, Roger Mallinson and Roger Chapman, were aboard the *Pisces III* helping to lay a transatlantic telephone cable between England and Canada. Capt. Leonard Edwards of the mother ship, the *Vickers Voyager*, explained what happened. ''While still on the surface, the hawser apparently tore off an atmosphere hatch, water poured into the flood compartment and the submarine sank. When it was 170 feet down, the hawser snapped and the submarine dropped to the bottom.''

It was there that *Pisces* rested, mired in the sediment of the ocean floor, 1375 feet below the surface and with only a 72-hour oxygen supply. To make matters worse, gale force winds, heavy seas, and poor visibility on the sea floor hampered rescuers looking for the 20-foot-long craft. About the only thing in the trapped men's favor was that they were able to maintain radio contact with the surface. They used valuable oxygen to stay on the radio in order to guide sonar toward the craft by singing Irish sea chanties and the Beatles' song ''Yellow Submarine.'' Mallinson spent his thirty-fifth birthday in *Pisces*, literally singing for his life.

The rescue was made three and a half days after the ordeal began. *Pisces II* and *Pisces V*, sister ships of the trapped sub, and an unmanned submersible were able to guide lifelines around *Pisces III*, which had been resting at a 70° cant. When Mallinson and Chapman finally reached the surface, there were only about 90 minutes of air left in the sub.

Songs from the Seabed

In 1973, off Key West, Florida, the hulk of a scuttled World War II destroyer was being considered as an artificial reef that might attract sea life for study. To investigate the feasibility, the submersible Johnson-Sea Link (shown here on another occasion), operated by the Smithsonian Institution, went down to take what should have been a leisurely look. But the small submersible got caught in the maze of lines and masts of the submerged wreck, and before it could be freed, two of its crewmen died of carbon dioxide poisoning and cold.

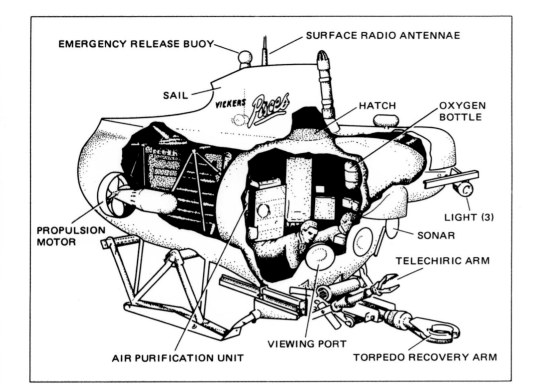

This cutaway drawing shows the inside of the submersible Pisces III that was trapped nearly 1400 feet down in the Atlantic off Ireland while laying a cable in 1973. The sub and her two crewmen were rescued without harm after about 70 hours.

Ready for Rescue

Since 1920 more than 1000 men have died in 29 United States submarines accidentally sunk in peacetime at depths below their hull-collapse limits. Officials of the U.S. Navy have estimated that all of these disasters offered at least the possibility of crew rescue if only there had been a rescue submarine capable of very deep dives. Today such a vessel exists.

The first of the navy's Deep Submergence Rescue Vehicles (DSRV) was launched in 1970. Measuring nearly 50 feet in length, the DSRV is designed to carry a three-man crew and have the capacity for retrieving 24 survivors at a time from a disabled submarine. Submarines of the United States Navy carry from 100 to 160 men. Two DSRVs can complete a rescue in two hours. Their maximum operating depth is 5000 feet, well below the collapse depth of any existing military submarine.

DSRV can only attach itself to standard-size hatches, such as those adopted today by all submarines of the United States and most of those of the members of the North Atlantic Treaty Organization. Unfortunately, none of the existing exploration submersibles has been equipped with such hatches.

Solo Circumnavigation

One of the greatest lures of the sea is the challenge of circumnavigating the globe. Every sailor knows that all it takes to accomplish this are a seaworthy craft, food and water, and knowledge about wind, water, and navigation. Yet it was only within the last century that a lone man accomplished this feat.

The lone navigator in this case was Captain Joshua Slocum, a native of Nova Scotia who became a United States citizen. The trip around the world was no foolhardy venture, for Slocum had considerable experience piloting fishing craft and cargo vessels. He planned his journey well, even going so far as to build his own boat, the 35-foot-long *Spray*, using the wreck of a 100-year-old oyster boat as a base to build on. Slocum shaped his vessel with new wood, patience, hard work, and an unheard-of caulking made of cotton and oakum.

At the age of 51, this daredevil mariner set sail from Boston on April 24, 1895, and returned three years and two months later.

During his journey, Slocum discovered the problems that beset solitary navigators, such as the immense loneliness that even his store of books couldn't always alleviate. And he quickly realized, after eating foul food and

jettisoning his supply of plums, that if he were ill, there was no one to man the ship.

The intrepid Slocum had covered more than 46,000 miles on his journey and was none the worse for wear, as he explained it. "I had profited in many ways," he later wrote. "I had even gained flesh and actually weighed a pound more than when I sailed from Boston. . . . And so, I was at least 10 years younger than the day I felled the first tree for the construction of the *Spray*."

His conclusion was not one of fear and loathing of the dangers he had encountered and the harrowing experiences he had been through. Rather, he wrote, "The sea has been much maligned . . . and the *Spray* made the discovery that even the worst sea is not so terrible to a well-appointed ship."

Surviving at Sea

I witnessed one of the oddest battles of man against the sea. We had *Calypso* in antarctic waters during the austral summer of December and January of 1972–73 and had been alerted to be on the lookout for the *Ice Bird*, a small vessel that was in distress. The only person aboard was a David Lewis, a medical doctor from Australia. The chances against spotting another ship on the high seas are as great as finding the proverbial needle in a haystack, perhaps greater.

On the morning of January 29, I awoke about 4 A.M.—it was broad daylight then—to find a small boat alongside the *Calypso*. I hurried topside—for we didn't want some wreck to damage the *Calypso*—where we were startled to see a man appear. His hair and beard were messy, his face reflected his ordeal. His spirits were good, however, despite the fact that his clothes were still wet from the time two months earlier that he had capsized, losing his mast. Shortly before, he had set out from Hobart, Tasmania, to circumnavigate Antarctica. He had managed to proceed 2500 miles with a makeshift mast, but was without radio or heater and was forced to eat only cold tinned food and biscuits since everything else had been wet or lost.

Dr. Lewis came aboard *Calypso* and told his story. He asked us to relay a message to his two daughters, aged 10 and 12, back home. It read: "Sorry, the boat is a bit broken, but I will mend it here. Hope you were not too worried. I miss you both so much. Was Christmas nice? I shared mine with the little ice birds. They had broken biscuits. There are thousands of penguins. Love my little girls more than anything in the whole world. Love, Daddy."

Dr. Lewis told us that as soon as he repaired *Ice Bird* he wanted to sail around the Cape of Good Hope. Incredible, but true.

To Prove a Point

A French physician named Alain Bombard had a theory that one could survive for long periods with only the nourishment that the sea itself could provide. In 1952 he set out in a rubber raft named *L'Hérétique* to prove his point. He planned to sail from Casablanca across the Atlantic to the West Indies. During the entire voyage he would eat only fish and use their juices and rainwater to provide the necessary liquids. Stowed in the raft were containers of emergency food and water rations. If the containers reached their destination with the seals unbroken, the expedition would be a success.

The first few days of the solitary voyage went well. "There were plenty of fish," he wrote. "Little flyingfish struck against my sail and fell in the raft." He varied his diet with plankton that he caught in a fine-mesh net. "It tasted like lobster, at times like shrimp, at times like some vegetable."

For 23 days he had no rainwater, but fish juices quenched his thirst. Still, he longed for great quantities of some liquid. "I dreamed of beer," he said.

Bombard discovered that the real difficulty was in what he called "the terror of the open sea." There was "no sound of fellow human beings or the familiar noises of the land. Only the rushing of the wind, the watery hiss of the breaking waves, the nervous flutter of the sail. And the face of the shark following you patiently, relentlessly—sometimes rubbing his sand-paper back under the raft."

Finally, after two months on the open Atlantic, *L'Hérétique* landed at Barbados. The emergency rations had not been opened.

In 1969 Thor Heyerdahl set out to prove the theory that ancient Egyptians had crossed the Atlantic to America in papyrus boats using ocean currents as navigational guides. Heyerdahl and six companions attempted to duplicate the voyage in a reed vessel named Ra, *in honor of the ancient Egyptian sun god.* Ra I *(shown here) was overcome by high seas, and some ten months later* Ra II *was built. Redesigned,* Ra II *successfully completed the transoceanic voyage from Safi to Barbados in just 57 days.*

OVERLEAF:
Antarctic ice of the type that crushed Ernest Shackelton's ship in the Weddell Sea; then began one of the most heroic struggles for survival in the history of polar exploration.

The jugs discovered at the site of the ancient shipwreck off Grand Congloué helped reconstruct the ship's last journey and dated the wreck as the oldest one found to that time.

Hero of the Antarctic

British explorer Ernest Shackleton's fourth expedition to Antarctica in 1914 intended to make the first crossing of the continent—from the Weddell Sea to the Ross Sea—but it never got underway. The ship was caught in the ice of the Weddell and crushed. Shackleton and his men found themselves stranded in the most remote of frozen seas with no way of telling the world of their plight.

By sled and sail, Shackleton and his men made their way to the nearest point of solid land—a hitherto untrod rock called Elephant Island. Leaving 22 men behind while he and five others went for help, they set out on one of the most daring voyages ever undertaken—a journey in a 20-foot open whaleboat across nearly 1000 miles of Antarctic Ocean in the remote hope of reaching a tiny inhabited speck in the South Atlantic called South Georgia Island.

The 16 days they spent in the *James Caird* were, in Shackleton's words, a tale of "supreme strife amid heaving waters." Once clear of the dangerous ice packs, they were hit by almost constant gales. Cramped in their narrow quarters and continually wet by the spray, they suffered severely from cold throughout the journey. There were no dry places in the boat, and bailing was a constant occupation. The incessant motion of the boat made rest impossible.

The meals were the bright beacons of their days. Breakfast consisted of "a pannikin of hot hoosh made from Bovril sledging ration, two biscuits, and some lumps of sugar." Lunch comprised the same sledging ration eaten raw and a pannikin of hot milk for each man. Tea consisted of the same menu. Then at night they had hot milk again.

A thousand times it seemed as if their little boat would be engulfed by the seas, but each time she survived. Spray froze on the boat and gave everything a heavy coat of ice that constantly had to be picked and chipped at. The men were frostbitten and had large blisters on their fingers and hands.

At midnight of the eleventh day Shackleton was at the tiller when he noticed a line that appeared to be clear sky between the south and southwest. A moment later he realized that what he had seen was not a break in the clouds but the white crest of the most gigantic wave he had ever seen in his 20 years at sea. When the wave struck, Shackleton wrote, "We felt our boat lifted and flung forward like a cork in breaking surf. We were in a seething chaos of tortured water."

Their hopes were buoyed when they saw two shags sitting on a mound of kelp. "Those birds," Shackleton wrote, "are as sure an indication of land as a lighthouse is, for they never venture far to sea." The next morning they were hit by one of the worst hurricanes any of the men had ever experienced. "The wind simply shrieked as it tore the tops off the waves and converted the whole seascape into a haze of driving spray." The seas drove them toward reefs where great glaciers ran down to the sea. Then, miraculously, the wind changed, and they were able to find a landing in a sheltered bay.

But their journey was not over. They still had to make the first trek over what had been thought to be impassable glaciers in order to reach the whaling station on the other side of the island. But they succeeded, and soon after the men who had been left behind on Elephant Island were rescued.

DIVING TO THE PAST

Many of the answers to our questions about man's past lie hidden beneath the sea. They are in ships that have sunk and in the cargoes they carried. They are in harbors or entire cities that either sank beneath the sea or were inundated by rising sea levels.

The future of marine archaeology is closely tied to advances in diving and underwater technology and also to education and to protection by governments of underwater sites. Looting has been practiced for centuries, but in recent years, with the enormous increase in the number of sports divers, it has become rampant. It is probably safe to say that there are no visible wrecks on the Spanish, French, or Italian coasts under less than 150 feet of water that have not been ransacked and almost totally destroyed. The challenge of marine archaeology is not to the diving scientist alone but to the value every man puts on rational inquiry.

The Archaeologist Joins the Diver

At the time of Odysseus's epic voyage, the Phoenicians had already begun their famous trade throughout the Mediterranean Sea, but no one knew this until marine archaeology told us. The excavation of a Bronze Age shipwreck off the southwest coast of Turkey in 1960 provided the evidence that the Phoenicians traded by sea as early as 1200 B.C. It was the first underwater excavation carried out methodically and to completion, and it was the first time that an archaeologist played an important role as a diver. He was Dr. George F. Bass of the University of Pennsylvania Museum.

A free diver made the first of a number of photographic montages of the site, and the positions of visible artifacts were plotted. The entire cargo was raised in coralline masses and excavated on land. Divers with hammers and chisels cut massive lumps free from the seabed and then they were winched to a boat on the surface. Pieces of concretion containing parts of the wooden boat were raised to the surface with the aid of two plastic lifting bags. The wood was fragmentary, but it contained pieces of planks with tree nails fitted into bored holes. The interior of the hull was lined with brushwood dunnage, as described by Homer, and the bark was still well preserved.

There finally emerged a picture of a small sailing vessel, about 35 feet long, that carried more than a ton of metal cargo. The cargo consisted largely of copper ingots, many of them stamped with signs in a language that still has not been deciphered—Cypro-Minoan. There were also broken implements packed into wicker baskets and pieces of casting waste. The cargo, then, was a load of scrap metal that had been destined to be forged into new weapons or tools.

Items were also found that evidently belonged to the ship itself or to its crew. They included a seal for stamping official documents, five scarabs, balance-pan weights, traces of food, including olive pits, stone maceheads, a razor, whetstones for sharpening tools, and an oil lamp.

In 1952 we discovered a wreck at a desolate rock called Grand Congloué, near Marseilles. Excavating that wreck would shed light on the life of men who lived 2100 years ago. We decided that we'd tackle it. We didn't realize that it would take five years and be the most extensive underwater archaeological operation yet attempted.

The wreck was a big one, approximately 110 feet long and with a cargo capacity of 350 tons. More than 7000 amphoras were recovered. The jars were of two types—squat Greek ones and slender Italian ones. They were used to carry wine, oil, or grains. As we unloaded them, we carefully noted their location in the wreck. It would help us to piece together the ship's last journey. Many of the Greek jars bore a seal with a trident and the letters SES. Professor Fernand Benoît was able to establish with this clue that the ship had been owned by a Roman named Marcus Sestius, who had lived on the Greek island of Delos at the time of the sinking.

To retrace the final voyage, we took *Calypso* to Delos. There we found in mosaic the letters SES worked into a design with a trident. The mosaic was to have been the floor of an unfinished villa. In our imaginations we can picture the powerful Marcus Sestius sending out his ship laden with wine in the fat amphoras of Delos. Stopping in Syracuse, it loaded more wine in Italian jars. At Naples it took on more than 7000 pieces of fine Campanian dinnerware. But perhaps a mistral was blowing as the ship approached Massalia, and it was tossed against the rock. And with the loss of his ship and his fortune, perhaps Marcus Sestius could not afford to complete his villa.

The Overdue Cargo

This 1864 Currier and Ives lithograph of the Union warship Monitor *and the Confederate* Merrimac *depicts one of the most memorable battles of the American Civil War and one of the most significant in naval history. Fought between ironclad vessels and won by the* Monitor, *it marked the end of the era of the wooden fighting ship.*

The clipper ship was the fastest sailing vessel ever built and, during the 1800s, helped bring continents closer together. These majestic ships were eventually replaced by steam-driven vessels. Clippers with their considerable speed were useful in the mid-nineteenth century in carrying East Coast passengers to the Gold Rush in California.

THE SEA THAT DIVIDES MEN

A Clash of Ironclads

On a Sunday in March 1862, a year after the American Civil War had begun, two ships clashed at Hampton Roads, Virginia, in a battle that made every navy afloat obsolete. Steam and armor plate had been combined in a new kind of warship, the ironclad.

The *Merrimac* had been scuttled by the Union when its forces had to abandon the navy yard at Norfolk, but she was raised by the Confederates and overhauled. Her masts and almost all the superstructure were removed. She was given a sloping roof covered with iron thick enough to deflect any cannonball that could be fired at her. The funnel rose out of the roof like a chimney, and ten square openings were cut in the armor as ports for the ten big guns—four on each side, one in the bow, and one in the stern. A sharp iron beak was fastened to the bow just beneath the waterline. It would prove to be a very effective ram.

Meanwhile, a Swedish engineer in the service of the Union was also working to combine the virtues of steam and iron in a new kind of fighting ship. The *Monitor*'s deck was also heavily armored, and her pointed bow was strongly reinforced to serve as a ram. A funnel rose from the low deck aft, and forward was a pilothouse made of iron beams. Amidships was a round gun turret with eight- and nine-inch iron plates. The swiveling gun turret represented yet another great stride in naval design and thus in naval tactics.

While the *Merrimac* was destroying one Union ship and putting four others out of action, the *Monitor* was steaming to meet her. The next day it was ironclad against ironclad. For hours their guns fired away at each other, but neither did the other any damage. At noon the *Merrimac* withdrew. No one had been killed or even seriously injured, but the battle had been one of the most decisive ever fought, for it was clear when it was over that the age of wooden warships was over.

Underwater Death

In the years immediately preceding World War II, the Japanese, like all the other great naval powers, still considered the battleship to be the backbone of sea power. They had built two monstrous ships, the *Yamato* and the *Musashi*, armed with nine 18.1-inch guns. Both of them were destroyed, not by other battleships but by naval aircraft. Appreciating then that it was not the battleship but naval air power that was to be supreme in the struggle for the seas, the Japanese hastily converted the still unfinished hull of a sister ship, the *Shinano*, from a battleship to an aircraft carrier.

On November 28, 1944, the *Archerfish*, an American submarine, was on patrol south of Tokyo. Its primary mission was to act as a "lifeguard" for any American plane that might be shot down on its way to or from a bombing mission over the Japanese mainland, but when the *Shinano* hove into view, its mission became to destroy the enemy. Four destroyer escorts barred every approach. Throughout the night and into the next day the submarine followed its prey.

On the sonar the pulsating rhythm of the carrier's propellers grew steadily louder. Finally, at 7000 yards, it came into view through the submarine's periscope. Conveniently, the nearest escort moved out of station to take a message flashed by the carrier. In doing so, she unmasked the giant ship at precisely the right moment. And a moment later the carrier made an alteration of course that left her a sitting duck for the sub. It was 3:17 A.M. Six torpedoes fired away at short intervals, and the submarine plunged to the depths. The first of the torpedoes ripped into the carrier's stern with a blinding explosion. Five other blasts followed one upon another. Each torpedo had found its mark.

The shock of the attack stunned the Japanese. The escorts loosed 14 depth charges but none of them found the submarine. Their shock was hardly felt aboard *Archerfish* compared to the jolt of the great ship breaking up on her way to the bottom.

The *Archerfish* never sank another ship, and perhaps that was appropriate, for she stood at the end of one era and the beginning of another. Naval air supremacy gave way to the age of the submarine.

THE SEA THAT UNITES MEN

Man's first boats were of three basic types. One was the dugout canoe—a log hollowed by fire or ax. Another was the raft, made from wood or reeds. A third was the skin-covered boat that probably developed from a bundle of reeds covered with an animal skin. •

These primitive boats gradually changed. The rafts became saucer-shaped, which helped keep their crews dry. Then they became longer, which

made them faster and easier to steer. Wooden paddles were found to be more efficient than hands. A long wooden paddle was used at the stern and thus the rudder was invented.

The Egyptians, who had invented the sail and the keel and who for centuries dominated the eastern Mediterranean in their ships made of long cedar planks, lost their lead to the Phoenicians, who developed a shorter, broader ship with one large square sail and with it became the great traders of ancient times. The next advance was probably the most important of all—the oarsman, usually a slave, was replaced due to developments in the use of sail and in ship design. Then for hundreds of years the wind did all the work.

In the early years of the nineteenth century the steamship was invented. It did not immediately replace sail. In fact, steam was at first used only for entering and leaving harbors and in times of calm. Even when the much more efficient propeller replaced the paddle wheel on the steamship, sail advocates remained staunch, particularly when the fast clipper ships arrived on the scene.

The *Cutty Sark* was one of the most famous of these "greyhounds of the sea." Built in Scotland in 1868, she first carried tea from China, often getting home 10 days before her rivals. Later the *Cutty Sark* entered the Australian wool trade and once came home in only 69 days, when the average voyage took 100 days.

There were disadvantages to the clippers, however. The sleek streamlined ships had very limited cargo capacity. To make the speed for which they were famous, an enormous amount of canvas had to be spread, day and night, fair weather or foul. That took a large, rough crew, and a captain had to be a superman to command them. Then cheap coal became available and steam engines were built that were more practical. The opening of the Suez Canal in 1869 shortened the route from Europe to the Far East, and for European sailors the day of the clipper was over. They lasted longer in the United States, where they were modified to hold more cargo and require smaller crews.

Coaling stations were established all over the world, and the tramp steamer came to dominate the sea lanes. It could carry any kind of cargo. With the great increase in world trade, specialized cargo ships were built. The most important of them was the tanker.

A supertanker is defined as any ship used for the transport of oil that weighs over 100,000 dead-weight tons. There are hundreds of such ships in operation today. Supertankers require superports. There are about 50 such ports around the world, and new ones are being constructed.

The glamor of ships passed to the ocean liner in the twentieth century. Great ships like the *Mauritania,* the *Normandie,* and the *Queen Mary* were luxurious floating hotels. Today the airplane carries passengers, but not freight, more efficiently than a ship, and since World War II the oceans have become more important than ever before as avenues of trade.

Riding high in the water, an empty oil tanker reveals the depth needed to carry its enormous liquid cargoes. Special harbors deep enough to accommodate them have had to be found or created in various parts of the world.

CHALLENGES OF THE FUTURE

One of the greatest challenges of the future might be, very simply, to find out what the sea is and what its relationship with mankind should be.

For most men the sea is alien. They are not consciously affected by it. They give it no emotional thought. It is essential that they do. We are threatened on many fronts and the sea can save us. The sea can provide some of the food to feed our multiplying populations. It can provide the energy source to power our homes and industries. It can provide some of the minerals that we will soon deplete in their terrestrial deposits. It can inspire peace.

Antiques of the Future

In 1956 a civil engineer named Wilhelm Prölss, who had spent most of his

Project for Sea City, a marine community designed for a self-sufficient population of 30,000. This is an example of architectural planning that considers the sea as part of the new urban environment.

professional career on aircraft design, put his mind to work on sailing ships. He turned out a design for a vessel he called the Dyna-Ship, and took it to the Schiffbau Research Center at the University of Hamburg for testing. The innovative design called for four or more masts 200 feet high. Sails would roll out from the masts on horizontal tracks on the curved stainless steel yards. Yards could be fixed, sails set, and masts rotated by a single man pressing a button. Planned for service on the North Atlantic, the vessel would be capable of 12 to 16 knots average speed, compared to the 10 to 15 knots averaged by diesel-powered ships, and in strong wind conditions the Dyna-Ship might hit a top speed of more than 20 knots. In a calm sea, an auxiliary diesel could be used.

During six years of testing and refining the design of the Dyna-Ship, it was proved that with modern materials and the navigational and weather forecasting aids available, a wind-powered ship would be as able to main-

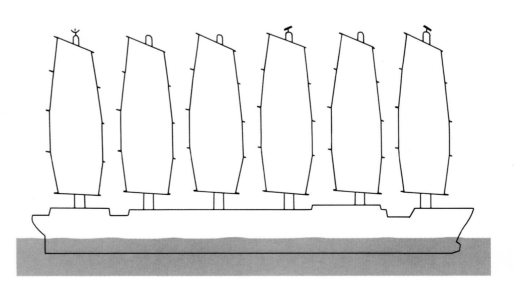

Totally automatic and computerized, this hypothetical sailing ship of the future has a small crew and uses satellites for navigation. Its sails and masts rotate to make full use of the wind.

406

tain schedule as the fuel-powered vessels. The savings realized in manpower and fuel would be considerable. Development of the Dyna-Ship awaits financing.

The Human Pipeline

It is possible that the man of the future will be able to travel by train from New York to Paris in an hour.

The human pipeline is not a pipedream, but a sound engineering possibility. The train will move silently through a huge 4000-mile-long vacuum tube laid on the bottom of the sea as a transatlantic cable. It will be powered by a linear electric motor with no gears.

Metal guideways will "steer" the train, but the train will never touch them. Instead, it will speed along about one foot above the rails, being supported, guided, and propelled by powerful electromagnetic forces. Such systems have successfully been tested in a number of countries throughout the world.

During the first half of such a trip from New York to Paris, passengers will face forward as the train accelerates, and the only sensation of speed they will have is of being pressed back into their chairs. In a vacuum tube there is no practical limit on speed. Halfway to the destination, the train will begin to decelerate, and the passengers will be automatically turned to face the rear of the train. They will once again feel as if they are being pressed into their chairs.

At an acceleration rate one-tenth as great as the earth's gravity, the train would be traveling at 120 knots after one minute. After 30 minutes the speed of the train would reach 3600 knots.

The human pipeline would be extremely costly to build, but after the initial cost the running expenses would be low and the convenience and nonpolluting factors would be considerable. Building such pipelines could only be efficient if operated between widely spaced cities.

Building on Water

Some of the most valuable and sought-after land in the world is that bordering on water, but there simply isn't enough waterfrontage to fill all the demands. In order to produce artificial offshore property, two groups of projects have been proposed: artificial islands constructed on piles, and enormous floating structures.

Sea City, proposed by British architects and engineers, would be an offshore island built on piles of concrete. It would be capable of housing 30,000 people. A 16-story amphitheater would surround a large lagoon with clusters of islands. The city would provide all of the amenities of a seaside resort. Plans are well advanced for a number of other offshore structures. They will be built to play diverse roles: resort hotels, deepwater ports, airports, nuclear energy plants, or oceanographic observation laboratories.

Energy from the Strait

Many years ago it was proposed that a dam be built across the Strait of Gibraltar. This idea had great drawbacks but there is a way we could take advantage of the strong currents that flow through the strait. It would require no technological breakthrough, but huge investments to install low pressure turbines, an aquatic replica of wind mills.

Water flowing through the strait at the surface moves in an easterly direction toward the hot, arid lands of the Middle East. Desert winds and hot sun evaporate the seawater, making the remaining water more saline. This saltier, heavier water sinks and sets up a countercurrent that moves west. The flow of water is reliable and strong. Hundreds of turbines could be placed in both currents in the strait to produce fuel-free, nonpolluting electric power. It would not interfere with shipping, since it could be installed below the draft depth of the largest ships. The topography at Gibraltar is compatible with such a complex generating plant.

THE
SEA
IN
DANGER

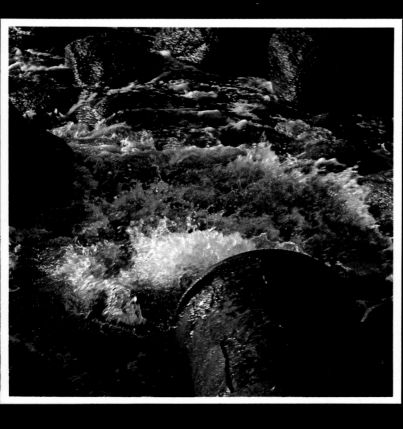

CHAPTER

THE PLANET IS ON FIRE

Flying in a private plane over any continent today is a terrifying experience. In summer Africa is burning solid. At night, from an altitude of 10,000 feet, you may count as many as 25 simultaneous fires, some of them enormous. The myth of the "beneficent" brush fire has spread worldwide and has brought about the disastrous proliferation of arid areas and such catastrophic droughts as the recent one in Nigeria. During daytime, the plane has to climb above 12,000 feet to find clear air. Below is a thick, dark, ochre coating of smoke that screens the sun's rays—and that humans have to breathe. Corsica, the Riviera, Florida, Canada, Australia, with individual variances, are also on fire, shrouded almost permanently in a thick, visible layer of choking, smoking air. Forests shrink, fertile soil is washed away. Recent research has demonstrated that the Sahara was covered with trees as recently as 6000 B.C., and that it was turned into a desert by nomadic tribes that burned the trees to provide grazing areas for their herds. From our little airplane, the thought is obvious: we are today in the process of turning the entire planet into a global Sahara.

Another alarming experience is flying by night above oil fields—in the Middle East, in Louisiana, Texas, or offshore in the Persian or Mexican gulfs. Everywhere burning torches light up the dark sky. Gas is burned just to get rid of it, because it would be uneconomical to store it. Flames dance sarcastically above every single refinery plant—a symbol of waste, of carelessness, and of man's contempt for nature.

In the sea, the reckless waste is shameless. Ninety-four percent of our whales have been slaughtered. The quantity of fish in all oceans has decreased by more than 30 percent. Half the shorelines of the world are dying. The bottom life of the continental shelves is ravaged by heavy trawlnets. The coral reefs are sick everywhere, probably from pollution. Seabirds are less than 50 percent of their number at the turn of the century. Generally speaking, more than 1000 species have been eradicated by man since 1900.

It is only recently that we are becoming aware of how severely we are plundering our planet. This can be explained by the suddenness with which our growth has entered the *explosive mode*. For at least one million years the human species has struggled for survival with a very weak set of defensive and offensive weapons: no shield, no carapace, no fur, no blubber, no fangs, no claws. The naked man had a brain, had agile hands and an articulated voice. He had to use all the tricks he could devise to compensate for his weakness, and survival became equivalent to fighting nature permanently. But with access to the sun's power, stored in coal, oil, and natural gas, man suddenly became, in the short time span of four generations, the undisputed ruler of the earth. This dramatic change was so abrupt that man has not yet realized that his role has changed, that his survival—and that of the world —no longer requires him to do battle with nature. Man must now become nature's protector.

The evolution of man had been slow, very slow. From a population of maybe tens of thousands one million years ago, he had reached a very few million by the birth of Christ. Today there are four billion humans, and by the year 2000 there will probably be nine billion or more. All aspects of the development of human destiny have entered the explosive mode: population, power per capita, mineral output, technological and scientific progress, nuclear bombs. As the outcome of any explosion is destruction, we do not need to be prophets to fear that a global explosion may turn into global destruction.

Some argue that pollution always existed, causing for example the medieval epidemics, or that species come and go, that we should be no more upset by the extinction of the bald eagle than of dinosaurs. This ignores the basic fact that we are no longer in a slow evolutive process but in a violent explosive one. No comparison is possible. There are no precedents. We have to face the danger as a new kind of man-made peril that only man-made measures can remedy. And as we have demonstrated in this book that the life cycle and the water cycle are inseparable, we must save the oceans if we want to save mankind.

Pesticide being sprayed

The world's seawater has kinetic, chemical, and physical properties that enable it to act upon some of man's wastes. It dilutes foreign substances and then begins to disperse them by means of its vast circulation system. Of course, dispersion can only handle as much material as the seas as a whole can bear. The limits could be reached, at the present pace, in a few years.

Another manner in which the sea disposes of unwanted substances is displayed in the Atlantic Ocean. The process is not fully understood, but it appears that excess carbon dioxide in the atmosphere is dissolved in the northern hemisphere and transported south, until it is "exhaled" in the less polluted air above the South Atlantic. This "breathing" of the ocean then helps reduce the concentration of carbon dioxide in the industrialized areas of Europe and North America in the northern hemisphere.

One of the most complex, yet very fundamental, aspects of the sea is its buffering action. This is its ability to neutralize excess concentrations of *both* acids and bases. Since man began burning large amounts of fossil fuel—coal, oil, and natural gas—with the coming of industrialization, there has been a disruption of the normal carbonate cycle in the sea as well as in the atmosphere. Many, if not most, marine animals are very sensitive to the alkalinity of the water, and any severe changes in that characteristic would have dramatic consequences.

In many ways life in the ocean is far easier than it is on land. Water is everywhere and this, along with oxygen, is basic for all animal life. A given volume of water, however, holds 40 times less oxygen than an equivalent volume of air and therefore animals living in the sea must adapt to this scarcity. Warm water cannot hold as much oxygen as cold water, and some tropical waters are so scarce in life that they are called "deserts." Oxygen-consuming bacteria may also deplete the gas' supply. With complete disregard for the environment, man has created many additional biological wastelands in the sea. Both bacterial growth stimulated by the dumping of sewage sludge and the heated effluent of power generation plants have reduced oxygen levels.

The shoreline is like a boundary, crowded on each side with life—terrestrial creatures feeding and breeding on one side; marine organisms occupying the other side so densely that the great commercial fisheries of the world usually operate in the waters above the continental shelf. Not only does man overfish the coastal waters, but he also uses the shoreline for transportation, recreation, and dumping grounds. Just where ocean life is most abundant, so is man most active and destructive.

The surface area of the water is an extremely vulnerable zone because that is where a great deal of the life process of the sea is carried on. Not only do the floating plants bloom, but many, many animals pass their larval stages as zooplankton in the photic zone. As adults, these organisms may not be directly affected by surface pollution; but such contamination may mean they will never get the chance to become mature. Even thousands of miles away from civilization, the open sea is collecting refuse from land carried by currents, oil slicks and garbage from ships, and toxic dirt from the air washed down by rain.

If all pollutants eventually reach the sea, then many come to rest on the sea floor. Heavy substances and particulate matter settle out of the water and even many of the dissolved deadly chemicals, such as mercury and lead, that have been incorporated into the tissues of animals and plants sooner or later sink within their dead bodies and enter the sediments.

In our search for the fabled chambered nautilus off New Caledonia, we saw miles and miles of dead and dying coral reefs—choked and killed by nickel-refining waste. Corals are extremely sensitive to any kind of sediment, even if it is nontoxic. Having evolved in very clean tropical water, the coral polyps are easily smothered. They are immobile, and their only

A tangle of dead fish demonstrates that the water they swam in has been polluted far beyond the tolerable level.

defense against suffocation is to secrete copious amounts of mucus to entrap the particles and then to beat the water with their hairlike cilia so that the dregs are carried away. When the coral is overwhelmed and dies, the erosion processes prevail and the reef with all its life forms dies as well.

SOWING LESS THAN WE REAP

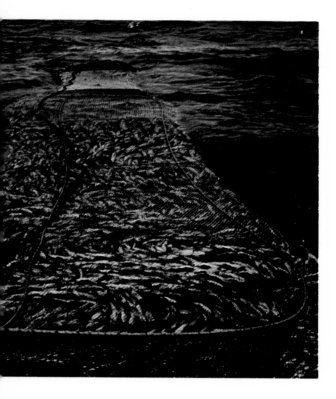

In a single haul, a commercial fishing operation has pulled in 58,000 pounds of Pacific hake.

For thousands of years the scenario has remained unchanged: each day hardy fishermen go to the sea in pursuit of fish, take their haul, and return home. No one entered the sea with the thought that no fish would be there. There always had been, and there always would be. But they were wrong.

About 60 million tons of fish a year are extracted from the world's water. The largest amount taken are the herrings, sardines, and anchovies, which account for more than one-third of the total catch; the next largest group are the cods, haddocks, hakes, and similar varieties, followed by shellfish of all kinds. There are areas that could be said to be completely overfished, particularly the North Atlantic, where some varieties are scarce because fishermen remove them faster than they can reproduce.

One of the problems of repopulation, of course, is the growth and reproductive rate of the organisms involved. Most marine life grows more slowly in the colder water ranges of their habitats than in the warmer waters. And since so many of the fish live in colder currents or move in the colder bottom waters to feed, their growth is slow. The cods and their commercially popular relatives like hake and haddock prefer water no warmer than 55° F. Sexual maturity is not usually reached until at least their fourth winter and often later. Most of the fish caught commercially are between five and ten years of age, which shows the difficulty the species has in surviving intensive fishing.

One of the reasons why overfishing is a problem is that it involves so much waste. Regional and national preferences cause some fish to be more popular than others to the point where perfectly good fish are thrown away or ground up and used for pet food, while the heavy trawlnets spread devastation on the bottom.

The French have a delightful recipe for manta ray cooked in butter, and my own village of St. André de Cubzac has a lamprey festival each year, celebrating a fish that is considered an undesirable intruder in the Great Lakes. In California some residents are alarmed at the rise in the sea urchin population, yet if they would learn to enjoy these little morsels the way other people do, that would be one way of solving the problem.

WASTE OF THE WORLD

Pure water is perhaps a misnomer, for even the most pristine stream has dissolved salts or minerals in it. Just because there is something in your water besides hydrogen and oxygen does not mean it is polluted. Pollution occurs only when the dissolved or foreign substance is undesirable or injurious to organisms. The contaminants may be inorganic compounds of metals like lead, mercury, cadmium, or chromium. Or they could be living matter like bacteria, viruses, or other microorganisms. Pollutants are sometimes pesticides, detergent, phosphates, petrochemicals, or radioactive or industrial chemical wastes. These pollutants are often linked with disease and result from improper water treatment and effluent discharge.

A recent Federal Water Quality Association report stated that seven percent of the sewered communities in the United States have no treatment at all, but dump raw sewage into the nearest body of water. And 60 percent of the treatment operations were found to be inadequate. Thus, man's own waste is the pollution most harmful to himself.

Adding Woes

Dredge spoils—the result of keeping deep harbors open to shipping—

A cannery at Monterey, California, sits idle after the area's fish populations dwindled as a result of overfishing.

constitute about 85 percent of the barge-delivered wastes disposed of in the oceans surrounding the United States. In most cases, the dredge spoil is deposited only a few miles from its original location.

The extraordinarily rapid buildup of sediment in the dump site may result in the destruction of spawning sites for marine organisms, a reduction in the food supply and the vegetation cover, and the trapping of organic matter. In addition, there is an increase in the turbidity of the water at the dump site which is detrimental to marine life in many ways.

Another by-product of man's misuse of technology is detergent pollution. Detergents work mainly by breaking the surface tension of water to allow it to penetrate clothing faster. They also allow water to wash away grease and oil much more easily, especially in heavy industrial operations.

The use of detergents to clean up oil spills is even more damaging than the oil itself. Detergents employed to disperse the oil from the *Torrey Canyon* disaster in England killed crabs and small fish that fed upon whelks. In the absence of their predators, the population of whelks soared until there was not enough food for their numbers and they starved to death. The detergents had made oil permeable to skin and thus multiplied its detrimental effects. The respiratory mechanisms of all marine organisms function in direct contact with water and their chemical systems are dependent on the characteristics of clean water. These interactions are affected by detergents and cause breakdowns of the vital respiratory membranes which can kill or severely incapacitate the animal. In addition, the reproductive products and delicate larvae of many marine creatures are unprotected against such chemicals and can easily be killed.

Social Waste of the Sea

In the sea, man found—he thought—a dump site that could swallow up his garbage and keep swallowing it forever.

The scope of waste dumping is increasing. In the first two decades after World War II, waste disposal at sea increased fourfold in the United States. A 1971 report prepared by the Environmental Protection Agency's Office of Solid Waste Management predicted: "It is almost certain that under prevailing conditions there will be increasing pressures to use the sea for disposal of the municipal and industrial wastes generated in rapidly expanding coastal zone metropolitan and industrial areas." In other words, look for more of the same on a grander scale.

West Virginia, a landlocked state in Appalachia whose mountain valleys are pockmarked with chemical plants, recently asked for the right to dispose of chemical wastes in the Gulf of Mexico, hundreds of miles down the Kanawha, Ohio, and Mississippi rivers.

The United States Navy, in an apparent attempt to kill two birds with one stone, has found it advantageous to get rid of its obsolete military explosives and chemical warfare agents by loading them aboard old ships and then scuttling the vessels. When faced with the question of the effects of the disposal of various chemical and biological warfare agents, a committee of the National Academy of Sciences admitted that "we have no information regarding possible deleterious effects of these operations on the ecosphere of the seas."

The ultimate effect of dumping all this junk in the ocean is still unknown. The construction and demolition debris dumped by New York City, for example, is so isolated that no one has studied the effects, harmful, beneficial, or otherwise. On the other hand, the military explosives and chemical and biological warfare agents constitute a potentially greater threat of unknown dimensions.

It is too easy to criticize governments, cities, and industries without criticizing ourselves. Every Sunday, millions of yachtsmen the world over throw bottles, plastic containers, and trash overboard. The result is incredible: almost everywhere in the ocean, a bathyscaphe diver will land within sight of modern man's litter to such an extent that it becomes difficult to keep these trivial objects out of camera range!

The Atomic Age   Radioactivity is potentially one of the greatest threats to the environment, because it is practically irreversible. Nuclear reactors in themselves and waste/storage systems are extremely vulnerable to natural and man-made

A rich foam atop water indicates the dangerous presence of detergent chemicals that poison fish life as in this river flowing through Tokyo to the sea.

disasters. We have already experienced a number of storage-tank leaks that have introduced dangerous levels of radioactivity in some areas. Even more of a hazard are the industrial nuclear reactors, because they are more difficult to control.

Fallout from the H-bomb test of 1954, as well as from the huge Russian and American experiments of 1962–63, spread over a great area of the Pacific. Like pesticides, the radioactive debris passed rapidly up through the food chains, increasing in concentration at successive levels.

Carried around the globe by wind, radioactive materials are spread everywhere, over the oceans and over continents alike, some halfway around the world from the test site. Strontium 90 can return to earth in rainwater and remain dangerous both on the land and in the sea for centuries to come. Strontium 90 reaches us directly through leafy vegetables and secondhand through milk and other dairy products. It is deposited in bone tissue and can affect blood cell production. Cesium 137 becomes even more concentrated by the time it reaches humans because it passes through many different organisms.

Radioactivity is a particularly pernicious pollutant because it has no natural outlet—it just piles up in our environment. The high-energy particles penetrate and damage all tissues. They disrupt genetic material—the very essence of our being. Some of these elements will be with us for a very long time, since it takes a few thousand years to break them down by half. Any miscalculation would be irreparable.

FROM AND ON THE LAND

Pollutants and contaminants find their way into the ocean in many, many ways. The first and foremost is river-carried matter.

Once man started to industrialize his refuse became complex, especially as rivers became natural sewage systems that cleaned ore excavated from surface mines or washed coal used to heat homes. Then came the dyes and bleaches. Farmers began to poison their fields selectively, trying to destroy insects—which they called pests—first with mercury and arsenic compounds, then with the more deadly and unfortunately stable pesticides developed from crude oil. These agricultural poisons, along with churned topsoil, began washing off the continents into the rivers. And man built not only his homes on the banks of the river but also his factories, not just water-powered grist mills but putrefying paper mills and steel mills and textile mills.

Natural Nursery

Estuaries, salt marshes, and wetlands play an essential part in the interaction between land and water. Though these areas are small in size, they are large in productivity. The protected waters, the flushing tides, the balanced community of living organisms make such coastal areas natural breeding grounds.

Among the commercially valuable marine animals that live in estuaries at critical times in their life cycles are salmon, clams, crabs, oysters, shrimp, menhaden, and flounder. Yet the destruction of wetlands continues under the justification that more land is needed for residential and industrial use.

The coastal wetlands are important not only for the marine creatures and waterfowl that live and breed in the area, but also for the vegetation that grows there. These plants trap solid particles in the water, and also provide surfaces for microorganisms that can metabolize organic waste. This is where the tides are so important, for they regularly provide the new water and fresh air that enable the cycle to continue. Thus not only is land saved from being washed out to sea, but water is also purified by the natural "sewage treatment" that is constantly going on.

In Florida's Biscayne Bay, oily surface slicks with pesticide concentrations 10,000 times greater than the surrounding waters have been found. Since plankton in the surface layers absorbs great quantities of pesticides, this greatly increases the process of bioconcentration in the food chain. Researchers at Woods Hole, Massachusetts, have shown that 10 parts per billion of DDT, endrin, or dieldrin slowed growth of plankton cocolithophores and that 100 ppb (parts per billion) harmed two species of diatoms. In contrast, a harmful dinoflagellate was not affected by 1000 ppb of any of the three compounds. The imbalances that are reached in plankton populations have profound effects on all sea life because they form the base of the food chain. Such an imbalance would be immediately felt by organisms that fed on a specific group of plankton.

The Mortal Lakes

Every lake is a terminal case, for there is little that can be done to prevent death due to silting, evaporation, increased salinity, and eutrophication. Lakes have nothing to compare with the ocean's wide-scale ability to redistribute sediment, both terrestrial and biogenic, or to circulate dissolved salts and minerals until a balance is achieved. Rather, lakes are condemned to fill up from the bottom as clays and muds filter out, and they become saltier in a way that reflects the type of lands their feeding rivers have coursed. In times of drought, lakes suffer loss of water through evaporation, but are affected to an even greater extent by the diminished flow of water from streams and springs. These are the mechanical ways in which lakes die.

Eutrophication, on the other hand, is a biochemical way of death. It occurs naturally when the algal bloom exceeds the food requirements of the aquatic organisms, and the algae die. The algae are then decomposed by oxygen-consuming bacteria. As the process continues, there is less and less oxygen for the animals and the lake becomes devoid of higher forms of life.

A polluted river at Sacramento, California, pours contaminated water into the world's hydrologic cycle. The river yields its waters to the oceans, which give it up to the sky where it is held until it falls as rain, perhaps thousands of miles away.

An oil-covered western grebe lies dying ashore—a victim of the Santa Barbara disaster.

The limited oxygen supply is further depleted because the algae, even though they are plants that produce oxygen through photosynthesis during daylight hours, change life-style at night and use oxygen during respiration.

As the oxygen supply in a lake decreases, a new type of bacteria begins ascending to dominance. These are the anaerobic microorganisms that can break down organic material in the absence of oxygen. Unfortunately the primary products of anaerobic bacterial action are hydrogen sulfide, which smells like rotten eggs, and mercaptans, which have a fouler odor.

Eutrophication is a natural process that has been accelerated by man's activity. There are a number of man-made agents that accomplish this, but the most important are phosphate and the sulfonic detergents. Because of their chemical composition, the phosphates and sulfonates are also plant nutrients and thus contribute to widespread algal bloom, much more than the natural processes can cope with, and the eutrophication cycle takes a great leap forward.

As man changes the composition of the waters and the nature of the habitat, he is also changing the type of animal or organism capable of living in a particular body of water.

An example of this is shown in the migration of the sea lamprey. Originally a North Atlantic resident that bred in freshwater streams, the lamprey was able to establish itself in the Great Lakes with the construction of the Welland Ship Canal, which provided access to Lake Erie from Lake Ontario by bypassing Niagara Falls. By the 1940s they were in Lake Superior and breeding in many of the streams that drain the upper Midwest. The invasion of lampreys caused a great stir because they preyed upon the commercially valuable lake trout and whitefish.

Thermal Pros and Cons

Heat and hot water are by-products of many manufacturing processes and especially of fossil-fuel or nuclear electric plants.

Marine life is extremely sensitive to changes in water temperature. For many animals a rise of just a few degrees is a signal that spawning should begin. If the signal is a false one, and the animals are induced to spawn too early, their premature offspring will be subjected to harsh weather conditions and they will not be able to find their natural food.

Warm water holds other hazards for fish. At higher temperatures they become increasingly more susceptible to toxins in the water. Warm water cannot hold much oxygen, but living in it means that fish require more than they would in cold water. They also have higher metabolic rates, require more food, and grow and mature more rapidly.

The changing needs of marine creatures subjected to such thermal pollution can lead to population imbalances. The Chesapeake Bay opossum shrimp, an important part of the food chain, are quickly damaged by warm water, while an undesirable species, the sea nettle, a jellyfish, can tolerate water that is 18° F. above normal. Quite frequently in artificially warmed areas it is found that one species will thrive at the expense of all others and eventually destroy itself.

POLITICAL WATERS

A disastrous oil spill off the California coast at Santa Barbara left a state park beach looking like a scene of desolation.

If pollution of the oceans is going to be stopped, it will take more than the efforts of conscientious individuals everywhere and some strict regulations by well-intentioned nations.

The single biggest problem in reaching international agreement on a topic lies in convincing sovereign nations with different goals, opposing political systems, and fluctuating positions in day-to-day politics that their interests all lie in the same direction.

Territorial limits, for example, can trigger disputes. Claims can range from three to 200 miles. In such broad areas, these nations are, in effect, asserting exclusive fishing rights in the waters around their lands since the major part of all commercial fishing is done above the continental margins, which rarely extend 200 miles from shore. There are occasional flare-ups between nations over fishing rights, with seizure of fishing craft and displays of gunboats.

The problem of dumping wastes on the high seas is certainly a matter of international concern, yet there is really no effective mechanism for making the ships of any one nation cease the practice. The only way such actions could be controlled today is through international regulations backed by inspection at the ports of loading, perhaps by an agency like the International Maritime Consultative Organization, which has played a role in securing agreements regarding oil spills and in the regulations concerning dumping of heavy metals in the sea.

Fortunately, we are today technically capable of tracing the sources of oil slicks, even if they occur on the high seas. Oil from each of the world's wells has its own identity, characteristics that show up on spectrophotographs taken from cameras mounted on satellites. If we seriously wanted to stop polluting the seas with oil, we could set up a global "Big Brother" system that could nail offending vessels with indisputable evidence.

Whatever their problems and differences may be, nations must inevitably get together to work problems out, for at some point the result will be mutually beneficial. Peru, for example, which wants to protect its anchovy fisheries as much as possible, does not want to alienate the United States to such a degree that this northern neighbor will no longer purchase the anchovy meal for cattle and chicken feed.

A comprehensive agreement—based perhaps on the principle of the Antarctic Treaty—has often been discussed and is the dream of many statesmen and nations. Yet past breaches and violations of the many voluntary agreements raise doubts about the effectiveness of the new pacts. A series of conferences had been held in an attempt to adopt a new set of international sea laws, but a meaningful treaty will be difficult to achieve in the near future as long as some nations continue to be shortsighted and narrow-minded in protecting their interests in the seas. Perhaps some calamity is needed to alter such thinking.

In the meantime the sea is protected only by the various multinational and bilateral agreements, however flimsy or contradictory, already in existence. Or by the occasional unilateral decisions, such as that by Ecuador that proclaimed virtually the entire Galápagos Archipelago a national park. One hopes the other nations of the world will respect the sovereignty of these islands and the truly unique creatures that live there.

Filling the Sea with Junk

When Nations Get Together

Refuse on the floor of the ocean, here at the Antarctic, is not only ugly, but can be dangerous. Metal cans and other scrap eventually break down chemically into relatively harmless substances, and could provide temporary shelter for fish. But they are rarely discarded clean and often still contain harmful chemicals.

STUDY OF THE SEA IN DANGER

Sewage treatment plants are the answer to polluting rivers and oceans. This unique facility in Calcutta combines sewage treatment, aquaculture, and agriculture. In the final stage after liquification of sewage, purified and nutrient-rich water is aerated and passed to ponds in which carp are raised.

There is no single science that covers water pollution, for there are many disciplines involved, including chemistry, physics, marine biology, oceanography, and limnology. And if remedies are to be offered and solutions pursued, the expert advice of lawyers is also required. In addition to the traditional tools of oceanography, which measure temperature, salinity, dissolved gases, or the rate of diffusion of liquids in the sea, advanced space-age technology is being used to aid in the detection of pollution.

Information on pollution has been gathered by NASA's Earth Resources Technology Satellite. The ERTS program was launched in 1972 and the first results were compiled within a year. Among the subjects investigated were sedimentation and erosion, municipal refuse, agricultural and industrial waste discharge, oil spills and ocean dumpings, solid waste disposal, and mobile and stationary sources of air pollution.

By studying photographs and other information gathered by the orbiting satellite, scientists were able to detect such offenses as chemical discharge from a paper mill into Lake Champlain. This information was used as evidence in a court suit to obtain a cease-and-desist order against the delinquent plant.

The groundwork for monitoring the oceans could be done with a global network of buoys. These buoys would be anchored in the open sea as well as on coastlines and would carry sensors all the way down to the bottom. Their information could be picked up by satellites and fed directly into computers. The surface data from the areas between buoys would be obtained by super satellites of the ERTS filiation and by space stations such as Skylab.

A Case in Point

There are so many loosely related aspects to land and sea endangerment that it is valuable to examine one of the most momentous cases in some detail to see the complex interaction of forces which result in the condition we label pollution.

New York City and its surrounding communities encompass a population of about 25 million people; in 350 years pristine wilderness has become a most endangered area. The city itself has approximately 535 miles of shoreline, all but 35 of which are classified as polluted. This certainly is a subject for a case history of neglect and ignorance of nature. It is an ecological disaster.

The single biggest source of contamination of New York waters results from sewage and disposable waste which are delivered by barges to the offshore areas of the New York Bight. About 4.5 million cubic yards of sludge a year from sixteen metropolitan sewage treatment plants have been dumped there for the past 40 years. In addition, 360 million gallons of raw sewage is contributed each day by New York City. On the New Jersey side of the Hudson Gorge, six million cubic yards of dredging spoils are dumped each year. Bottom-grab samples retrieve cigarette filtertips, Band-Aids, and aluminum foil. The oxygen concentration is less than one part per million. Lead is found in the water column at concentrations of 151 ppm, copper at 60 ppm, chromium at 40 ppm and DDT and DDE at 150 ppm. The sediments contained 338 ppm of copper, 197 ppm chromium, and 249 ppm of lead. Nematodes and other worms that normally thrive on pollution are absent.

In times past, garbage and refuse were dumped at sea, but could be seen floating back into New York Harbor, often within a matter of days, and soon the practice was stopped. Today, however, much of the waste deposited even further out has been shown to drift back toward the seashore, carried mainly by deepwater currents.

The air pollution of the New York metropolitan area is under close scrutiny and may be the first aspect of the area's pollution problems to be alleviated. In 1973 the mayor of New York jokingly remarked "I don't like to breathe air I can't see." But there have been efforts to reduce the level of air pollution, which stems mainly from petroleum plants in neighboring New Jersey.

Throughout this environmental crisis, the city of New York has survived. The quality of life has been somewhat diminished, but there were people who believed the situation was not hopeless and have persevered. They had few scientific studies to back them up and not much cooperation from businessmen. But they had hope, which is what may lead—and must lead—all of us into action to help save our environment.

The problem of the sea in danger is serious, deadly serious. On every front we see confusion, conflicting evidence, and continued pollution. The most serious problem is political: the rich nations are concerned with convenience, the poor nations with survival. A rift is developing between the rich

HOPE OR ELSE

Industrial wastes from a cannery in California kill millions of organisms each year, as evidenced in the bleak shoreline of this once-flourishing bay.

and poor countries, a dramatic and potentially explosive dichotomy.

The traditional thinking of economy and profit must be altered to accommodate these differences, for who profits when a man won't deliver his bread to a starving man who has no money? Looking at economy from another point of view, if regulations were such that an industrial plant was required to circulate its effluent through its own drinking supply and cafeteria before discharging them into the river, the water would certainly be free of contaminants. The cost of eliminating poisons becomes very relative. Technology can provide all the answers; people are likely to accept the solution. All that is needed is internationally controlled regulations and a strong, sincere, determination of the main governments.

Now the crisis is at hand. This is not the raving of a placard-carrying doomsayer, but the observation of thousands of learned and concerned individuals. I am 100 percent pessimistic when I predict some sort of a disaster, for it will surely come. But it is in our power to reduce dramatically its seriousness and its consequences. Yet I am also 100 percent optimistic for recovery after the disaster. After the deluge, the sun will shine, and men will hope again that the golden age will come. Most of the misconceptions prevailing in the minds of the generation in power originated in ignorance. Schools are now emphasizing our dependence on nature. There will soon be a generation with a new philosophy.

Although New York is not the largest city in the world either in population or area, it seems to concentrate all the problems of urban centers, including a polluted waterfront.

Some Alternatives

Many varied and remarkable things have been done with treated sewage. True, most of these programs are small and several are still in the experimental stage, but they point in the right direction.

In one experiment conducted by the Woods Hole Oceanographic Institution in Massachusetts, oysters, seaweed, sandworms, and flounders were developed in an environment utilizing triple-treated sewage and seawater.

425

This experiment is only one of several that used waste products in unusual ways. In one case, refuse was chemically deactivated, the liquid removed, and the remaining material compacted into blocks for use in building and construction. On the island of Taiwan, diluted city sewage is pumped into fish ponds every three days. Almost nine tons of milkfish and tilapia are harvested each year from these enriched ponds which cover only 2.5 acres of land. Such attempts may not advance beyond the experimental stage because of cost, resistance of the public to the new products, or problems in implementing the processes on a large scale. But they do show a real concern for the pernicious polluting properties of our domestic waste.

Clean Water and the Problems in Getting It

There are several ways of cleaning up water. The first attack on polluted water must be a frontal one—we must stop dumping wastes directly into the sea. Municipal sewage treatment plants are one way. Another is to control watercraft which dispose oily and human waste directly in the sea. In response to this problem, the Environmental Protection Agency proposed a standard for marine sanitary devices.

The EPA has also been active on another front in its effort to clean up the water. The agency has patented a process for purifying water that can be used by paper mills and other private polluters and that hopefully can be adapted for use by inland municipal sewage treatment plants. In this process, called FACET, an acronym for fine activated carbon effluent treatment, the discharged water, instead of being channeled directly into a stream, is circulated through a series of tanks containing a slurry of finely ground activated carbon. The carbon, which can readily remove organic material from the water, is made by charring wood or coal. After the effluent is circulated through these tanks, the water is pure enough to be used again, which means that the system is self-contained and does not constantly draw on a river or lake for new water supplies.

The Environmental Protection Agency hopes eventually that the process could become compact enough so that it could be used in individual homes as a primary treatment stage before household wastes enter the municipal sewage system.

No Bugs, No Chemicals

It is possible to control weeds and crop pests without taking recourse to chemical agents. In Florida, a weed control program was put into effect which utilized neither chemical nor mechanical destruction in freshwater lakes that had become overgrown. The agent of control was a Siberian fish, the white amur. Scientists estimated that 100 of these fish, ranging in size from two to 20 pounds, could eat a ton of weeds a day and still not upset the other creatures in the lake. The weeds had become a problem because they clogged irrigation ditches and channels used for drainage. Manatees, the vegetarian sea cows, could also be used to clear the channels of the invading water hyacinth.

There are many natural substances that can be used as herbicides and pesticides, but the problem is that in large-scale farming—which is as cost conscious as big business—these alternatives are too costly or too laborious to use. But they are ingenious. Males of some insect species have been treated with X-rays and made sterile. Introduction of screw worms treated this way has largely eradicated this pest after a few seasons.

Synthetic hormones have been sprayed on infested areas to cause premature emergence of larvae, which are unable to survive. Bacteria harmful only to specific species have been successfully used. Sonic devices that kill insects that come within range have been used to protect grain. Techniques are being explored to break down natural defenses and render pests vulnerable to predators.

Approaching the problem from the opposite direction, scientists have developed 22 varieties of wheat that are resistant to the Hessian fly. The fly population has been so reduced in California through use of resistant wheat that it is now possible to plant nonresistant wheat again. Resistant strains of other food crops are being developed.

One such strain is a fast-growing cotton plant that reaches maturity earlier in the growing season, before the deadly boll weevil reproduces in large numbers. Cotton farming was a natural place to begin looking for alternatives to chemical pesticides because, of all the major crops in the United States, no other is as besieged by pests.

Though cotton farmers in the United States work only 1.5 percent of the cultivated land in the country, they use almost 50 percent of the agricultural pesticides sold in America. The potential for developing the short-season cotton strain—which is smaller than the traditional type but which can be planted more closely in the field and yields the same or more per acre—has been known for many years. But the research was not pressed until the use of DDT was banned.

It would also be wise to consider changing our farming methods from single-crop operations to polyculture. With a more balanced mixture of plants, there would be more ecological stability and a diminished need for the excessive chemical controls we use today, for it often happens that agricultural pests develop an immunity to chemical poisons.

A THREATENED OASIS

Planet earth's rare gift of water essential for life must be preserved for the future.

In our solar system, the earth is the only planet with an appreciable supply of liquid water. This rare gift is essential for life and, consequently, as the only intelligent and conscious species, mankind should consider the protection of the water system—rivers, lakes, seas, and oceans—as the first condition for survival. This has unfortunately not been the case, and we have very little time left to reverse the trend if we are to hand over a healthy earth to future generations.

GLOSSARY

Abyss. The region of the ocean basin that lies below the continental slope, generally deeper than 6000 feet. It occupies three-quarters of the sea floor area.

Acanthodii. The most primitive group of jawed fish, often called spiny sharks. They predate the placoderms but left a poor fossil record.

Actinopterygii. A subclass of the bony fish, which includes most modern "normal-looking" fish. They have ray fins, in contrast to the crossopterygians which have lobe fins.

Adaptation. The physical, behavioral, and genetic changes a group of organisms undergoes to best succeed in its environment.

Adenosine triphosphate (ATP). An organic compound in living tissues that acts as an energy storage and transfer agent in metabolism.

Adsorption. The process of collecting substances, usually as liquids, on the surface of materials. This property is present in activated charcoal, certain clay minerals, and many other materials.

Aerosol. Particles suspended in a gas. In a sense, seawater suspended in air.

Agnatha. Fishlike creatures without jaws. This class of animals includes the lampreys, sea hags, and the extinct ostracoderms.

Alcyonarians. Members of a subclass of Anthozoans (corals and anemones) that includes the sea pens, sea fans, sea pansies, whip corals, and pipe corals. Alcyonarian corals are characterized by eight tentacles that are pinnate—that is, they possess featherlike side branches.

Algae. A large group of marine plants that contain chlorophyll but lack true roots, stems, and leaves. They may be single-celled or multicelled and may live alone or in colonies. Microscopic algae, called phytoplankton, are the base of the ocean's food web and carry out most of the photosynthesis that occurs in the sea.

Allantois. An embryonic membrane that forms in reptiles and birds, where it functions as bladder and lung. In mammals, the blood vessels in the allantois carry nutrient material from the placenta to the embryo.

Ambergris. A waxy, gray, strong-smelling substance that forms in the digestive tracts of sperm whales and is found in no other living creature. It is used as a perfume fixative.

Ammonoidea. A group of extinct cephalopods with shells commonly in flat spirals. Their name is derived from Ammon (Jupiter), a god who is often pictured with a spiraling ram's horn.

Amphibian. Any tetrapod vertebrate that spends portions of its life in and out of water and whose reproductive phase is entirely dependent upon a return to water. The larval forms usually have gills, which are not present in adults, as with frogs.

Amphipod. A member of the class Crustacea, order Amphipoda. They have no carapace, and their body is laterally compressed, giving them a shrimplike appearance. Amphipods range in size from one or two millimeters to the giant of the order, *Alicella gigantea,* 14 centimeters long. Most amphipods are marine, a few are semiterrestrial, and some are freshwater. Beach fleas are amphipods.

Ampullae of Lorenzini. Sensory organs in the head region of sharks and rays that are exposed to the exterior through small, round, porelike openings in the skin; they are very prevalent in the snout region. The function of the ampullae is not fully understood, but they are sensitive to changes in pressure, salinity, and electric fields.

Anaerobic. In the absence of oxygen; in contrast to *aerobic,* in the presence of oxygen. Both terms are commonly used in reference to respiration.

Annelid. The phylum that includes segmented worms possessing kidney and heartlike structures. Among the taxa are polychaetes, earthworms, feather-duster worms, and parchment worms.

Anoxia. A condition in which not enough oxygen is being carried by the blood to permit the cells of the body to carry on normal respiration. When prolonged, it can result in death. The cells of the nervous system and brain are the most sensitive to anoxia and may suffer permanent damage.

Anthozoa. The class of coelenterates that includes the anemones and corals. Anthozoans are exclusively polypoid and often have calcareous skeletons.

Anticline. A folded layer of rocks in which the center of the structure is elevated relative to the lateral sides.

Aqualung. The first automatic self-contained breathing apparatus for divers. It was coinvented by Emile Gagnan and Jacques-Yves Cousteau.

Arachnid. A class of arthropods that includes the spiders, mites, ticks, and scorpions.

Arthrodire. A group of placoderms which were rather efficient predators.

Arthropod. The phylum of animals that includes the largest number of species. Arthropods have reached the peak of invertebrate evolution, exhibiting bilateral symmetry, a coelomic cavity, and well-developed organ systems—all characteristics of higher forms of life. They have segmented bodies covered with a hard exoskeleton and jointed appendages that give them a great deal of mobility. During the course of evolution they were the first great group to make the transition from water to land, where they have adapted to a wide range of niches. Insects, arachnids, and crustaceans are the best-known members of this phylum.

Ascidians. Members of the subphylum Urochordata, known as tunicates (sea squirts). They are common vertebrates found throughout the world and an important link in the evolutionary chain. Their larvae are bilateral and have a notochord; thus they are precursors of the vertebrate body plan.

Asexual reproduction. A form of reproduction that requires a single parent; often the mode used by simple invertebrates. The techniques include budding and fission.

Asthenosphere. The upper, plastic portion of the earth's mantle. It is the region where convective motions take place.

Atoll. A ring-shaped coral reef, with an enclosed lagoon, formed by the subsidence of a volcanic island around which the coral reef had originally formed. Carbonate sand may fill in some of the shallow reef areas, forming atoll islands.

ATP. *See* Adenosine triphosphate.

Aurora australis. The Southern Lights. An atmospheric phenomenon resulting in bright bands of light flashing across the sky in the southern hemisphere. The *Aurora borealis* is a similar phenomenon in the northern hemisphere.

Autotrophs. Organisms that manufacture their own food, usually by photosynthesis.

Bacteriophage. A virus or possibly an unknown microorganism that inhabits and destroys bacteria.

Balance of nature. The dynamic equilibrium between organisms, including predators and prey, that can be damaged or destroyed by the removal of one or more animals or plants in an ecosystem.

Baleen. The long flexible plates made of material similar to that of human fingernails that hang down from the upper jaw of many species of whales, including the rorqual and blue. Whales swim with their mouths open, and the baleen functions as a strainer, separating krill and other microscopic animals from seawater.

Barbels. Small feelerlike extensions of the underjaws of certain fish, such as cod. They contain sensory receptors for taste (or smell) and touch.

Bathyscaphe. A free-diving device that functions in water much as a dirigible balloon does in air. It consists of a large cylindrical tank, filled with an incompressible lighter-than-water liquid, to which a passenger observation compartment is attached. Bathyscaphe means "deep boat" in Greek.

Bathythermograph (BT). A device used by physical oceanographers to

measure the temperature of ocean water as a function of depth. As the instrument descends, temperature as related to depth is automatically plotted on a smoked-glass slide within the BT.

Benthic. Pertaining to the sea bottom. Animals inhabiting this region are termed *benthos*. Also appears as *benthonic*.

Berm. A feature characteristic of most beaches. The berm is the backshore part of the beach that runs from the sand cliffs or dunes seaward to the berm crest, from where the beach slopes more drastically to the water's edge.

Binary fission. A means of asexual reproduction in which the parent organism duplicates its components and then splits into two daughter cells.

Bioconcentration. The process by which predatory animals concentrate substances derived from the tissues of their prey. Many contaminants are concentrated via the food chains in this fashion.

Bioluminescence. The emission of cold light by living organisms; common in deep-sea creatures. The source of light is either certain associated bacteria or photophores (luminescent cells) in the animals. Used by organisms to differentiate species and sexes, predators and prey, in a world where no sunlight penetrates.

Biosphere. The living world; all the life on earth. This encompasses all the habitable regions of the earth including those in the atmosphere, hydrosphere, and the like.

Bivalve. Refers to molluscs with two shells. Clams, mussels, scallops, oysters, and cockles are examples. Also referred to as pelecypods or lamellibranchs.

Blowhole. The hole in a cetacean's head through which it breathes and its spout passes.

Blubber. The thick layer of insulating fat found on warm-blooded marine animals.

Brachiopod. Any member of the phylum of bivalved animals that closely resemble bivalve molluscs in outward appearance, but that are more primitive in organization. Shells are dorsoventrally oriented. They are commonly found as fossils. *Lingula*, a living brachiopod, dates back to the Cambrian period, some 570 million years ago.

Brownian motion. The constant zigzag movement of colloidal particles suspended in a liquid; caused by the collision of the particles with rapidly moving molecules of the liquid medium. This phenomenon is named after Robert Brown (1773–1858), who first described it.

Bryozoans. A phylum of animals commonly called moss animals (from the Greek *bryon,* or "moss"). They are microscopic and form colonies that resemble thin, matlike, algaelike growths on shells, rocks, etc. Colonies can gain considerable size; some float free, often confused with seaweed. They are very important as fossils of the Paleozoic era.

Budding. A mechanism of asexual reproduction in which the parent creates one to several daughters as appendages. These daughter cells can detach themselves and exist separately; some, however, may remain attached and thereby produce colonies.

Buoyancy. The property of being suspended by a difference in density with the surrounding medium.

Buttress zone. The outer (seaward) limit of a reef system. The base generally slants steeply and forms a "wall" against which the waves crash.

Byssus. Strong, threadlike material secreted by some marine creatures (mussels and some other bivalves) that helps them adhere to submerged surfaces. The fabled Golden Fleece.

Calcification. The process of adding calcium to a material. It may result in the stiffening of a skeleton or shell or in the creation of a calcium framework.

Carapace. A hard protective shell that covers part, or all, of the upper surface of an animal. Common in arthropods; we speak of the carapace of the lobster or crab.

Carbonization. One of the fossilizing processes which occurs due to the distillation of organic substances in the tissues, resulting in a carbon film.

Carnivore. An organism that eats the tissues or cells of animals.

Cartilaginous fish. A fish that lacks bony structures. The internal skeleton is composed of cartilage instead of bone. Examples include sharks, rays, and many extinct fish.

Cell. The basic organizational unit of life. It is composed of organelles or undifferentiated protoplasm and is usually bounded by a membrane or wall.

Cephalopod. Any of a class of molluscs with tentacles extending from the head. They are usually free-swimming and have well-developed eyes. Shell may be internal or external. Examples are octopods, squids, cuttlefish, and nautiloids.

Cephalotoxin. Poisonous substances found in the beaks of several cephalopods, notably the octopus.

Cetacean. Any of an order of aquatic, mostly marine, mammals (whale, dolphin, and porpoise). Appendages reduced, tail expanded into flukes.

Chlorophyll. The green pigment found in photoautotrophic organisms that catalyzes the photosynthetic process.

Chondrichthyes. *See* Cartilaginous fish.

Chondrostean. A fish belonging to the most primitive group of the ray-finned fish. The sturgeon is a living example.

Chordate. An animal with a notochord or vertebral column, gill slits at some stage of its life, and a dorsal hollow nerve cord. Mammals, birds, fish, and *Amphixous* are all chordates.

Chorion. One of the embryonic membranes surrounding a developing fetus; its function is protective and nutritive. In mammals with internal development, the chorion forms part of the placental connection with the mother's tissue, through which the embryo receives nourishment.

Chromatophore. A cell in a fish or other organism that contains a colored substance. By controlling the distribution of pigment in its chromatophores, a fish can change its color patterns and camouflage itself against a variety of different backgrounds.

Chromosome. A threadlike strand within the nucleus of a cell that contains the genes; the bearers of the genetic code, governing all aspects of cellular development and reproduction. Chromosomes are constant in number within a given species. (*See* DNA.)

Cnidaria. *See* Coelenterate.

Coacervate. A concentrate of molecules in suspension, held together by ionic forces; usually formed on a solid surface such as clay.

Coccolithophore. A tiny, spherical, planktonic plant covered with calcareous plates. Coccoliths (the plates) are one of the major sources of calcareous sediments on the ocean floor.

Coelacanth. A member of a group of lobe-finned fishes that may be ancestors of the land vertebrates. This group was thought to have been extinct since the Mesozoic era, but several specimens of the genus *Latimeria* have been caught during the past 40 years off Madagascar.

Coelenterate. A member of the phylum Cnidaria (often called Coelenterata); characterized by a body cavity with both digestive and circulatory functions, by stinging cells (nematocysts), and two distinct tissue layers. Examples of coelenterates are corals, jellyfish, and hydroids.

Coelom. An embryonic cavity from which the other body cavities are derived; found in most multicellular organisms. In adult animals, the coelom houses the digestive tract and various other organs associated with it.

Commensalism. A relationship between animals and/or plants, in which there are advantages for some of the partners while none are adversely affected. Pilot fish and sharks are an example.

Continental drift. The theory, currently in favor, that the continents on earth have drifted into their current positions after once having together formed one or more landmasses.

Continental shelf. The portion of the sea floor that extends from the low-tide mark to the abrupt break in slope at a depth of approximately 600 feet (200 meters). The seaward edges of the shelves mark the true continental boundaries.

Continental slope. The steep slope that extends from the continental shelf seaward to the abyss. In some places it is broken by the *continen-tal rise* (a change in slope or actual elevation) before it reaches the sea bottom.

Convergent evolution. Similarity in form occurring in two separate evolutionary lineages, such as in the tuna and mako shark.

Copepod. Any of a large subclass of small freshwater and marine crustaceans that are a major constituent of zooplankton.

Coral. A solitary or colonial anthrozoan coelenterate that characteristically secretes a calcareous skeleton. These skeletons may form *coral reefs*.

Countershading. Color patterns exhibited by fish and cetaceans. These animals are darker on top and lighter below; thus they appear an even gray when strong sunlight shines from above.

Crinoid. Stalked pelmatozoan echinoderms, very common as fossils from the Carboniferous period. Surviving crinoids are referred to today as sea lilies—flower-shaped marine animals that are anchored to the substrate by a stalk opposite the mouth, common on reefs in the Indo-Pacific.

Crossopterygii. Lobe-finned fish, including the coelacanths, that may have been the ancestors of the land vertebrates.

Crustacean. Any member of the largest group of marine and freshwater arthropods, including such diverse forms as lobster, copepods, and barnacles.

Current. Movement in water masses caused by density differences, winds, the Coriolis force, and other energy sources.

Cytoplasm. The material that makes up the living cell. A specialized form of *protoplasm* (which is material that makes up any living substance).

Deep scattering layers (DSL). A series of layers formed by certain marine organisms that reflect or scatter sound. Most of these organisms rise toward the surface at night and may range as far down as 2000 feet in the daytime. Some layers are unaffected by the diurnal cycle.

Detritus. In the biological sense, organic particles that settle to the sea bottom, where some types of benthic organisms can use them as a source of food.

Diatom. A single-celled planktonic plant of the subphylum Chrysophyta. They are characterized by a siliceous test and are among the most abundant of all marine phytoplankton.

Dinoflagellate. A single-celled planktonic plant of the subphylum Pyrrophyta. They are characterized by two flagella that they use for locomotion. When environmental conditions are favorable, explosive population increases occur, causing in some cases a dangerous red tide that can result in extensive fish kills and the contamination of shellfish in the area. Many dinoflagellates are bioluminescent.

Divergent evolution. Evolution in very different directions from a common ancestor, resulting in closely related forms looking very different. A good example is the dissimilar appearance of sea urchins and sand dollars.

DNA (deoxyribonucleic acid). The complex organic compound in the genes that controls the replication and activities of living tissue. These helical molecules, along with certain proteins, constitute chromosomes. *RNA* (ribonucleic acid) acts as a messenger in conjunction with DNA.

Doldrums. A region of calm winds along the equator, caused by the rising of warm air in that area.

Dolphin. Cetaceans of the family Delphinidae, generally possessing a beaklike snout and numerous teeth. They are not to be confused with the fish also called dolphins.

Doppler effect. The apparent change in frequency of sound or light waves caused by differences in the direction and velocity of the sound or light source and of the observer.

Echinoderm. Any of a phylum of radially symmetrical coelomate marine animals generally having a high level of organization. Although adults are radially symmetrical, larvae are bilateral. They are an important step forward in the course of evolution, as all higher forms of life are bilateral. This phylum includes the starfish, sea urchin, and sea cucumber.

Echo sounding. A technique of determining the depth of the sea bottom by the use of sound reflection; the results are produced graphically (*echogram*). The device used is termed an *echo sounder, depth sounder,* or *depth reader.* A similar technique for locating objects is called *echolocation.*

Ecosystem. The totality of organisms in a community; sometimes refers to physical attributes of an organic community. It may be considered in equilibrium if the community is stable for a finite period.

Electrolyte. A liquid solution capable of conducting electric currents because its dissolved components can dissociate into their ionic form. Salt (NaCl) in ionized form (dissolved—Na^+ and Cl^-) is a typical example.

Elver. A young eel recently metamorphosed from its larval stage.

Embryology. The science dealing with the earliest development of an organism.

Empiricism. A search for knowledge by observation; sometimes used in contrast to *experimentalism* (the pursuit of knowledge through direct testing in controlled conditions).

Enzyme. An organic catalyst; something that affects the rate of a reaction without being permanently altered itself.

Epifaunal organisms. *See* Infaunal organisms.

Epoch. A division of geologic time, commonly used for the subdivision of a *period* of the earth's history.

Equilibrium. A state of balance; in biology the steady state of undisturbed interactions; in chemistry a reversible reaction remaining unchanged.

Era. The largest subdivision of geologic time, e.g., Paleozoic, Mesozoic, Cenozoic.

Estuary. A semiclosed body of water with free access to the open ocean, wherein seawater is measurably diluted by freshwater runoff from land and where the tides of the ocean affect water movements. Most estuaries are drowned river valleys or valleys cut by glaciers (fjords).

Eucaryote. An organism, one- or many-celled, that has its genetic material bound up in a nuclear membrane. This is in contrast to *procaryotes* which have their genetic material dispersed through their cytoplasm. Procaryotes include bacteria and blue-green algae. All higher forms of life are eucaryotes.

Euphausiids. Pelagic, shrimplike crustaceans averaging one inch in length. All members of the order Euphausiacea are marine, and most are filter feeders; a few are predaceous. An important part of the polar food chain; whalers called them krill.

Euryterid. One of the group of extinct "sea scorpions" that lived in the early Paleozoic era.

Eutrophication. The natural aging process by which the dissolved oxygen in a standing water body, usually a lake, is reduced to the point where most forms of life cannot survive.

Evolution. A gradual process of change. Usually used in the sense of "organic evolution," which is the change in living organisms, often in the direction of specialization.

Extinction. The destruction or termination of something; usually refers to a group of organisms.

Fathometer. A sonar device used for measuring ocean depth; also called a *depth sounder.*

Fault. A fracture in the earth that involves movement of the rocks on either side of it.

Fertilization. Initiating the development of offspring by means of the introduction of genetic material in sexual reproduction; combining of an egg and sperm. It may be *external* (occurring in the water) or *internal* (occurring within the female).

Flagellata. A group of flagellate protozoa in the subphylum Mastigophora.

Flagellum. A threadlike or whiplike extension of the protoplasm of some cells, often used in locomotion such as the tail of a sperm cell.

Flukes. The caudal extensions of a cetacean; they form the forked, horizontal fishlike tail.

Fluorescence. The property of emitting light under stimulation by an outside energy source such as X or ultraviolet rays.

Food chain. The succession of predation from the phytoplankton to the zooplankton, to small fish and larvae through various levels, to the largest predators such as tuna and man.

Foraminifera. Any of an order of marine protozoans with chambered shells; so named because the shells usually have holes (or foramina) through which pseudopods extend. They are abundant planktonic and benthonic animals.

Fossil. Any preserved remain or trace of a dead organism; generally older than the recent epoch.

Fossil fuels. The energy sources man uses which derive from fossil organisms, including coal, natural gas, and petroleum.

Gamete. The reproductive cell of sexually reproducing organisms, possessing one-half the full number of chromosomes (haploid).

Gastropod. Any of a large class of molluscs, often with a univalve shell. Examples are snails, nudibranchs, and limpets.

Gene. A replicative unit in the chromosome, composed of DNA.

Geotaxis. An innate response by a living organism to the gravitational forces of the earth.

Gestation. The period of time in the development of an individual from conception to birth.

Gill. The complex of breathing organs in an aquatic animal, composed of osmotic membranes, supports, clefts, and vascular tissue.

Glacier. A mass of compacted snow and ice that forms when the rate of melting is less than the rate of snow accumulation.

Gonad. The reproduction organ of an animal, usually the ovary or testis.

Gondwanaland. The supercontinent that may have once incorporated all the southern landmasses before continental drift took place more than 60 million years ago.

Graptolite. An extinct protochordate planktonic animal, possibly related to the living pterobranch.

Great Barrier Reef. A 1250-mile-long coral reef off the coast of Australia; the largest coral reef in existence today.

Greenhouse effect. The superabundance of atmospheric carbon dioxide over many industrialized areas mimics the glass of a greenhouse and acts as a heat trap. This phenomenon may someday warm the earth enough to melt the ice caps and precipitate a glacial advance or "ice age."

Guano. The feces of birds and bats, commercially important as fertilizer.

Gulf Stream. The warm ocean current, approximately 50 miles wide, flowing from the Gulf of Mexico along the eastern U.S. coast, then turning toward Europe at the Grand Banks.

Guyot. A flat-topped submarine peak. Guyots are indicative of past, lower sea levels and/or submergence of the seamount.

Hadal zone. The oceanic trenches.

Herbivore. An organism that subsists on plant nutrients; a plant feeder.

Hermaphroditic. Having the sex organs or reproductive aspects of both male and female. Many invertebrates and some fish show this capability.

Heterotroph. An organism that must depend on food outside itself; usually a form that directly or indirectly eats plants. Opposed to *autotroph.*

Holograph. A photograph taken with laser light to produce three-dimensional images. The process requires no lenses and works by interference phenomena in the split laser beams.

Holoplankton. Members of the planktonic community that spend their entire life cycles as part of the plankton. This is in contrast to *meroplankton:* planktonic forms that spend only part of their life as plankton. Examples are fish eggs and larvae. Nanoplankton is a size classification: very small forms, ranging from 5 to 60 microns. (One micron = 10^{-4} cm or 3.94 x 10^{-5} inches.)

Holosteans. Any of a largely extinct group of fish characteristic of the Mesozoic era. Living members include the gar pike and bowfin.

Holothurians. A class of echinoderms known as sea cucumbers. They are bottom dwellers, living in sand and mud. They have retractable tentacles near their mouth with which they capture food.

Hydrodynamics. The branch of physics dealing with the motion and forces of water.

Hydroid. A polypoid form of coelenterate in form like the genus *Hydra*.

Hydrophone. A device for detecting and pinpointing underwater sounds.

Ichthyologist. A specialist in the study of fish, their anatomy, classification, and life history.

Igneous rock. A rock that has solidified from molten magma, either within the earth or on the surface in the form of volcanic lava.

Inbreeding. The process of continually mating individuals from closely related or identical ancestors.

Infaunal organisms. Animals that live in the ground within a certain area. We speak of the infaunal organisms of a mud flat, etc. This is in contrast to *epifaunal organisms* that live above ground.

Instinct. An inborn tendency to behave in a certain way that is characteristic of a species.

Internal wave. An ocean wave generated along a horizontal density gradient between two water masses beneath the surface. Unlike wind-generated waves, they are not visible on the surface. Internal waves can be a hazard to submarines.

Intertidal zone. The portion of the sea and sea bottom included between the low- and high-tidal marks. Most obviously seen on rocky coasts.

Invertebrates. Those animals that lack a spinal column or notochord; that is, all animals except fishes, amphibians, reptiles, birds, and mammals.

Isopods. Any of the second largest order of crustaceans next to the Decapoda (lobsters, crabs, shrimps). Most isopods are marine, a few are freshwater, some are terrestrial, and a few are even parasitic. They have a characteristically flat body and lack a carapace. Most isopods are 5 to 15 mm long. The giant of the group is a 14-inch isopod that lives in the deep sea *(Bathynomus giganteus).*

Isostasy. Approximate balance of areas of the earth's crust which tend to "float" on the semifluid mantle below. Continents consist of material lighter than that of the ocean floor and thus they float higher than the sea floor.

Jacklighting. Drawing fish to a net or fishing area by use of torches or lights.

Kelp. Common name for large brown seaweeds belonging to the brown algae.

Kraken. A Norwegian sea god and name for giant squid or octopods.

Labyrinthodont. A long-extinct group of primitive amphibians. Fossil labyrinthodonts are common in deposits from the Carboniferous period, which are 300 million years old.

Larvae. The immature form of any animal that changes structure (usually by metamorphosis) as it becomes an adult. Usually free-swimming.

Lateral-line organ. A series of sensitive hairlike cells imbedded in a matrix that responds to changes in pressure. These sensory receptors run laterally down the body of many fish. They can detect water pressure, direction and rate of flow of water, and low-frequency sound.

Laurasia. A hypothetical continent composed of North America and Eurasia that supposedly separated in the late Paleozoic era.

Lithosphere. The solid portion of the earth; the earth's crust, as opposed to the atmosphere.

Littoral. Referring to the shore or coast; the environment of flora and fauna from the shore to a depth of about 600 feet (200 meters).

Lysosomes. Cellular organelles, saclike in shape, enclosing enzymes from the rest of the cytoplasm. The enzymes, thus contained, break down cellular macromolecules, or the organism itself after death.

Mammal. Any of a class of vertebrates usually characterized by homeostatic body temperature, live birth, and mammary glands.

Mantle. Layer of earth between the *crust* (lithosphere) and *core;* also, the fleshy fold of tissue that secretes the shell of the mollusc.

Mariculture. The technique of artificially cultivating and culturing marine plants and animals for human consumption. It is most successful, at this point, with shellfish.

Meiosis. Cell division in which the chromosome number in the resulting cells is haploid, or half the usual number (reduction division). Meiosis results in the formation of cells for sexual reproduction; when two sex cells unite, the normal diploid number of chromosomes is restored. This is the opposite of *mitosis,* in which the resulting cell is diploid. Mitosis is the process that almost all the cells of our body use to replicate themselves.

Melanophore. A specialized type of chromatophore responsible for dark pigmentation; found in cephalopods, crustaceans, and fish.

Mesoglea. Jelly-containing layer of tissue between the ectoderm and endoderm of coelenterates.

Metabolism. Processes by which food is assimilated into tissue (anabolism) and those by which accumulated matter is broken down (catabolism) to release energy in living organisms.

Metamorphosis. Process by which an immature form of an animal undergoes radical structural change into the adult form.

Metazoans. Term used in some systems of taxonomy to refer to all animals whose bodies are composed of complex, differentiated cells.

Mid-Atlantic Ridge. Great ridge-and-valley system rising in the center of the Atlantic, running north and south throughout the ocean. One of several oceanic ridge systems found throughout the world, it is the site of sea-floor spreading.

Migration. Instinctive response to reproductive or food needs that triggers repeated movement of entire species to a new area. *Vertical migration* refers to the diurnal movement of plankton from deep to shallow depths in the sea.

Mimicry. The close resemblance in behavior, coloration, or physical appearance of one organism to another. It usually is protection from predation.

Mitochondria. Cytoplasmic organelle that serves as an enzyme storehouse and is vital to the metabolism of the cell.

Mitosis. *See* Meiosis.

Mohorovicic Discontinuity. Whenever there is an earthquake or nuclear blast, for example, shock waves travel through the earth. When these waves meet areas of different densities, they speed up or slow down. Waves abruptly increase in speed below 50 km, indicating a change in material. This area of change became known as the Mohorovicic discontinuity, named after the Yugoslavian seismologist who discovered it. Called Moho for short, it is the boundary between the earth's crust and mantle.

Molecule. Smallest particle of substance or compound that can exist in the free state while still maintaining the properties of the material.

Molluscs. Any member of the phylum Mollusca; generally characterized by a soft, nonsegmented body with gills, enclosed in a mantle and shell.

Mouthbreeders. Any of the freshwater and saltwater species of fish that take fertilized eggs into the mouth for incubation. Common marine varieties include the catfish, cardinalfish, and jawfish.

Mutation. Any change, artificially or naturally induced, in the genetic makeup of an animal that will be incorporated into the chromosomal material of the succeeding generation.

Mutualism. A symbiotic relationship between two species of organisms that is mutually beneficial.

Nacre. The pearly inner layer in the shell of several molluscs. Often called mother-of-pearl, it has been used for centuries as an ornamental material.

Nautiloids. Any of the subgroup of cephalopods; the only living member is the nautilus.

433

Nekton. All free-swimming organisms whose movements are not governed by currents and tides.

Nematocysts. Stinging cells of coelenterates.

Nephridia. Kidneylike structures that perform excretory functions in many invertebrate groups. The marine worm *Nereis* has a pair of nephridia in each of its body segments.

Neurons. Nerve cells; the structural and functional units of the vertebrate nervous system. The squid, an invertebrate, has a giant neuron, which makes it a valuable animal to research scientists.

Nitrogen narcosis. A mental disorder resulting from breathing compressed air below 100 feet. Often called "rapture of the deep," it results in double images, exaggeration of some senses and attenuation of others, feelings of euphoria and terror, and poor mental judgment. A sufficient explanation for this disorder has not been found, but it is related to the presence of nitrogen in the circulatory system.

Notochord. An elongated rod that acts as a supporting structure in lower chordates. It appears in the embryonic stages of higher vertebrates but is later replaced by the vertebral column becoming the centrum of the vertebrae.

Nucleic acids. A group of acids that carry genetic information. The two kinds are DNA and RNA. They can be found in various cell structures as well as in the cytoplasm. (*See* DNA.)

Olfactory receptors. Sensory elements that detect chemicals dissolved in water or in air. Touch is not involved, as it is in taste.

Ontogeny. The development of an individual.

Ooze. A general name given to sediment deposits on the floor of the deep ocean that were derived from biological sources.

Osteichthyes. In some taxonomies of living things, the class to which all bony fish belong. The other large class of fish surviving today is the Chondrichthyes, or cartilaginous fish.

Ostraciiform movement. A type of locomotion in fish where the body remains rigid and only the tail fin flexes, providing all the forward propulsion. It is used by boxfish and trunkfish.

Ostracoderm. Primitive fish that roamed the seas 400 million years ago. They lacked jaws, and their bodies were covered by thick bone or bonelike plates. Ostracoderm means shell-skinned.

Outfall. Discharge from treatment plants, factories, power plants, and the like that is released into the aquatic environment.

Oviparous. *See* Viviparous.

Ovoviviparous. *See* Viviparous.

Oxidation. The combination of a substance with elemental oxygen. In more general terms, a substance is said to be oxidized if it loses one or more electrons in a chemical reaction.

Pangaea. The supercontinent believed to have existed 200 million years ago. Pangaea broke up, and the pieces began to drift apart, forming the continents as we know them today.

Parasitism. A relationship in which one organism obtains food and other benefits at the expense of the other; the host organism is usually harmed.

Parthenogenesis. Unisexual reproduction in which the egg begins to develop and grow without ever having been penetrated by a sperm cell or accepting its nuclear material.

Pelagic zone. All ocean waters covering the deep-sea benthic province (bottom). The pelagic zone is divided into an open sea (oceanic) province and an inshore (neritic) province.

Permafrost. Permanently frozen subsoil found in the higher latitudes of the earth.

Petrification. A process in which organic matter is replaced by silica, lime, or some other mineral deposit to form a stony substance.

Pheromone. A substance secreted by an organism that stimulates a behavior or physiological response in another individual of the same species.

Phosphorescence. The emission of visible light without the production of any noticeable heat. This is a common phenomenon among deep-sea creatures. A synonym for this term is bioluminescence.

Photoautotrophs. Organisms that depend upon light as an energy source (for photosynthesis) and which use carbon dioxide as their principle source of carbon.

Photon. A unit, or quantum, of electromagnetic energy. The energy of light is carried by photons.

Photosynthesis. A biological process in which plants convert carbon dioxide and water into usable carbohydrates. Solar energy, captured by the green pigment chlorophyll, supplies the power to carry out the many complex chemical reactions that are involved in the photosynthetic process.

Phylogeny. The evolutionary development, or lines of descent, of a plant or animal species.

Phytoplankton. The microscopic plants that inhabit the oceans of the world. They are responsible for most of the photosynthesis that goes on in the ocean and are the basis of the oceanic food web.

Pingos. Ice intrusions in arctic waters that resemble reef formations.

Pinnepeds. A group of marine carnivores that includes all the seals and walruses. Characteristically the digits at the end of each limb are connected and covered with a thick web of skin. Most species possess claws.

Placenta. A structural link between developing embryo and mother through which the embryo receives nourishment and rids itself of waste products. It is a mammalian characteristic.

Placoderms. Any of a group of archaic jawed fish that became extinct by the end of the Paleozoic era, about 200 million years ago. Placoderms, like ostracoderms, were covered, to a degree, by armor plates.

Plankton. In general, any organism, large or small, that floats or drifts with the movements of the sea. Plants are called phytoplankton, and animals zooplankton. Most organisms that spend their entire life as part of the plankton are microscopic in size.

Plate tectonics. The study of the earth's crustal plates and the forces that cause them to "drift" over denser mantle rocks.

Platyhelminthes. The phylum to which the flatworms belong. Two classes, the Cestoda (flukes) and Trematoda (tapeworms), are parasitic. A third class, Turbellaria, is free-living. The flatworms are the most primitive bilaterally symmetric animals and are an important evolutionary link.

Polymerization. A process by which two or more molecules are united to form a more complex molecule. The new molecule has different physical and chemical properties from those of its components.

Polyp. The body form of various coelenterates that remain attached to the substrate. A polyp has a hydralike form; it is a fleshy stalk with tentacles. Polyps can occur singly or as part of a colony. The polyp stage of a coelenterate's life cycle may be contrasted to the medusa or free-swimming stage.

Preadaptation. A quality possessed by a living organism that is not essential for its survival today but might be vitally important to its existence at some future stage of development.

Predator. A creature that survives by capturing and feeding on other living organisms.

Pressure drag. A resistance to forward motion that results from turbulent forces generated as an object or animal moves through its medium. A fish with a fusiform shape produces little turbulence and therefore reduces the pressure drag holding it back.

Primary production. The amount of organic material (represented by carbon) produced per given unit of seawater by photosynthetic autotrophs.

Procaryotic cell. A cell that lacks a membrane-bound nucleus. Nuclear materials are found within the cytoplasm.

Protochordates. A group of primitive animals possessing a notochord during some part of their life cycle (acorn worms, pterobranchs, and tunicates).

Protozoa. Microscopic, nonphotosynthetic, eucaryotic, unicellular animals. Major groups include the ciliates, amoeba, flagellates, and sporozoans.

Pterobranchs. Tiny marine animals that form plantlike colonies. They are considered the oldest protochordates as they show primitive gill development and a food-filtering system from which more typical chordate features could have been derived.

Pycnogonid. An arthropod commonly known as a sea spider. Almost all legs and no body, these unique creatures have successfully adapted, through evolutionary processes, to life in the abyss as well as shallow waters.

Radioactivity. A phenomenon exhibited by certain elements wherein various forms of radiation are emitted as a result of changes in the nuclei of atoms.

Radiolaria. One-celled marine animals (Protozoa) that are encased in a spiny coat of silica.

Radula. A tonguelike structure found in many molluscs. It has a rough surface and is used to rasp or file organic material from the surfaces of rocks or plants or to bore holes in the shells of other animals.

Reclamation. A process in which wasteland is converted into usable space. Inundated shorelines are often recovered by diking back the sea.

Red tide. A phenomenon produced by the explosive growth of certain dinoflagellates, causing the sea to turn red. Toxins produced by these animals can kill fish on a large scale and poison shellfish, making them unsafe for human consumption.

Reef. An underwater structure built by carbonate-secreting organisms like coral and encrusting algae. Fringing reefs are connected to shore and run parallel to the coast. Barrier reefs also parallel the coast, but they are separated from shore by a lagoon.

Refraction. The bending of light ray or sound wave as it passes through one medium and then through another medium of different density. Ocean waves are also refracted or bent by submerged or shoreline features.

Regeneration. The natural replacement of lost tissue or an entire body part.

Rift. A fault in the earth's crust where the movement of material is away from the fault in a lateral direction. The line of submerged mountains that bisects the Atlantic Ocean in a north-south direction is often referred to as the Mid-Atlantic Rift.

"Ring of fire." Active volcano and earthquake activity around the margin of the Pacific Ocean, a result of large-scale movements of the earth's crust.

Rip currents. Nearshore, wave-induced water movements that push water toward the beach. In the surf zone, the water moves parallel to the beach. When these longshore currents combine, they produce jetlike streams of water, a few tens of meters across, which move seaward through the surf. Such rip currents have swept many a bather out to sea.

Salinity. A measure of the amount of salt (usually determined by chloride ions) dissolved in water; expressed in parts per thousand.

Sarcopterygii. The lobe-finned, air-breathing fish.

Sargasso Sea. The region of the North Atlantic between the West Indies and the Azores, characterized by an abundance of the drifting seaweed *Sargassum* and by almost motionless water inside the circulating North Atlantic gyre.

Scyphozoan. A coelenterate which features a well-developed medusoid stage. Jellyfish are examples of this class, in essence a swimming, upside-down anemone.

Sediment. A particle or bed of particles that is the result of the weathering and erosion of preexisting rocks or organic materials. When the sediments become lithified (compressed or cemented into rock), they become *sedimentary rocks.*

Seine. A type of fishing net that hangs vertically in the water, its bottom edge weighted and its upper edge supported by floats. Seines trap fish by enclosing them in a confined area. Varieties are purse seines and haul seines. Other fishing apparatus include *long lines* (with many hooks), *trawls* (which are dragged along behind a ship), and *gill nets* (a wall of netting that hangs suspended in the water, trapping fish within its mesh).

Sial. The rocks, rich in aluminum, sodium, and potassium, that compose the upper layers of the earth's crust.

Silicoflagellate. Flagellated, planktonic organisms with siliceous shells; common in most of the colder parts of the ocean.

Sima. The rocks of the lower part of the earth's crust, which are rich in magnesium, calcium, and iron minerals.

Sirenians. A group of herbivorous marine mammals that includes the manatee and dugong; distantly related to the land elephant.

Sonar. A system of echolocation used to detect objects underwater; acronym for SOund NAvigation Ranging. Applicable to man's and dolphin's systems.

Spawn. The eggs of aquatic animals, including fish, invertebrates, and amphibians. The larval stage of many bivalves is referred to as *spat.*

Species. The most precise unit of taxonomy; a group of organisms that can interbreed with members of no other group.

Spectrum. The ordered arrangements of light and radiant energy, usually classified by wavelengths. The device to measure this distribution is called a *spectrophotometer*, and produces a *spectrogram.*

Spermatozoa. The male gamete in a sexually reproducing animal; usually a tadpole-shaped motile structure with a long whiplike tail.

Spicule. A needlelike supporting structure, which may be composed of calcium carbonate, silica, or chitin; characteristically found in sponges.

Statocyst. A balance organ found in many organisms; also called an *otocyst.* It consists of a fluid-filled sac and may contain a *statolith* or *otolith* (a small concretion that moves, under the influence of gravity, with the fluid and tells the animal which end is up or down).

Stomiatoid. Any of a group of bathypelagic fish commonly called viperfish or dragonfish, which may possess fanglike teeth and luminous barbels.

Storm surge. A change in sea level associated with a storm. They occasionally cause catastrophes in low-lying areas.

Stratosphere. The portion of the earth's atmosphere that is relatively uniform in temperature, above the *troposphere* and below the *mesosphere*, beginning about seven miles up.

The upper boundary of the troposphere reaches about five miles above the poles and ten miles above the equator. The stratosphere extends ten to 15 miles higher, and the mesosphere goes up to the 50-mile level. The next region is the *ionosphere*, which extends to an altitude of 350 to 600 miles. The outermost atmospheric layer is the *exosphere*—900 miles of thinly dispersed helium surrounded by a hydrogen layer that extends 4000 miles before it tapers off into the void of space. Atoms and molecules in the exosphere are so far apart they seldom collide and some are lost forever to space.

Stridulate. To produce by friction rapid vibrations sounding like chirping; frequently done by many marine invertebrates and some fish and mammals.

Stroboscope. A device to make moving objects appear stationary by emitting bright light impulses at variable intervals. Stroboscopy is used to analyze movements and to measure revolutions per minute.

Stromatolite. A mat of algae and sediment which may harden and fossilize. Some stromatolites are three billion years old.

Sublittoral zone. A division of the benthic province extending from approximately a depth of 150 feet (50 meters) to the edge of the continental shelf, a depth of about 600 feet (200 meters).

Submersible. A controlled, free-diving underwater vessel that does not have ballast carried externally (as in a submarine). Generally submersibles are small and used for research.

Surf. Waves breaking in a coastal area.

Swell. Waves that have traveled away from the site of generation, usually in a regular pattern.

Swim bladder. A gas-filled sac found in many fish, which helps them maintain buoyancy and sometimes aids them in respiration. For hearing and sound production it may act as a resonating chamber.

Symbiosis. Any relationship between two or more organisms. Sometimes confused with mutualism. Encompasses the subdivisions of commensalism, parasitism, and mutualism.

Syncline. A folded layer or layers of rock with the center depressed relative to the sides and with the youngest rocks exposed in the center after erosion levels the surface.

Synergism. In chemistry, the process by which two or more substances reinforce each other's effects, or in combination produce a new effect. Many poisons show this property.

Synthesis. The creation of a substance, usually by an organism or experimental process.

Taxonomy. That branch of science dealing with the systematic classification of living things.

Tectonics. Movements and mountain-building forces within the earth; also the study of those forces and the changes they produce.

Teleostei. A superorder within the class Osteichthys (bony fishes) to which modern bony fish belong. Living bony fish *not* belonging to this group include the sturgeon, paddlefish, garpike, and bowfin.

Thermal inversion. An atmospheric phenomenon that occurs when a warm air mass overrides a cooler one. Pollutants from below are trapped beneath the warm air in a horizontal layer.

Thermistor. A solid-state electronic device used to measure water temperature. The thermistor's electrical resistance decreases as the water temperature increases. This change in resistance is monitored and converted into temperature differences.

Thermocline. A layer of water between warm surface water and cool deep water where temperature decreases rapidly with depth. A thermocline can act as a barrier to the transmission of sound in the sea, as well as to vertical migrations of fish.

Thermohaline circulation. Movement of water caused by differences in salinity and temperature. In the antarctic region, for example, cold salty water, which has a high density, sinks and flows northward along the bottom to the equator.

Thermoregulation. Maintenance of body temperature at a specific level by means of heat production and various methods designed to conserve heat. Thermoregulation is characteristic of warm-blooded animals and promotes homeostasis.

Tide pool. A pocket of seawater isolated by the receding tide. It is an unusual environment; salinity and temperature can fluctuate greatly before the next tide comes in.

Till. Sediment carried or deposited by a glacier. *Tillite* is a sedimentary rock composed of this.

Topography. The surface features of an area, including relief, lakes, rivers, and so on. Oceanographers are interested in the topography of the ocean bottom and spend a great deal of time charting the depths of the sea.

Trade winds. Persistent winds that blow from about 30° N. latitude toward the equator from the northeast, and from 30° S. latitude toward the equator, from the southeast. These winds drive the North and South Equatorial Currents in a westward direction and indirectly the Equatorial Countercurrent in an eastward direction.

Trawl. A specific type of oceanographic sampling and commercial fishing gear, most often towed behind a moving ship. A trawl can be moved over the bottom, where it collects benthic invertebrates and fish, or at a midwater depth, where it can sample pelagic fish life with some success.

Trenches. Deep fissures in the ocean floor where old crust is destroyed and forced back into the interior of the earth.

Trilobite. Any of a primitive group of arthropods that became extinct about 230 million years ago.

Tsunami. A seismic sea wave produced by a sudden movement of the ocean bottom. It can race undetected across the ocean at hundreds of miles per hour, but when it reaches shallower shore areas, it can grow to heights of 100 feet or more, wreaking havoc on these far-distant shores.

Tundra. A vast, nearly level, treeless plain characteristic of the arctic region.

Turbidity current. A dense, sediment-laden current of water that usually flows downward through less dense water along the slope of a continent. Some submarine canyons may have been cut by turbidity currents. Few have actually been observed; their dynamics and effects on submarine erosion are poorly understood.

Ultrasonic frequencies. Mechanical vibrations (sound) above the range of human detectability, which is 16 to 16,000 cycles per second. Whales and dolphins have hearing particularly sensitive to ultrasonic sound.

Undertow. A current of water moving seaward under the breaking surf. More generally, it can be any current of water moving beneath the surface water in a different direction.

Upwelling. An oceanic phenomenon wherein deep water is drawn to the surface. Upwelling can be induced by wind and surface currents. Bottom waters are rich in nutrients, and areas of upwelling are areas of high biological productivity.

Vacuole. A relatively clear, fluid-filled cavity within a cell. Vacuoles serve a variety of functions, from discharge of water and waste products to food and enzyme storage.

Vertebrate. Any animal possessing a backbone and a cranium. Vertebrata, the large subphylum of chordate animals, includes all mammals, birds, reptiles, amphibians, and fish.

Virus. Ultramicroscopic particles considered by some to be alive and by others to be complex proteins that sometimes include nucleic acids and enzymes. Viruses can multiply only in connection with a living cell.

Viscosity. The internal friction of a fluid, caused by attractions of the molecules for each other, which makes it resistant to flow.

Viviparous. The retention of the egg within the mother during development and subsequent live birth, seen in mammals and some fish. In contrast to *oviparous,* in which the egg is laid and development takes place externally, seen in birds and some reptiles, and *ovoviviparous,* in which development is internal but no placental connection ever forms between mother and embryo, seen in sharks and some fish.

Wavelength. If one pictures ripples emanating from a pebble dropped in a pond, the wavelength is the distance between one wave crest and the next; or between one wave trough and the next; or between any point on one wave and the corresponding point on the next one. For sound, a wavelength is the distance between one zone of compression or refraction and another.

Xiphosura. A nearly extinct subclass of arthropods whose fossil record dates back to the Ordovician period, 450 million years ago. Three genera and five species of Xiphosurans exist today. The most common representative is *Limulus,* the horseshoe crab, a true living fossil.

Yolk sac. A sac containing yolk, a nutrient for developing embryos that consists mainly of fat and protein. Commonly seen in fish embryos.

Zooplankton. Tiny animals, rarely more than a few millimeters in length, that are unable to counter the movements of water in the oceans and are carried away, drifting at the mercy of the current.

Zooxanthellae. Symbiotic algae that live in the tissue of most reef coral, as well as other species. These dinoflagellate relatives require sunlight for photosynthesis, a factor that may restrict reefs to sunlit waters. The coral animals utilize the oxygen that the algae produce.

Zygote. The fertilized ovum in plants and animals; more technically, it is the diploid cell resulting from the fusion of male and female gametes.

BIBLIOGRAPHY

GENERAL OCEAN SCIENCE

Carson, Rachel L. *The Sea Around Us*. 2nd ed., rev. New York: Oxford University Press, 1961; NAL, 1954 (paper).

————. *The Edge of the Sea*. Boston: Houghton Mifflin, 1955; NAL, 1971 (paper).

Coker, R. E. *The Great and Wide Sea*. New York: Harper & Row, 1962 (paper).

Deacon, G. E. R. *Oceans*. London: Hamlyn, 1968.

Dietrich, Gunther. *General Oceanography: An Introduction*. Translated by Deodor Ostapoff. New York: Wiley, 1963.

Dugan, James, et al. *World Beneath the Sea*. Washington, D.C.: National Geographic Society, 1967.

Engel, Leonard. *The Sea*. New York: Time–Life, 1969.

Fairbridge, Rhodes W., ed. *The Encyclopedia of Oceanography*. New York: Halsted, 1975.

Firth, Frank E., ed. *The Encyclopedia of Marine Resources*. New York: Van Nostrand, Reinhold, 1969.

Goldberg, E. D., *The Health of the Oceans*. Paris: UNESCO, 1976 (paper).

Heberlein, Hermann. *Le Monde sous-marin*. Zurich: BEA, 1959.

Herring, Peter J., and Malcolm R. Clarke. *Deep Oceans*. London: Arthur Barker, 1971.

Laurie, Alec. *The Living Oceans*. Garden City: Doubleday, 1973.

McCormick, J. M. and J. Thiruvathukal. *Elements of Oceanography*. Philadelphia: W. B. Saunders, 1976.

Olschki, Alessandro, ed. *SUB, enciclopedia del subacqueo*. Florence: Sadea/Sansoni, 1968.

Outhwaite, Leonard. *The Ocean*. London: Constable, 1961.

Ross, David A. *Introduction to Oceanography*. Englewood Cliffs: Prentice-Hall, 1970.

Scientific American. *Oceanography*. San Francisco: W. H. Freeman, 1971.

Sverdrup, H. U., Martin W. Johnson, and Richard H. Fleming. *The Oceans, Their Physics, Chemistry and General Biology*. New York: Prentice-Hall, 1942.

Weyl, Peter K. *Oceanography*. New York: Wiley, 1970.

THE POLES AND THE ABYSS

Bruun, Anton Frederic, et al. *The Galathea Deep Sea Expedition*. Translated by Reginald Spink. London: Allen & Unwin, 1956.

Heezen, Bruce C., and Charles D. Hollister. *The Face of the Deep*. New York and London: Oxford University Press, 1971.

Holdgate, M. W., ed. *Antarctic Ecology*. London and New York: Academic Press, 1970.

Idyll, C. P., ed. *Abyss, the Deep Sea and the Creatures That Live in It*. 3rd rev. ed. New York: Apollo, 1976.

Menzies, Robert J., Robert Y. George, and Gilbert T. Rowe. *Abyssal Environment and Ecology of the World Oceans*. New York: Wiley, 1973.

Stefansson, Vilhjalmur. *Hunters of the Great North*. New York: AMS, 1922.

EARTH AND OCEAN SCIENCES

Badgley, Peter C., Leatha Miloy, and L. Childs, eds. *Oceans from Space*. Houston: Gulf Publishing, 1969.

Bascom, Willard. *Waves and Beaches*. Garden City: Doubleday, 1964.

Bouteloup, Jacques. *Vagues, marées, courants marins*. Paris: Presses Universitaires de France, 1968.

Defant, Albert. *Ebb and Flow: Tides of the Earth, Air and Water*. Ann Arbor: University of Michigan Press, 1958 (paper).

King, Cuchlaine A. M. *Beaches and Coasts*. London: Edward Arnold, 1959.

Kummel, Bernhard. *History of the Earth*. 2nd ed. San Francisco: W. H. Freeman, 1970.

Lacombe, Henri. *Les Energies de la mer*. Paris: Presses Universitaires de France, 1968.

————. *Les Mouvements de la mer: courants, vagues et houle, marées*. Paris: Doin, 1971.

Menard, H. W. *Marine Geology of the Pacific*. New York: McGraw-Hill, 1964.

Pettersson, Hans. *The Ocean Floor*. Riverside, N.J.: Hafner, 1969, repr. of 1954 ed.

Shepard, Francis P. *The Earth Beneath the Sea*. 2nd ed. Baltimore: Johns Hopkins, 1967.

Shepard, Francis P., and Harold R. Wanless. *Our Changing Coastlines*. New York: McGraw-Hill, 1971.

Smith, F. G. Walton. *The Seas in Motion*. New York: Thomas Y. Crowell, 1973.

Trewartha, Glenn T. *An Introduction to Climate*. 4th ed. New York: McGraw-Hill, 1968.

EVOLUTION

Beerbower, James R. *Search for the Past*. 2nd ed. Englewood Cliffs: Prentice-Hall, 1968 (paper).

Buettner-Janusch, John. *Origins of Man*. New York, London, and Sydney: John Wiley, 1966.

Colbert, Edwin H. *Evolution of the Vertebrates*. 2nd ed. New York: John Wiley, 1955.

Darwin, Charles E. *The Origin of Species*. New York: NAL (paper); Cambridge, Mass: Harvard University Press, 1975 (paper facsimile of the 1st ed.).

Fenton, Carroll L., and M. A. Fenton. *The Fossil Book: A Record of Prehistoric Life*. Garden City: Doubleday, 1958.

Romer, Alfred S. *The Vertebrate Story*. rev. ed. Chicago and London: University of Chicago Press, 1971 (paper).

MARINE ARCHAEOLOGY

Bass, George F. *Archaeology Underwater*. New York: Praeger, 1966.

Bass, George F., ed. *A History of Seafaring*. New York: Walker, 1973.

Dumas, Frédéric. *Deep Water Archeology*. Translated by Honor Frost. Chester Springs, Pa.: DuFour Editions, 1962.

Marx, Robert F. *The Lure of Sunken Treasure*. New York: David McKay, 1973.

————. *Port Royal Rediscovered*. Garden City: Doubleday, 1973.

Peterson, Mendel. *History Under the Sea*. Washington, D.C.: Smithsonian Institution, 1954.

Potter, John, Jr. *The Treasure Diver's Guide*. Garden City: Doubleday, 1972.

Throckmorton, Peter. *Shipwrecks and Archaeology, The Unharvested Sea*. Boston: Little, Brown, 1970.

————. *The Lost Ships*. London: Jonathan Cape, 1965.

MAN AND THE SEA

Balder, A. P. *Complete Manual of Skin Diving*. New York: Macmillan, 1968.

Barnaby, K. C. *Some Ship Disasters and Their Causes*. Maritime Library Series. Cranbury, N.J.: A. S. Barnes, 1970.

Beebe, William. *Half Mile Down*. New York: Duell, Sloan and Pierce, 1955.

Clark, Eugenie. *The Lady and the Sharks*. New York: Harper & Row, 1977.

Cousteau, Jacques-Yves, and James Dugan. *The Living Sea*. New York: Harper & Row, 1963.

Cousteau, Jacques-Yves, and James Dugan, eds. *Captain Cousteau's Underwater Treasury*. New York: Harper & Row, 1959.

Cousteau, Jacques-Yves, and Frédéric Dumas. *The Silent World*. New York: Harper & Row, 1953.

Darwin, Charles. *The Voyage of the Beagle*. New York: Bantam Books, 1972.

Davis, Robert H. *Deep Diving and Submarine Operations*. 7th ed. Chessington, Surrey: Siebe, Gorman and Co., 1962.

Deacon, G. E. R., ed. *Seas, Maps, and Men*. London: Crescent Books, 1962.

Frey, Hank, and Shaney Frey. *130 Feet Down, A Handbook for Hydronauts*. New York: Harcourt Brace Jovanovich, 1961.

Goldberg, E. D. *Strategies for Marine Pollution Monitoring*. New York: Wiley, 1976.

Gordon, Bernard L., ed. *Man and the Sea*. Garden City: Doubleday, Natural History Press, 1972.

Groueff, Stéphane. *L'Homme et la mer*. Paris: Larousse, Paris-Match, 1973.

Groupe d'Etudes et de Recherches Sous-Marine. *La Plongée*. Paris: B. Arthaud, 1955. (Published in English under title *Complete Manual of Free Diving* [New York: Putnam, 1957])

Johnston, R., ed. *Marine Pollution*. New York: Academic, 1976.

Khan, M. A. Q., and J. P. Bederka, Jr., eds. *Survival in Toxic Environments*. New York: Academic, 1974.

Lee, Owen. *Skin Diver's Bible*. Garden City: Doubleday, 1968 (paper).

Lockwood, A. P. M., ed. *Effects of Pollutants on Aquatic Organisms*. New York: Cambridge University Press, 1976.

Miles, S. *Underwater Medicine*. Philadelphia: Lippincott, 1966.

Penzias, Walter, and M. W. Goodman. *Man Beneath the Sea*. New York, London, Sydney, and Toronto: Wiley-Interscience, 1973.

Piccard, Jacques, and Robert S. Dietz. *Seven Miles Down: The Story of the Bathyscaphe Trieste*. New York: Putnam, 1961.

Sagan, L. A., ed. *Human and Ecological Effects of Nuclear Power Plants*. Springfield, Ill.: Charles C. Thomas, 1974.

Schlee, Susan. *The Edge of an Unfamiliar World, A History of Oceanography*. New York: Dutton, 1973.

Shenton, E. H. *Exploring the Ocean Depths*. New York: Norton, 1968.

Titcombe, R. M. *Handbook for Professional Divers*. Philadelphia: Lippincott, 1973.

United States Department of the Navy. *U.S. Navy Diving Manual*. Washington, D.C.: U.S. Government Printing Office, 1963.

Vaissiere, Raymond. *L'Homme et le monde sous marin*. Paris: Larousse, 1969.

Villiers, Alan, et al. *Men, Ships, and the Sea*. Washington, D.C.: National Geographic Society, 1973.

Wilson, R., and W. J. Jones. *Ecology and Environment*. New York: Academic, 1974.

MAMMALS

Andersen, Harald T. *The Biology of Marine Mammals*. New York and London: Academic Press, 1968.

Burton, Robert. *The Life and Death of Whales*. New York: Universe Books, 1973.

Coffey, D. J. *Dolphins, Whales and Porpoises: An Encyclopedia of Sea Mammals*. New York: Macmillan, 1977.

Howell, A. Brazier. *Aquatic Mammals*. New York: Dover, 1970.

Kenyon, Karl W. *The Sea Otter in the Eastern Pacific Ocean*. 2nd ed. New York: Dover, 1975 (paper).

Magnolia, L. R., comp. *Whales, Whaling and Whale Research: A Selected Bibliography*. The Whaling Museum, 1977.

Norris, Kenneth S., ed. *Whales, Dolphins and Porpoises*. Berkeley: University of California Press, 1977.

Perry, Richard. *The World of the Polar Bear*. Seattle: University of Washington Press, 1966.

Rice, D. W. *A List of the Marine Mammals of the World*. 3rd ed. U.S. National Marine Fisheries Service, 1977.

Ridgeway, Sam H., ed. *Mammals of the Sea*. Springfield, Ill.: Charles C. Thomas, 1972.

Scheffer, Victor B. *The Year of the Whale*. New York: Scribner's, 1969.

Slijper, E. J. *Whales and Dolphins*. Ann Arbor: University of Michigan Press, 1976.

Small, George L. *The Blue Whale*. New York and London: Columbia University Press, 1973.

Stackpole, Edouard A. *The Sea-Hunters: New England Whalemen During Two Centuries, 1635–1835*. Philadelphia: Lippincott, 1953; Westport, Conn.: Greenwood, 1973 (repr. of 1953 ed.).

UNDERSEA RESOURCES

Bardach, John. *Harvest of the Sea*. London: Allen & Unwin, 1969.

Brandt, Andres von. *Fish Catching Methods of the World*. 2nd ed., rev. London: Fishing News, 1972.

Gulland, J. A., ed. *The Fish Resources of the Ocean*. London: Fishing News, 1971.

Idyll, C. P. *The Sea Against Hunger*. New York: Thomas Y. Crowell, 1970.

Iverson, E. S. *Farming the Edge of the Sea*. London: Fishing News, 1968.

Mero, John L. *The Mineral Resources of the Sea*. New York: American Elsevier, 1964.

Milne, P. H. *Fish and Shellfish Farming in Coastal Waters*. London: Fishing News, 1972.

Rounsefell, G. A. *Ecology, Utilization and Management of Marine Fisheries*. St. Louis: Mosby, 1975.

Simmons, I. G. *The Ecology of Natural Resources*. New York: Halsted, 1974 (paper).

Ulrich, Heinz. *America's Best Bay, Surf and Shoreline Fishing*. New York: Barnes, 1960.

Walford, Lionel A. *Living Resources of the Sea*. New York: Ronald, 1958.

NATURAL HISTORY

Abbott, I. A., and G. J. Hollenberg. *Marine Algae of California*. Stanford: Stanford University Press, 1976.

Abbott, R. Tucker. *Sea Shells of the World: A Guide to the Better-Known Species*. Edited by H. S. Zim. New York: Golden Press, 1962.

Alexander, W. B. *Birds of the Ocean*. New York: Putnam, 1963.

Barnes, Robert D. *Invertebrate Zoology*. Philadelphia and London: Saunders, 1963.

Boney, A. D. *A Biology of Marine Algae*. London: Hutchinson Educational, 1966.

Buchsbaum, Ralph. *Animals Without Backbones*. 2nd ed. rev. Chicago and London: University of Chicago Press, 1976.

Bustard, Robert. *Sea Turtles*. New York: Taplinger, 1972.

Carr, Archie. *So Excellent a Fishe: A Natural History of Sea Turtles*. Garden City: Doubleday, Natural History Press, 1967.

Chapman, V. J., ed. *Wet Coastal Ecosystems*. New York: Elsevier, 1977.

Cobb, J. S., and M. M. Harlin, eds. *Marine Ecology: Selected Readings*. Baltimore: University Park Press, 1976.

Coppelson, V. M. *Shark Attack*. London: Angus & Robinson, 1958.

Curtis, Brian. *The Life Story of the Fish*. New York: Dover, 1961.

Dance, S. P., ed. *The Collector's Encyclopedia of Shells*. 2nd ed. New York: McGraw-Hill, 1976.

Easton, W. H. *Invertebrate Paleontology*. New York: Harper Bros., 1960.

Eisenberg, J. F., and William S. Dillon, eds. *Man and Beast: Comparative Social Behavior*. Washington, D.C.: Smithsonian Institution, 1971.

Eibl-Eibesfeldt, Irenaus. *Ethology, The Biology of Behavior*. New York: Holt, Rinehart, Winston, 1970.

Ellis, R. *The Book of Sharks*. New York: Grosset & Dunlap, 1975.

Fox, H. Munro, and Gwynne Vevers. *The Nature of Animal Colours*. London: Sidgwick and Jackson, 1960.

Frings, H., and M. Frings. *Animal Communication*. 2nd ed. rev. Norman: University of Oklahoma Press, 1977.

Gillett, Keith, and Frank McNeill. *The Great Barrier Reef and Adjacent Islands*. Sydney: Coral, 1959.

Gordon, B. L. *The Secret Lives of Fishes*. New York: Grosset & Dunlap, 1977.

Halstead, B. W. *Poisonous and Venomous Marine Animals of the World*. 3 vols. Washington, D.C.: U.S. Government Printing Office, 1965.

Hardy, A. C. *The Open Sea: Its Natural History—Fish and Fisheries*. London: Collins, 1959.

————. *The Open Sea: Its Natural History—The World of Plankton*. London: Collins, 1956; Boston: Houghton Mifflin, 1959.

Herald, Earl S. *Living Fishes of the World*. Garden City: Doubleday, 1971.

Hickman, Cleveland P. *Integrated Principles of Zoology*. St. Louis: Mosby, 1961.

Hyman, L. M. *Invertebrates*. 6 vols. New York: McGraw-Hill, 1940–1967.

Johnson, M. E., and Harry James Snook. *Seashore Animals of the Pacific*. New York: Dover, 1967.

Jones, O. A., and R. Endean, eds. *Biology and Geology of Coral Reefs*. 3 vols. New York: Academic, 1976.

Lagler, Karl F., John E. Bardach, and Robert R. Miller. *Ichthyology*. Ann Arbor: University of Michigan Press, 1962.

Lane, Frank W. *The Kingdom of the Octopus: The Life History of the Cephalopoda*. New York: Sheridan, 1960.

Lanyon, W. E., and W. N. Tavolga. *Animal Sounds and Communication*. Monticello: Lubrecht & Cramer, 1960.

Le Danois, Edouard. *Fishes of the World*. London: Harrap, 1957.

Mackal, R. P. *The Monsters of Loch Ness*. Chicago: Swallow, 1976.

Marshall, N. B. *The Life of Fishes*. New York: Universe Books, 1966.

————. *Explorations in the Life of Fishes*. Cambridge, Mass.: Harvard University Press, 1971.

McConnaughey, B. H. *Introduction to Marine Biology*. 2nd ed. St. Louis: Mosby, 1974.

Migdalski, E. C., and G. S. Fichter. *The Fresh and Salt Water Fishes of the World*. New York: Knopf, 1976.

Najarian, H. H. *Sex Lives of Animals Without Backbones*. New York: Scribner's, 1976.

Nelson, J. S. *Fishes of the World*. New York: Wiley, 1976.

Neugebauer, W. *Marine Aquarium Fish Identifier*. New York: Sterling, 1975.

Newell, G. E., and R. C. Newell. *Marine Plankton: A Practical Guide*. London: Hutchinson Education, 1966 (paper).

Nichol, J. A. Colin. *The Biology of Marine Animals*. London: Pitman, 1960.

North, W. J. *Underwater California*. Berkeley: University of California Press, 1976.

Odum, Eugene P. *Fundamentals of Ecology*. Philadelphia: Saunders, 1971.

Ommanney, F. D., and the Editors of *Life* magazine. *The Fishes*. New York: Time Inc., 1963.

Orr, Robert T. *Animals in Migration*. New York: Macmillan, 1970.

Peres, Jean-Marie. *La Vie dans les mers*. Paris: Presses Universitaires de France, 1972.

————. *Clefs pour l'oceanographie*. Paris: Seghers, 1972.

Randall, John E. *Caribbean Reef Fishes*. Neptune City, N.J.: Harrowood Books, 1977.

Ricketts, Edward F., and Jack Calvin. *Between Pacific Tides*. 4th ed. Stanford: Stanford University Press, 1968.

Scheffer, V. B. *A Natural History of Marine Mammals*. New York: Scribner's, 1976.

Schmitt, Waldo L. *Crustaceans*. Ann Arbor: University of Michigan Press, 1965.

Smith, L. S. *Living Shores of the Pacific Northwest*. Seattle: Pacific Search, 1976.

Tavolga, W. N., ed. *Sound Production in Fishes*. New York: Halsted, 1977.

Taylor, H. *The Lobster: Its Life Cycle*. New York: Sterling, 1975.

Torchio, Menico. *The World Beneath the Sea*. New York: Crown, 1973.

Wilson, D. P. *Life of the Shore and Shallow Sea*. London: Nicholson and Watson, 1951.

Yonge, C. M., and T. E. Thompson. *Living Marine Molluscs*. London: Collins, 1976.

CONSERVATION

Carr, Donald E. *Death of the Sweet Waters*. New York: Norton, 1966.

Carson, Rachel. *Silent Spring*. rev. ed. New York: Fawcett, 1973 (paper).

Commoner, Barry. *The Closing Circle*. New York: Knopf, 1971.

Ehrlich, Paul R., and Anne M. Ehrlich. *Population, Resources, Environment*. 2nd ed. San Francisco: W. H. Freeman, 1972.

Kay, David A., and Eugene B. Skolnikoff. *World Eco-Crisis*. Madison: University of Wisconsin Press, 1972.

Marx, Wesley. *The Frail Ocean*. New York: Ballantine Books, 1967.

Moorcraft, Colin. *Must the Seas Die?* Boston: Gambit, 1973.

INDEX

Numbers in *italic* type refer to pages on which illustrations appear.

Sea fan, 185; *60, 184*
Sea flea, 359; *232*
Sea-floor spreading, theory of, 241
Sea gooseberry, 234
Sea hare, 18, 26, 117
Sea hawk, 196
Seahorse, 25, 34, 84, 91–92; *90*
Seal, 196, 200, 201, 203, 206, 208, 278, 343, 349, 354, 360; *3, 169, 194, 201, 204–5, 294;* dives, 43, 209; evolution, 198; intelligence, 166–67; sound production, 142; teeth, 200. *See also* specific names
Sealab I, 262
Sealab II, 262–63; *264*
Sea lamprey, 421
Sea lancelet, 311; *310*
Sea lion, 98, 196, 198, 203; dives, 43, 209; ears, 201; evolution, 198; flippers, 202; intelligence, 166–67; locomotion, 93; sound production, 142; teeth, 200; temperature regulation, 206
Sea Lions, The (Melville), 289
Seamounts, 236, 237
Sea otter, 39, 54, 73, 166, 196, 199, 202, 203, 206–7, 209, 343, 377; *54, 104, 167*
Sea pen, 230; *220*
Sea robin, 47; *47*
Sea scorpion, 310
Sea Shadow (Arpino), 291; *270*
Sea slug, 120–21; *18*
Sea snail, 358, 359; *159*
Sea snake, 25, 86, 96, 108; *87*
Sea, songs of, 290–91
Sea spider, 300; *299*
Sea squirt, 312; *190*
Sea star. *See* Starfish
Sea urchin, 41, 48, 162, 163, 179, 236, 391, 413; *237*
Seawall, 328
Seawater, 82, 203, 222–24, 304, 330, 412
Seaweed, 114, 231, 378
Seaworm, 307
Sedentary feeders, 47–48
Sediment, 191, 230, 236, 327, 424
Seismic-wave-warning system, 328
Seismology. *See* Earthquakes
Sei whale, 197, 202, 209
Sessile animals, 120, 358
Sewage, 416, 423–24, 425–26; *423–25*
Sex cells. *See* Gametes
Sexual reproduction, 23–35
Shackleton, Ernest, 400–1; *398–99*
Shad, 390
Shadow zones, 330
Shark, 39–40, 56–57, 82, 84, 85, 106–7, 125, 138, 234, 278, 313, 315, 376; *5, 46;* ampullae of Lorenzini, 139; camouflage, 111; claspers, 35; commensalism, 56–57; eyes, 56, 76; *77;* feeding, 39, 57, 190; reproduction, 25, 34, 35; senses, 43, 46, 47, 56, 76; teeth, 51, 107
Sharptail eel, *84*
Shearwater, 197, 355
Shellfish, 42, 228
Shells, 116, 310, 373; *372*
Shinano (ship), 403–4

Ships, 85, 377, 404, 405–7; *394–400, 402–4, 406. See also* specific names
Shipworm, 121, 328; *328*
Shipwrecks, 336–37, 392–94, 402; *400*
Shore life, 159, 203
Shoreline, 229, 412; *227*
Shrimp, 25, 114, 234, 298, 418; *130;* farming, 380; locomotion, 93; reef scavenging, 190; statocyst, 135
Sial, 240, 241; *238*
Sight. *See* Vision
Silent World, The (Cousteau film), 64
Silica, 234, 240
Silicon, 240
Silicoflagellate, 114
Silversides, *157*
Sima, 240, 241; *238*
Singing seal. *See* Ross seal
Siphonophore, 234
Sirenians, 196, 198, 201, 209. *See also* Dugong; Manatee
Skate, 84, 96, 138, 279
Skate (submarine), 326
Skeleton: coral, 174–75; exoskeleton, 114; phytoplankton, 231
Skimmers, 197
Skirmsl, 282
Skua, 356
Slocum, Joshua, 395–96
Slug, 120–21
Snail, 54, 114, 116, 137, 158, 190, 191. *See also* Nudibranch
Snailfish, 358
Snake, 25, 39, 86, 96, 108; *87*
Snowshoe hare, 360
Sodium chloride. *See* Salt
SOFAR (Sound Fixing and Ranging) channel, 330
Soft coral. *See* Alcyonacean
Sole, 20, 380
Sonar, 145, 146, 336, 370; *331*
Sonic sound, 143
Sonnenberg, Robert, 262
Sound, underwater, 140–47, 330–31; acoustic techniques, 330–31; as a defense, 121; channels, 330; echo, 330; remote sensors, 331, 332–33
Sound Fixing and Ranging channel. *See* SOFAR
Sounding, 145, 330–32
Spacecraft, oceanographic uses, *332–35*
Spadefish, 84
Spat, 379
Spawning: damselfish, 153; grunion, 33; oyster, 42; salmon, 20; shad, 390; wrasse, 152
Spearfishing, 384
Speleology, 247
Sperm, 23, 24, 25, 26
Spermaceti, 212, 213
Sperm oil, 215
Sperm whale, 42, 43, 52, 125, 146, 147, 197, 198, 200, 201, 202, 206, 209, 212–13; *208, 211*
Spider, 25
Spider crab, 298
Spinal cord, 131
Spiny lobster, 141; *143*

Splitfin flashlight fish, *70–71*
Sponge, 114, 130, 136, 176, 178, 179, 189, 190, 247, 248, 306, 307, 385; *49*
Sporophyte, 23
Sports, 383–84; *344–45*
Spotfin butterflyfish, *114*
Sprat, 234
Spray (boat), 395–96
Spring tides, 228
Spyhopping of gray whale, 99
Squaw, 197
Squid, 31, 43, 67, 113, 117, 197, 231, 236, 359; *24, 68–69;* eyes, 43; feeding, 190, 234, 236; locomotion, 81, 82, 93, 100, 101; symbiosis, 55
Squirrelfish, 337; *85, 228*
Stage decompression, 257
Starfish (sea star), 42, 43, 48, 94–95, 116, 122, 162, 190, 230, 301, 311, 312, 358; *43, 163*
Starfish House, 63, 260; *262*
Starling, 391
Statocyst, 135
Statolith, 135
Steamship, 405
Steller's sea cow, 211
Stenuit, Robert, 260
Stickleback, 32
Stinging cell. *See* Nematocyst
Stinging coral, 185, 189; *183*
Stingray, 109, 117; *85*
Stomach, 207–8
Stomiatoids, *53*
Stonefish, 109
Stork, 203
Storm petrel, 203
Storm surge, 227
Strabo, 322
Striped sea catfish, 43; *45*
Stromatolite, 178, 179
Strontium, 382, 418
Sturgeon, 46, 279, 315; *316*
Sub-Igloo, 359
Submarines, 85, 101, 145, 231; *101, 325; Archerfish,* 404; *Argyronete,* 266–67; *Nautilus,* 325–26; *Nekton,* 338; *337*
Submerged decompression chamber (SDC), 266
Submersible, unmanned, 332, 394
Suez, Gulf of, *240*
Sulfide ore, 370
Sunfish, 299
Sunlight, 64–66; *67. See also* Light
Supertanker, 405
Surface area (water), 412
Surface tension, 222
Surf clam, 230
Surfing, 383–84
Surgeonfish, 117, 154
Surtsey Island, 243
Swamp, 159
Swan, 216; *217*
Sweeper, *157*
Swim bladder, 82, 90, 140, 315
Swordfish, 98, 315
Symbiosis, 55; algae—coral, 55–56; clownfish—sea anemone, 155; hermit crab—sea anemone, 161; mushroom coral—worm, 185; pearlfish, 162

Symmetry, 311

Table reef, 175
Tabular icebergs, 348
Tail fins. *See* Caudal fins
Tails, 88, 90; flukes, 201–2
Tang, 117
Tanker, oil, 405; *404*
Tapetum lucidum, 76
Tapeworm, 307
Tektite II mission, 337–38
Teleosts, 315
Television, underwater, 336–38; *65*
Temperature, arctic, 344–45; body, 206; North Pole, 326; water, 19, 158–59
Teredo navalis. See Shipworm
Tern, 177, 197, 199
Territorial claims of nations, 421–22
Territoriality of animals, 31, 123; damselfish, 154, 155
Tetrodotoxin, 382–83
Thermal pollution, 421
"Thermistor chain," 329
Thermocline, 234
Thermohaline circulation, 225
Thomson, Elihu, 259
Thomson, Wyville, 323
Thornback skate, 138
Threnody (Anthony), 291
Tidal power, 370–71
Tidal time, 392
Tide pool, 228
Tides, 174, 178, 227, 228; *160*
Tilapia, 33, 426
Time, sense of, 391–92
Tin, 369
Titanic (ship), 348, 393
Toadfish, 140; *141*
Toothed whales, 145, 198, 201. *See also* Dolphin; Porpoise; Sperm Whale
Tornaria larva, 301
Torpedo ray, 52, 53, 137, 138; *52, 138*
Torrey Canyon (ship), 416
Tortoise, 17
Total pressure, 251
Toxins, 382–83
Trade winds, 225
Train, undersea, 407
Tramp steamer, 405
Transitional flow of water, 83
Traveling-wave propulsion, 86
Treasure, sunken, 368
Trenches, deep-sea, 229, 238, 241
Trieste (bathyscaphe), 238, 252
Triggerfish, 100, 117, 189
Trilobite, 73, 310, 359; *178, 310*
Tripodfish, 236
Triton, 189; *115*
Trocophore larva, 301
True's porpoise, 197
Trumpeter swan, 216; *217*
Trumpetfish, 49, 111; *84, 111*
Trunkfish, 86; *89*
Tsunami, 226–27
Tuamotu Archipelago, 249; *176*
Tube-nose birds, 354
Tube worms, 174
Tuffy (porpoise), 263
Tuna, 81, 82, 85, 100, 190, 202, 234; evolution, 299; fishing, 377;

PHOTOGRAPH CREDITS

Numbers refer to the pages on which the illustrations appear.

Heather Angel, England: 92, 178 top, 310 (2x); Fred Bruemmer, Montreal: 217; A. A. Carter, Japan: 417, 427; Cousteau Society: 4–5, 13, 62, 63 (3x), 65 below, 166 above, 250, 258, 260–63, 264–65, 336, 339, 380, 396, 400 (2x); Danish National Tourist Office, N.Y. (Photo Sv. Thoby): 286; Exxon Corporation: 363; General Electric Co., Dave Woodward: 266; Global Marine, Inc.: 326; Richard W. Grigg, Univ. of Hawaii: 174 below, 239; Robert Harding Associates, Inc., London, Brian Hawkes: 354; G. Renne, 362; Thor Heyerdahl: 397; Hirmer Verlag, Munich: 277 below; Information Canada Photothèque, Ottawa, D. Wilkinson: 360; C. Scott Johnson, Naval Undersea Center, San Diego: 124; Robert Glenn Ketchum, L.A.: 16, 162, 163 above, 373, 378, 379, 423; Herman Landshoff, N.Y.: 365; Dr. N. A. Locket, Royal Society, London: 317 ; Louvre, Paris: 392; Marineland of Florida: 34; Mariners Museum, Newport News, Va.: 256; Herbert Migdoll, N.Y.: 270; Howard Morris, N.Y.: 276; Museum of the City of New York: 402; Tom Myers, Sacramento: 242, 384, 419, 420, 421, 424; NASA: 176 above, 222, 240, 242, 332, 333 (3x), 334–35, 428; National Marine Fisheries, Wm. L. High: 413; National Oceanic and Atmospheric Administration (NOAA): 319, 330, 337; Naval Undersea Center: 394; New Bedford Whaling Museum: 287; New York Convention and Visitors Bureau: 425; Ann Parker, Brookfield, Mass.: 274; Peabody Museum of Salem: 322; Philadelphia Museum of Art: 280; Phillips Petroleum Co.: 371; Photri, Alexandria: 315, 325, 374–75 (Yamada), 375 below (Yamada), 410 (Jack Novack), 412; Photo Circle, Inc., Leo Touchet: 352–53; Pilkington Bros., St. Helens, Lancashire, Eng.: 406; Private Collection: 273; Radius Photo Service: 341, 348, 350; Scripps Institute of Oceanography: 264, 329; Sea Library: Michael Abbey, 232–33; Bob Barbour, 147, 200, 207; J. Boland, 38, 47, 57, 88, 143; John Bright, 219; Peter Capen, 89, 282, 283, 311; Jim and Cathy Church, 409; Mick Church, 65 above; Ron Church, 98, 130 below; Pat Colin, 298; Jim Cooluris, 25, 387; Ben Cropp, 54, 327; Jack Drafahl, 31, 42, 59, 65 right, 74, 75 above, 75 below, 92, 110A, 185, 372; Larry Dunmire, 269; Robert Evans, 11, 64 below, 67, 75 center, 118, 119, 126, 194, 197, 199, 294; Jill Fairchild, 196; H. Genthe, 79, 130 above, 159 above, 172, 176 below, 190 below, 220, 366, 372 below; Al Giddings, 312 above, 385; Charles Gilbert, 2–3, 376; Daniel W. Gotshall, 414–15; George Green, 226, 236; Howard Hall, 37; William L. High, 167; J. C. Hookelheim, 70, 90, 107, 131, 143 below, 145, 168, 314, 416; Hyperion Sewage Plant, 231 right, 300 (2x), 301; R. H. Johnson, 390; Stan Keiser, 24, 49, 60, 84, 85 center, 99, 128, 179, 180–81, 184, 198, 230, 246, 290, 368; Gina Kellogg, 33 above, 308–9, 328; G. L. Kooyman, 358; Chris Korody, 160; Peter E. Lake, 211 above left, 211 above right; Lingle, 97, 146; Ken Lucas, 316; Tom McHugh, 73, 77, 312 below; Bill Macdonald, 103, 134; J. A. Mattison, 104, 166 below; D. Nelson, 46; D. Odell, 212, 213; David C. Powell, 108; Jim Rakowsk, 18; D. Reed, 120; Randy Reeves, 211 below; G. A. Robilliard, 299, 356, 359, 422; Bruce Robinson, 71 above; Carl Roessler, 19, 30, 44–45, 48, 49, 50, 64 above, 71 below, 76, 80, 83, 84 left, right, center, 85 left, right, 87, 91, 94, 101, 109, 110B, 111, 112, 114, 115, 116, 123, 140, 149, 150, 152 (2x), 153, 169, 177, 178 below, 182, 183, 187, 188, 190, 201, 202, 210, 223, 228, 235, 248, 284, 288, 297, 298 above, 375 above; Eda Rogers, 7, 15, 26, 33 below, 43, 55, 106, 132–33, 165, 191, 237 above, 296; Ernie Rogers, 144; John Running, 159 below, 174 above; Dale J. Sarver, 22, 161 left, 163 below; Sea Library, 227; Patrick Shawn, 135; J. Sill, 68–69, 138; Alexander Smart, 156–57; Tom Stack, 40; Walter A. Starck, II, 121; Steinhart Aquarium, 52; Rick Tegeler, 93, 237 left below; N. K. Temnikow, 193, 203, 204–5, 342, 346, 355, 357, 398–99; Doug Toth, 51, 56, 137; W. E. Townsend, Jr., jacket, 35, 141, 154, 206, 214, 307; Bob Trelease, 96; Lewis Trusty, 164, 231 left; Paul Tzimoulis, 95, 117, 186, 245, 254–55, 388, 404; J. van der Walt, 142, 161 right; W. F. van Heukelem, 28–29; R. Waples, 293; Ozzie Wissell, 72; Woods Hole, 66; Woody Woodworth, 225; Secrist-Richesu, L.A.: 171, 175; Siebe Gorman & Co., Ltd.: 251; Staatliche Antikensammlung und Glyptothek, Munich: 275; Dennis Stock Magnum Photos: 21; U.S. Navy: 331; Herbert Warden, N.Y.: 403; Webb AgPhotos: Tom and Ceil Ramsay, 57 (2x); Jerry Stebbins, 351; Westinghouse Electric Corp., Oceanic Div.: 259; Wide World Photos, N.Y.: 224, 278 above, 395; Winnepeg Art Gallery, Canada, Ernest Mayer: 277.